II SCIENTIFIC-PRACTICAL FORUM "OIL CAPITAL"

IOP Conference Series: Materials Science and Engineering
Volume 921

Khanty-Mansiysk, Russia
21 - 22 February 2019

ISBN: 978-1-7138-3190-7
ISSN: 1757-8981

Printed from e-media with permission by:

Curran Associates, Inc.
57 Morehouse Lane
Red Hook, NY 12571

Some format issues inherent in the e-media version may also appear in this print version.

Institute of Physics
Dirac House, Temple Back
Bristol BS1 6BE UK

Phone: 44 1 17 929 7481
Fax: 44 1 17 920 0979

techtracking@iop.org

Additional copies of this publication are available from:

Curran Associates, Inc.
57 Morehouse Lane
Red Hook, NY 12571 USA
Phone: 845-758-0400
Fax: 845-758-2633
Email: curran@proceedings.com
Web: www.proceedings.com

TABLE OF CONTENTS

IOP Conf. Series: Materials Science and Engineering **921** (2020) 011001 doi:10.1088/1757-899X/921/1/011001

Introduction

From February 21 to 22, 2019, in Khanty-Mansiysk, the II SCIENTIFIC-PRACTICAL FORUM "OIL CAPITAL" was held by the government of the Khanty-Mansiysk Autonomous Okrug-Yugra with the support of the Russian National Committee of the World Petroleum Council (the RNC of WPC).

In addition to the key event - the International Youth Scientific and Practical Conference - the extensive programme of the Forum included a competition and a specialised exhibition of innovative projects, Round Table discussions on hot topic issues of the Fuel & Energy Complex, a presentation of the corporate standards of the "Personnel for the Fuel & Energy Complex" company, and a social and cultural programme for participants and guests. For the first time, the International Petroleum Academic Congress named after F.K. Salmanov was held as part of the Forum.

The organisers of the "Oil Capital" Forum highlight that the main tasks of the event are the capability enhancement for international cooperation between young scientists and specialists, and the promotion of innovative developments in the Fuel & Energy sector. At the same time, the Forum is considered as a catalyst for the optimisation of practical activities in the oil and gas industry of the Khanty-Mansi Autonomous Okrug-Yugra - a region that plays a key role in the oil and gas complex of Russia.

The forum brought together a wide range of participants - representatives of different countries and generations, students, scientists, experts. Representatives of federal authorities, the government of the Khanty-Mansi Autonomous Okrug-Yugra, and professional industry associations also participated in the professional discussion of hot topic industry issues.

Opening the Forum, the governor of the Khanty-Mansi Autonomous Okrug-Yugra Natalia Komarova emphasised the practical significance of the event. She cited several examples of introducing innovative solutions into hydrocarbon extraction, which were presented by young specialists a year ago - at the 1st "Oil Capital" Forum.

The delegation of the World Petroleum Council in Khanty-Mansyisk involved the Senior Vice-President of the World Petroleum Council Anatoly Zolotukhin, the Chairman of the Russian National Committee of the World Petroleum Council Valeriy Yazev, the Director of the RNC of WPC Vlada Streletskaya. Valeriy Yazev has made a presentation at the Plenary Session with the report "Oil and gas factor in geopolitics and sustainable development". Anotoly Zolotukhin held a Round Table "Global trends in the oil and gas industry: today and tomorrow," and Vlada Streletskaya presented a report revealing the importance and features of training a new generation of young professionals - industry leaders.

The presentation of the first issue of journals of the VI Youth Forum of the WPC "Future Leaders" was held at the central venue of the "Oil Capital" Forum. Participants of the Khanty-Mansyisk Forum had the opportunity to be among the first to get acquainted with the publications of these journals. The appearance of new editions produced by the RNC of WPC was greeted with interest and approval.

"We consider participation in the organisation and holding of events of the "Oil Capital" Forum as an important element in the preparation of the 6th Future Leaders Forum of the World Petroleum Council – VI Youth Forum of WPC, which will be held this summer in St. Petersburg," Vlada Streletskaya noted. "Both forums, which provide opportunities for professional communication and quality professional training of young specialists from different countries, are platforms for dialogue between generations, and focus on attracting young, self-motivated people to the industry. The experience of the "Oil Capital" Forum and the impressions of participating in this event will help to better prepare for the Future Leaders Forum".

IOP Conf. Series: Materials Science and Engineering **921** (2020) 011002 doi:10.1088/1757-899X/921/1/011002

Peer review declaration

All papers published in this volume of *IOP Conference Series: Materials Science and Engineering* have been peer reviewed through processes administered by the Editors. Reviews were conducted by expert referees to the professional and scientific standards expected of a proceedings journal published by IOP Publishing.

- **Type of peer review:** Open
- **Conference submission management system:** Program committee
- **Number of submissions received:** 31
- **Number of submissions sent for review:** 29
- **Number of submissions accepted:** 29
- **Acceptance Rate (Number of Submissions Accepted / Number of Submissions Received X 100):** 93,55%
- **Average number of reviews per paper:** 3
- **Total number of reviewers involved:** 43
- **Any additional info on review process:** -
- **Contact person for queries:**

  - Kaplan Mikhail Alexandrovich
  - E-mail: mishakaplan@yandex.ru / 79031927386@yandex.ru
  - Tel.: +79031927386
  - Forum Scientific Secretary

IOP Conf. Series: Materials Science and Engineering **921** (2020) 012001 doi:10.1088/1757-899X/921/1/012001

Application studies of organomineral fertilizer for restoration under the conditions of the Far North of disturbed illuvious-ferruginous podbur

D Yu Aladin, S M Sevostyanov, N F Deeva, D V Demin

142290, Russia, Pushchino, Moscow Region, 2 Institutskaya st.
Institute of Basic Biological Problems RAS.

E-mail: aladin-danila@rambler.ru

Abstract. The work is devoted to the use of organomineral fertilizer (OMF) based on food waste for sowing three mixtures of perennial cereals and the reclamation of an illuvial-ferruginous podbur in the Far North. Under the conditions of a model experiment, it was found that when OMF is introduced in fractions of 25 and 50%, the physicochemical properties of the soil improve. There is an increase in the green mass of plants relative to the control by 2.3 times. Adding 50% of the presented fertilizer leads to inhibition of cereals. The mass of plantings at this dose of fertilizer approaches the figures appearing in the experiment with 25% OMF. The presented organic-mineral fertilizer based on food waste can be used for the rehabilitation of disturbed soils of the illuvial-ferruginous podbur as fertilizer.
Keywords: organic-mineral fertilizer, perennial cereals, physicochemical properties of soils, podburs.

1. Introduction

In the Arctic, environmental degradation is taking on dangerous trends. Disturbance and destruction of natural ecosystems causes serious damage. Violation of the fragile Arctic nature may be irreversible. The natural landscapes of the Arctic have lesser stability, self-healing processes are carried out slowly. Anthropogenic impact leads to degradation of the Arctic lands, which in the long run may lead to irreversible environmental processes. Activation of the processes of soil deformation, thermokarst and thermoerosion is manifested especially intensively in the centers of industrial development and along linear structures (oil and gas pipelines, railways and highways, power lines, etc.). The annual growth of uncultivated disturbed lands is: in the oil industry - 5-6 thousand ha, in the gas industry - 2.5-3 thousand hectares, in the construction of pipelines - 0.4-0.5 thousand hectares [10].

Bioremediation is now recognized as possibly the most attractive clean-up approach for Polar Regions. Numerous contaminated sites in Polar Regions have legacy contamination that is often associated with weathered contaminants that tend to be recalcitrant to bioremediation. There is growing interest in cold-region landfarming because it is a relatively inexpensive and effective method for dealing with contaminated soils in inaccessible areas. This technique has proven to be effective in the Arctic. Reports of effective landfarming of higher-molecular weight compounds in cold climatesare limited and concerns remain about the effectiveness of this technology [12].

IOP Conf. Series: Materials Science and Engineering **921** (2020) 012001 doi:10.1088/1757-899X/921/1/012001

Podburs are formed in the flat and mountainous regions of the tundra, northern and middle taiga under conditions of good drainage on stony-fine-grained eluvium-deluvium of extrusive and metamorphic rocks and polymineral sandy - sandy loamy rocks rich in bases and ferruginous primary minerals. The profile of podbur consists of a peaty litter or a humus horizon with rare mineral grains clarified due to the removal of ferruginous films. Below lies an alpha-humus horizon of brown or buffy-brown color, formed as a result of the illuvial accumulation of aluminum-ferrous-humic compounds, the skeleton and grains of fine earth are covered with brown allochthonous films. The horizon brightens down and gradually passes into the parent rock. The substrates are characterized by an acidic and strongly acidic reaction of the entire profile with a decrease in acidity downward, base unsaturation, accumulative distribution of sludge, exchange bases and humus of fulvate composition with a predominance of mobile and aggressive fractions. The distribution of gross and oxalation-soluble forms of Fe and Al oxides is predominantly accumulative [5].

Polar soil environments differ from other soil environments in a number of ways that might affect hydrocarbon biodegradation. Polar soils have unique periglacial features, including permafrost and numerous types of patterns primarily due to freeze-thaw effects. In the Arctic tundra, an active soil zone, above the permafrost, thaws for a period of typically 1 ± 2 months in the summer. Summer temperatures in the active zone are highly dynamic and vary greatly, from near freezing at the permafrost interface to occasionally above 20°C at the surface. Permafrost restricts water movement and sometimes results in a saturated active zone. The active zone is where most hydrocarbon contaminants exist and is presumably the site of most biological activity. Little is known about the relationships between permafrost and hydrocarbon contaminants. Many Arctic tundra soils are low in organic content, a characteristic which can be expected to directly affect sorption of hydrocarbons to the soil and to indirectly affect biodegradation of hydrocarbons. Arctic tundra soil microbial communities have not been well characterized. There is presently no convincing evidence that the composition or structure of these communities is unique to Arctic tundra; although, the biomass may be low and the organisms more cold-adapted, relative to soils in temperate regions. Organic decomposition is slow in Arctic tundra soils, largely due to low temperatures. Some studies indicate the presence of hydrocarbon-degrading microorganisms in Arctic tundra soils and groundwater. Similarly, hydrocarbon-degrading microorganisms have been reported in Antarctic soils. Physical, chemical and biological factors have complex effects on hydrocarbon biodegradation in soil. For this reason, experts frequently recommend that soil bioremediation projects begin with treatability studies to empirically test the biodegradability of the hydrocarbon contaminants and to optimize treatment conditions. On the other hand, it is possible that the expense of such treatability studies could be avoided or minimized, if certain soil characteristics could be measured and used to predict the potential for bioremediation of a site, the kinetics of hydrocarbon removal or the optimal values for certain controllable treatment conditions. For example, certain co-contaminants such as heavy metals might preclude hydrocarbon bioremediation. Or, soil particle size distribution might partly dictate the potential rate and extent of hydrocarbon removal [13].

The presence of permafrost limits the development of a biologically active soil layer in the Far North. Due to harsh natural conditions and a very short vegetative period, restoration of the vegetation cover is very slow, also due to the development of solifluction processes and water erosion. As a result, the mining and other human activities inevitably leads to disruption of the soil cover [9]. The effectiveness of restoration of disturbed soil is substantially dependent on climatic conditions and, as a consequence, biological activity. With the natural restoration of soils after anthropogenic impact, the processes take a long time and are complicated by random factors [8].

When restoring disturbed soils in the Arctic conditions, it is supposed to create a fertile layer and sow perennial grasses that can form a stable turf cover [3, 6, 7]. To form this cover, additional resources of organic matter, mineral fertilizers and ameliorants are needed [4]. Moreover, the availability of these resources in the Arctic is limited, mainly due to the remoteness of disturbed objects.

A certain amount of organic waste is produced in settlements. Their processing into organic fertilizers will provide funds for the restoration of disturbed soils. One of the most promising areas of processing large volumes of food and other organic waste is composting, which is a recycling technology based on the biochemical decomposition of the organic fraction. Together with food waste, it is possible to process leaves, branches, mowed grass, waste from biochemical wastewater treatment, etc. To date, composting can be carried out both by private farms and enterprises on an industrial scale. On an industrial scale, composting can be carried out in special storage pits (long compost heaps). The processing technology is extremely simple: layers of food waste are strewn with soil, plant debris and undergo aerobic fermentation. Using high-level technologies, compost can mature in 2-3 weeks. Intensification of the process requires periodic aeration of the storage pits. The composting process is exothermic, a temperature regime from 50 to 70 °C is created inside the processed mass, which provides anthelmintic and antimicrobial action [11].

A technology has been developed for producing organic fertilizer based on food waste by solid-phase fermentation directly at the place of its formation.

Use of this technology can contribute to improving the environmental situation due to the processing of accumulated food waste and reducing the cost of biological reclamation of adjacent territories.

The aim of the work is to predict the possibility of using organomineral fertilizer based on food waste to restore disturbed soil cover using grass mixtures from perennial cereals in the Far North.

2. Objects and methods

The experiment was conducted in laboratory conditions with soil selected in the area of the city of Nadym of the Yamalo-Nenets Autonomous Okrug. The sampling depth is 5-30 cm. Sampling from a given depth is explained by the fact that, in places of intense anthropogenic load, these horizons go to the surface.

The soil is illuvial-ferruginous podbur, the Bh horizon (5-30 cm) is a brownish-yellow loam [5].

Organomineral fertilizer obtained by solid-phase fermentation from food waste from the village of Yubileiny, Yamalo-Nenets Autonomous Okrug. The composition introduced lime for deoxidation of the substrate and the formation of the structure.

Physicochemical analysis of OMF showed that it has an alkaline reaction (Table 1); according to the soil grouping by nitrifying ability, the content of mobile phosphorus, exchange potassium and humus, they belong to a very high class of security. The listed indicators (except for humus, the excess of which within the limits of this class) are exceeded several times [1], which confirms its high agrochemical effectiveness.

Table 1. Chemical and physical properties of soils, soil mixtures and organomineral fertilizers (OMF)

Measured indicators	OMF	Soil before OMF	After OMF 25% introduction
pH, aqueous extract	8,7	5,3	7,2
pH, salt extract	8,2	4,2	6,8
Organic matter, %	10,6	1,4	2,5
$N\text{-}NH_4$, mg/kg	1977,8	16,3	359,3
$N\text{-}NO_3$, mg/kg	112,8	5,1	63,3
P_2O_5, mg/kg	430	11	90
K_2O, mg/kg	8062	548,7	3299
Alphitite content (<0,01 mm), %	19,8 (sandy loam)	48,3 (heavy loam)	52,4 (light clay)

Mixtures of soil samples with organic fertilizer were prepared in a ratio of 3: 1 (75: 25%) and 1: 1

(50:50%) by dry weight. The determination of physico-chemical properties (Table 1).

Change in chemical parameters was carried out with one dose of application - 25% by dry weight. It was found that the chemical characteristics improved, the reaction of the medium changed from acidic to neutral, the content of nutrients (NPK) and organic matter increased significantly (Table 1).

In order to conduct the experiment, soil samples and mixtures with organic fertilizer content of 25% and 50% were placed in plastic vessels, with holes in the bottom in the shape of Mitcherlich's vessels and trays for collecting water. The surface area was 95 cm^2. Soil mixtures in all variants were equalized in moisture to ultimate field water capacity and maintained in such a process during the experiment. Each variant of the experiment was laid out in triplicate.

In each variant, 0.5 g of three cereal mixtures of herbs were sown on a vessel.

Mix No. 1: perennial ryegrass (*Lolium perenne*) - 25%; red fescue (*Festuca rubra*) - 47%; meadow bluegrass (*Poa praténsis*) - 13%; creeping bent (*Agróstis stolonífera*) - 15%.

Mix No. 2: red fescue altered (*Festuca rubra commutata*) - 15%; red fescue (*Festuca rubra*) - 35%; meadow bluegrass (*Poa praténsis*) - 10%; sheep fescue (*Festuca ovina*) - 5%; perennial ryegrass (*Lolium perenne*) - 35%.

Mix No. 3: couch grass (*Elytrígia répens*) - 20%; meadow fescue (*Festuca pratensis*) - 60%; perennial ryegrass (*Lolium perenne*) - 20%.

The presented perennial cereals were taken in the experiment based on the literature data and the principles of the availability of seeds and their resistance to adverse conditions of the Far North - frost resistance, resistance to waterlogging, and the ability to reproduce in the short growing season [2,9].

Similar mixtures were seeded in control variants. The containers were covered with glass and placed in a greenhouse.

In total, the experiment included nine variants: three - with the sowing of three mixtures of herbs in the original soil; three - soil with the introduction of 25% OMF with the sowing of herbs, similarly with the introduction of 50% OMF.

The experimental conditions are a greenhouse, which eliminated accidental effects on plants.

Duration of illumination is 12 hours a day, 12 hours a night, illumination 25000 lm/m^2, temperature $+ 20 \pm 2°C$, air humidity $60 \pm 5\%$.

After emergence (on the third day), the coverslips were removed. Irrigation was carried out every three days with distilled water with a pH of 6.3, at the same time the height of the plants (maximum, average, minimum) was measured in all variants of the experiment.

This is due to the fact that mixtures of herbs were sown, the seeds of which have different sizes, different rates of germination and differ in productivity (mass of one plant).

The time of the vegetation experiment is 21 days after germination. Further, in all variants of the experiment, the weight of the green mass in the wet and dried state was evaluated, and a comparison was made relative to the control.

3. Results and discussion

Throughout the experiment, in variants with control, there was a lag in the increase in the mass of cereal plants, which is associated with the initial physicochemical properties of the soil presented. In all variants with 50% fertilizer, there was a more intensive growth of plants at an early period of time, but by the end of the experiment their indicators were similar. Nevertheless, on day 18, the tips of leaf blades began to dry and blacken, which indicates an excessive concentration of nutrients when a given dose of fertilizer is applied, which leads to inhibition of plant growth.

The maximum yield with the addition of 25 and 50% OMF was noted for grass mixture No 2, consisting of *Festuca rubra commutata, Festuca rubra, Poa praténsis; Festuca ovina; Lolium perenne* (Table 2).

Table 2. The mass of plants grown in illuvial-ferruginous podbur

Variant	Grass mixture	Green mass weight (g)	The increase in control, %	The weight of the dried green mass (g)	The increase in control, %
Control	1	0,83±0,02		0,180± 0,002	
Control	2	0,84±0,02		0,192± 0,002	
Control	3	0,85±0,02		0,188±0,002	
soil+ 25% OMF	1	1,82±0,02	119	0,258 ± 0,002	43
soil + 50% OMF	1	1,80±0,02	117	0,258 ±0,002	43
soil+ 25% OMF	2	2,00±0,12	138	0,275±0,025	43
soil + 50% OMF	2	2,20± 0,15	162	0,275±0,025	43
soil+ 25% OMF	3	1,92±0,015	126	0,270±0,05	44
soil + 50% OMF	3	2,20± 0,15	159	0,268±0,02	43

The green mass of plants in relation to control increased by 2.3 times, and 1.4 times for dry weight. At a 50% content of this fertilizer in the soil mixture, the largest increase in the weight of plants in the wet state is noted. In the dried state, the mass of plant samples is the same for all variants of grass mixtures with fertilizer applied.

4. Conclusions

During the experiment, it was found that organomineral fertilizer, based on food waste, is characterized by a high content of nutrients NPK, which confirms its high agrochemical effectiveness.

Organomineral fertilizer has a positive effect on the physicochemical properties of the pickup, and prevents the drying of the surface soil layer.

When applying of 50% OMF, inhibition of plants is observed on day 18 of the experiment, which indicates an excessive concentration of nutrients.

Adding OMF in fractions of 25 and 50% leads to an increase in the yield of grass mixtures in relation to control by an average of 2.3 times.

During the vegetation of plants for their biomass, the best results were obtained with a mixture of red fescue (*Festuca rubra*) - 35%; perennial ryegrass (*Lolium perenne*) - 35%; altered red fescue (*Festuca rubra commutata*) - 15%; meadow bluegrass (*Poa praténsis*) - 10%; sheep fescue (*Festuca ovina*) - 5%.

The presented organic-mineral fertilizer based on food waste can be used for the rehabilitation of disturbed soils of the illuvial-ferruginous podbur as a fertilizer in the Arctic.

References

[1] Guidelines for the comprehensive monitoring of soil fertility of agricultural lands. M. "Rosinformagroteh", 2003, 240 p.

[2] Biological reclamation of disturbed lands in the Yamal Peninsula: Recommendations. - Novosibirsk, 1994, - 48 p.

[3] Likhanov A. I., Archegova I. B. Development of theoretical and practical aspects of the restoration process of disturbed lands in the north of the Komi Republic // Theoretical and Applied Ecology. 2014. No3. pp. 79-85

IOP Conf. Series: Materials Science and Engineering **921** (2020) 012001 doi:10.1088/1757-899X/921/1/012001

[4] Nazarova G.V., Ivanov V.V., Gavrilyeva L.D., Mironova S.I. On the issue of using waste in the restoration of disturbed lands // Successes in modern natural science. - 2012.-No 11. pp.135-136.
[5] National Atlas of Soils of the Russian Federation / Under the General Ed. corresponding member RAS S.A.Shoba. Faculty of Soil Science, Moscow State University M.V. Lomonosov. - M.: Astrel Publishing House, 2011, pp. 500-502.
[6] Iglovikov A.V., Denisov A.A. Potash regime of disturbed lands in the Far North at the biological stage of reclamation // Bulletin of the Altai State Agrarian University. 2019. No 12 (182). pp. 56-64.
[7] Pystina N.B., Unanyan K.L., Ilyakova E.E., Khokhlachev N.S., Luzhkov V.A. Improving the technology of landscape restoration on the slopes in the Arctic climatic conditions. // Scientific Bulletin of the Yamal-Nenets Autonomous Okrug. 2017. No. 3 (96). pp. 62-68.
[8] Kapelkina L.P., Sumina O.I., Lavrinenko I.A., Lavrinenko O.V., Tikhmenev E.A., Mironova S.I. Self-growth of disturbed lands of the North // VVM Publishing House LLC, St. Petersburg, - 2014. - p. 204.
[9] Tikhanovsky A.N. Biological land reclamation in the Far North // Agrarian science. - 2004. - № 8. – pp. 12-13.
[10] Agbalyan E.V. The state of the environment in the Arctic // Successes in modern science. 2011. No4., pp. 74-76
[11] Andreeva T.S., Maistrenko E.V., Ibragimova N.I. Ways to solve the problems of solid waste disposal on the territory of the Khanty-Mansiysk Autonomous Okrug - Ugra // Waste and Resources Online Journal, 2019 No. 3.
[12] Sanscartier D., Laing T., Reimer K., Zeeb B. 2009. Bioremediation of weathered petroleum hydrocarbon soil contamination in the Canadian High Arctic: Laboratory and field studies. Chemosphere 77:1121-1126.
[13] Mohn, W.W., Stewart, G.R., 2000. Limiting factors for hydrocarbon biodegradation at low temperature in Arctic soils. Soil Biol. Biochem. 32, 1161-1172.

IOP Conf. Series: Materials Science and Engineering **921** (2020) 012002 doi:10.1088/1757-899X/921/1/012002

Concentrated salt-resistant emulsions of the "oil in water" type used for the development of oil and gas wells

E V Belenko[1], A V Khvatov[2] and P A Sakharov[2]

[1]MIREA - Russian Technological University, 78 Vernadsky Avenue, Moscow, 119454 Russia
[2]Emanuel Institute of Biochemical Physics, Russian Academy of Sciences, Kosygin str., 4, Moscow,119334 Russia

E-mail: belkub@yandex.ru

Abstract. Currently, water-based and hydrocarbon-based oil and gas well silencing fluids are widely used. The most promising liquids for silencing and developing wells are direct emulsions, whose viscosity can be regulated by changing the concentration of the "oil" phase. Urgent task is to develop polymer-free structured emulsion compositions for developing and silencing low-temperature oil and gas wells that are stable when the water phase is saturated with sodium or potassium chlorides. For use in protected areas, the developed formulations should use vegetable oils and esters based on them as the oil phase, along with hydrocarbons.

Currently, water-based and hydrocarbon-based oil and gas well silencing fluids are widely used. The first group includes solutions of mineral salts, foams, fresh and reservoir waters, as well as direct emulsions. Hydrocarbon-based liquids include commercial or thickened oil and invert emulsions with an aqueous phase content of up to 80% vol. Among the disadvantages of hydrocarbon systems, we can distinguish their blocking effect due to capillary forces that are realized when mineralized reservoir water is displaced from the pore space, which can lead to the formation of persistent water-oil emulsions in the reservoir. In addition, oil-based compounds have increased combustibility and a low level of environmental safety for the environment.

Sodium chloride (halite), which has the optimal combination of low cost and high availability in comparison with other salts, is the most widely used as silencing fluids for low-temperature gas wells. Sodium chloride is usually used for the preparation of silencing liquids with a density of up to 1.18 g/cm^3. For the preparation of liquids with a density of 1.18 to 1.30 g/cm^3, calcium chloride is used, and for the preparation of silencing liquids with a density of more than 1.30 g/cm^3, potassium carbonate (potash), as well as other salts or mixtures thereof can be used. Water brines are thickened with polymer additives based on polysaccharide derivatives to reduce the level of filtration of the water phase into the productive layer. Polymers are limited in solubility in a saline medium, so their aqueous solutions are close to dispersions in properties. Water-polymer compositions may also contain a highly dispersed filler and surfactant for hydrophobizing the pores of a productive formation. The article [1] describes an oil and gas well silencing fluid containing a surfactant, a polymer emulsion, a highly dispersed hydrophobic material, and an aqueous solution of mineral salt. The presented composition has insufficient insulating properties and increases the water content of the extracted hydrocarbons. A polysaccharide gel for silencing wells containing mineralized water, a polysaccharide thickener, a boron crosslayer, diethanolamine, Quaternary ammonium compounds, and a mixture of non-ionic and anionactive surfactants has low sedimentation stability and complexity of preparation [2]. Polymer gels based on high-molecular polyacrylic derivatives, magnesium or calcium chloride and surfactants [3] reduce reservoir properties due to high filtration rate and significant adsorption activity of macromolecules of the thickener. As a result, the use of anion-active polymers and surfactants in the silencing liquid in combination with concentrated solutions of mineral salts leads to the formation of highly dispersed precipitation, as well as blocking adsorption layers that are impervious to hydrocarbons.

Replacement of polymer thickeners with oligomeric and low-molecular-weight alcohol-containing components (triethanolamine, polyglycols, aliphatic alcohols, etc.) [4, 5] in combination with saturated aqueous solutions of alkaline and alkaline earth metals provides the required level of inhibition of hydrate formation, but does not allow to effectively regulate the rheological, filtration and inhibitory characteristics of the system. For example, a silencing liquid based on triethanolamine and a solution of calcium chloride reduces the permeability of terrigenous zones as a result of increasing the volume of water-sensitive minerals and reducing the porosity of the formation [4]. The low viscosity of the silencing fluid, which does not contain polymer thickeners, makes it ineffective when silencing highly permeable reservoirs, since the absorption volumes increase and the time of well development increases. Included in many liquids the aqueous solution of calcium chloride causes the formation of high viscosity oil-water emulsion, coagulation of clay particles and precipitation of insoluble precipitation of inorganic salts in the channels of the filter layer, which leads to increased water saturation of the productive formation, reduce permeability of the reservoir hydrocarbon and increases the water content of hydrocarbons. Disadvantages of silencing fluids based on polyglycol [5] are the propensity to foam and the presence of sodium hydroxide in the polyglycol composition, which interacts with carbonate rocks to form water-insoluble salts, and also causes swelling and dispersion of clay minerals, reducing the permeability of the bottom-hole zone of wells and their productivity coefficient.

The most promising liquids for silencing and developing wells are direct emulsions, whose viscosity can be regulated by changing the concentration of the "oil" phase. To ensure the stability of such emulsions in mineralized water, low-viscosity derivatives of lignosulfonates, polysaccharides, and oligomers based on acrylate or sodium methylsilicate are used [6-8]. Direct emulsions contain up to 40 % of hydrocarbons (diesel fuel, gas condensate) and are stabilized mainly by oil-soluble emulsifiers. A common disadvantage of emulsion compositions is their low stability at a high concentration of mineral salts in the water phase, which is 10-25 %. This narrows the range of regulation of the specific gravity of the silencing fluid, and also does not allow you to completely abandon the use of anionic polymer stabilizers.

Thus, an urgent task is to develop polymer-free structured emulsion compositions for developing and silencing low-temperature oil and gas wells that are stable when the water phase is saturated with sodium or potassium chlorides. For use in protected areas, the developed formulations should use vegetable oils and esters based on them as the oil phase, along with hydrocarbons.

The stability of the direct emulsion, which is a microdispersed oil phase (OF) (with an acid number of 80 mg (KOH)/g) distributed in fresh water, reaches a maximum in the pH range = 9 - 9.5, corresponding to the highest degree of electrostatic protection. In the absence of an emulsifying surfactant, the stability of the OF emulsion dispersed in the aqueous phase is very low. The rate of coalescence of oil droplets is proportional to the difference in densities of dispersed phase and dispersion medium, droplets OF the radius and inversely proportional to the viscosity of the water phase. Therefore, drops of the oil phase in direct emulsions must be stabilized with the addition of non-ionic oxyethylated surfactants (OS). The combination of the electrostatic factor of stabilization of emulsions due to the presence of negatively charged carboxylate groups on the surface of micelles and the sorption-solvate protective effect of hydrated adsorption layers of surfactants allows creating a sufficient potential energy barrier of the order of 20*k*T, which increases the half-life of the oil drop to several years [9]. The structure of the protective barrier formed in a freshwater dispersion medium is shown in (Figure 1).

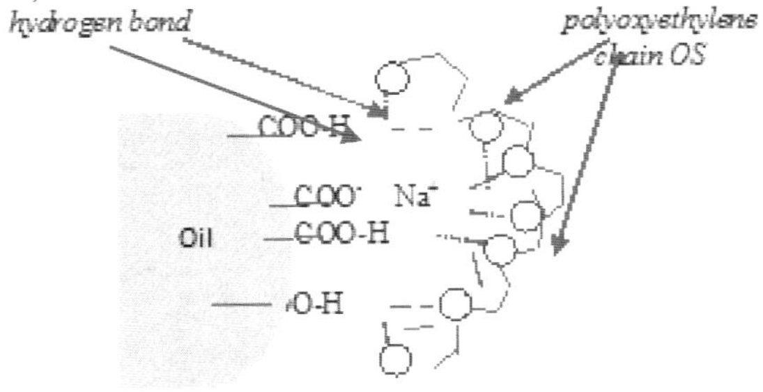

Figure 1. Formation of an adsorption layer of an oxyethylated surfactant on the surface of an oil drop.

Oxyethylated surfactants provide stability of the direct emulsion only if the dispersion medium is a good solvent for polyoxyethylene chains that are distributed near the drop surface and are linked to the carboxyl groups of OF by hydrogen bonds. Therefore, polyoxyethylene surfactant chains provide the implementation of a steric stability factor that occurs when entropy decreases when the adsorption layers interpenetrate when neighboring droplets converge.

Electrostatic stabilization of direct emulsion microdrops is not sufficient when the ionic strength of the water phase increases when the 1-1-valence electrolyte (NaCl) is dissolved. The destruction of the emulsion (flocculation consolidation of the droplets with the formation of the top layer of "cream") is due to two reasons: a critical lowering of the electrostatic repulsion between the drops in the transition to the isoelectric condition (coagulation concentration) and decrease the solubility of surfactants, which is accompanied by the degeneration of the entropy factor, which becomes the cause of the hydrophobic attraction dehydrated adsorption layers of adjacent particles OF.

To ensure the stability of the direct emulsion under conditions of saturation of the aqueous phase with electrolyte, it is necessary to introduce an additional structural and mechanical stability factor, which is realized as a result of interpolymer complexation between carboxylate groups on the surface of OF micelles and the substituent associations of divalent cations of alkaline earth or transition metals Me^{2+} (Ca^{2+}, Mg^{2+}, Cu^{2+}, Fe^{2+}) [10, 11] (Figure 2).

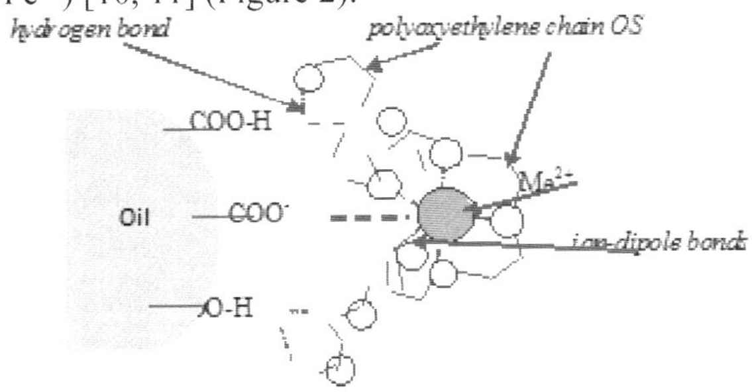

Figure 2. Formation of a structural-mechanical barrier on the surface of microemulsion droplets due to interpolyelectrolyte complexation of $Me^{2+}\cdot[EO]_5$ substituent associates with COO^--groups.

Podand poly-cationic associates are formed as a result of the coordination of divalent ions (Me^{2+}) within the crown-like five-membered oxy-ethylene cycles of surfactant molecules: $Me^{2+}\cdot[EO]_5$. At the same time, 1 structural unit of the substandard polycation $Me^{2+}\cdot[EO]_5$ electrostatically shields 2 COO^--groups, so the structural stability coefficient (K_{st}) of the emulsion is equal to the ratio of the amounts of substances of the components of the interpolymer complex (IPC), taking into account its stoichiometry. Accordingly, the steric stability factor of the emulsion (K_s) describes the entropic stabilization of the system and is determined by the ratio of five-membered oxyethylene cycles and carboxyl COOH- groups.

IOP Conf. Series: Materials Science and Engineering **921** (2020) 012002 doi:10.1088/1757-899X/921/1/012002

When the mass ratio between surfactant and OF changes, the stability and rheological properties of the direct emulsion undergo significant changes, which is associated with different levels of electrostatic and structural-mechanical stabilization of the system (Table 1).

Table 1. Colloidal-chemical properties of a direct emulsion with a different ratio of the oil phase (OF) and nonionic surfactant.

Colloidal-chemical properties of the emulsion	Mass ratio of OF: surfactant in emulsion (100 g)			
	80:20	65:35	50:50	30:70
Acid number OF, mg(KOH)/g	80	80	80	80
The mass ratio of the phases «water-in-oil»	85:15	85:15	85:15	85:15
NaCl concentration in the aqueous phase of the emulsion, % wt.	26	26	26	26
Degree of dissociation COOH-groups	39,5	39,5	39,5	39,5
Number (υ) of COOH-groups OF, mol	0,026	0,021	0,017	0,009
υ (COO^--groups) OF, mol	0,017	0,014	0,011	0,006
υ (surfactant), mol	0,005	0,009	0,012	0,017
υ $(EO)_5$, mol	0,009	0,015	0,022	0,031
υ Ca^{2+}, mol	0,0022	0,0022	0,0022	0,0022
υ (COO^-) as part of the IPC: $[Ca\text{-}(EO)_5{+}2{\cdot}COO^-]$, mol	0,0044	0,0044	0,0044	0,0044
K_{st} $(2{\cdot}\upsilon(Me^{2+})/\upsilon\ (COO^-))$	0,26	0,32	0,42	0,67
K_s $(\upsilon[EO]_5/\upsilon\ (COOH))$	0,17	0,50	1,04	2,96

Figure 3. Change in the coefficients of structural (Kst) and steric (Ks) stability of the emulsion with an increase in the mass fraction of oil in the surfactant composition.

As a result of lowering the stability coefficients, the stability of the direct emulsion with increasing oil concentration in the system also decreases, which can lead to the transition of the system to a state characterized by a deep secondary potential minimum. This is a fairly stable energy state in which the oil droplets are as close to each other as possible, but retain their integrity. In this case, the emulsion acquires high structural-mechanical and rheological properties, acquiring the features of a strong gel (Figure 4).

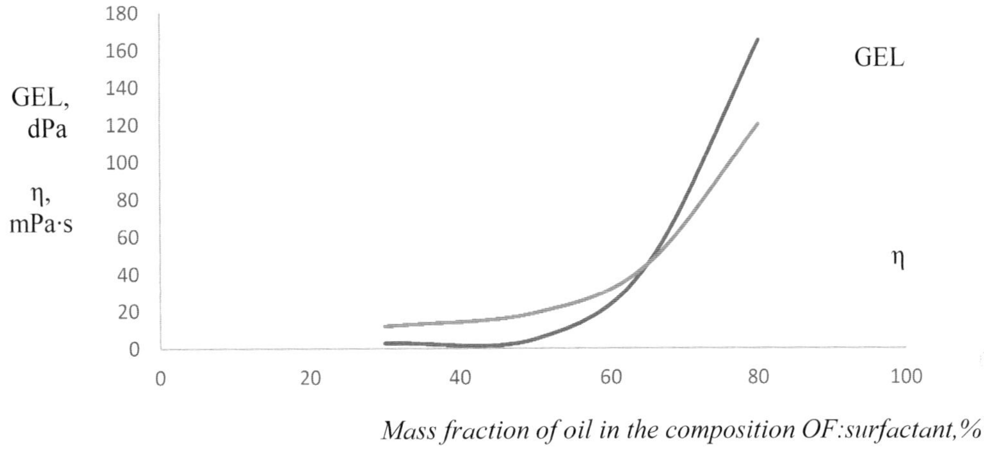

Figure 4. Change in the rheological properties of the direct oil emulsion in the composition OF: surfactant (total concentration of the complex OF + surfactant = 15 % by weight).

The increase of the static strength of the structure (GEL) and dynamic viscosity (η) with increasing concentration of the oil phase indicates the presence of long-range order interactions between the dispersed particles of the emulsion resulting in the development of hydrophobic complexation by overlapping adsorption layers of adjacent particles, between which there is an electrostatic barrier. Further convergence of the droplets is prevented by a structural-mechanical barrier due to the presence of IPC at the phase interface. On the contrary, when the oil concentration is low, the distance between the microdrops increases, which prevents hydrophobic interactions; the stability of the emulsion increases significantly, and the structural spatial grid of dispersed particles disappears. Increasing the oil concentration above 80%wt. it is impractical, since it leads to an increase in coagulation processes and stratification of the emulsion with an insufficient level of steric and structural-mechanical stabilization ($K_s < 0.1$; K_{st} Increase the static strength of the structure (GEL) and dynamic viscosity (η) with increasing concentration of the oil phase indicates the presence of long-range order interactions between the dispersed particles of the emulsion resulting in the development of hydrophobic complexation by overlapping adsorption layers of adjacent particles, between which there is an electrostatic barrier. Further convergence of the droplets is prevented by a structural-mechanical barrier due to the presence of IPC at the phase interface. On the contrary, when the oil concentration is low, the distance between the microdrops increases, which prevents hydrophobic interactions; the stability of the emulsion increases significantly, and the structural spatial grid of dispersed particles disappears. Increasing the oil concentration above 80%wt. it is impractical, since it leads to an increase in coagulation processes and stratification of the emulsion with an insufficient level of steric and structural-mechanical stabilization ($K_s < 0.1$; $K_{st} < 0.1$).

It is convenient to characterize the stability of emulsions by the phase inversion temperature (PIT), which is the transition temperature of the direct emulsion to the reverse. To determine the PIT investigated the direct emulsion was stirred at 8000 rpm with increasing temperature and identify PIT by the sharp drop in conductivity when added to aqueous dispersion medium (Figure 5, a) passes into a state with an oil dispersion medium. In the temperature range above PIT, after mixing is stopped, the emulsion usually breaks down to form a three-layer structure: the lower layer of water, the middle thin layer of the bicontinual emulsion, and the upper layer of oil (Figure 5, b).

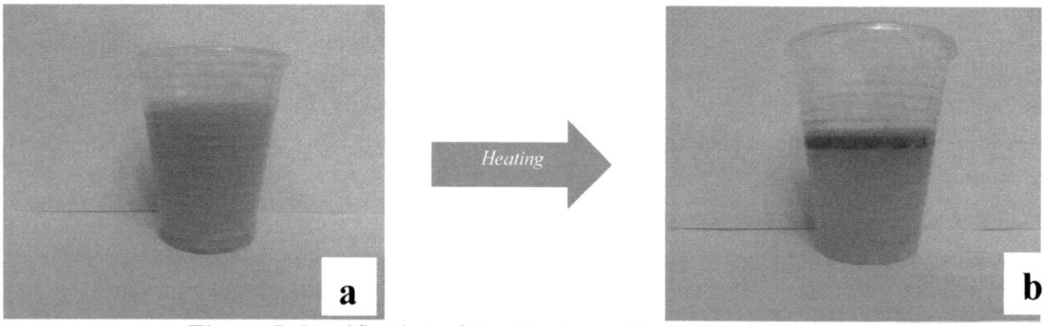

Figure 5. Stratification of the direct emulsion when heated above PIT.

The direct emulsion PIT decreases with increasing electrolyte concentration in the water phase, as well as with additional introduction to the system relative to the polar oil phase as a solubilizer. The more polar the oil is (it contains multiple bonds, phenyl groups, branches of the hydrocarbon skeleton, etc.), the lower the typhoid. In addition, PIT is lowered by additives that increase the polarity of the oil phase – fatty acids and alcohols. Additives less polar, but well soluble in water, significantly increase PIT. Figure 6 shows a change in the PIT of direct emulsions containing different ratios of OF: surfactants, with an increase in the concentration of sodium chloride.

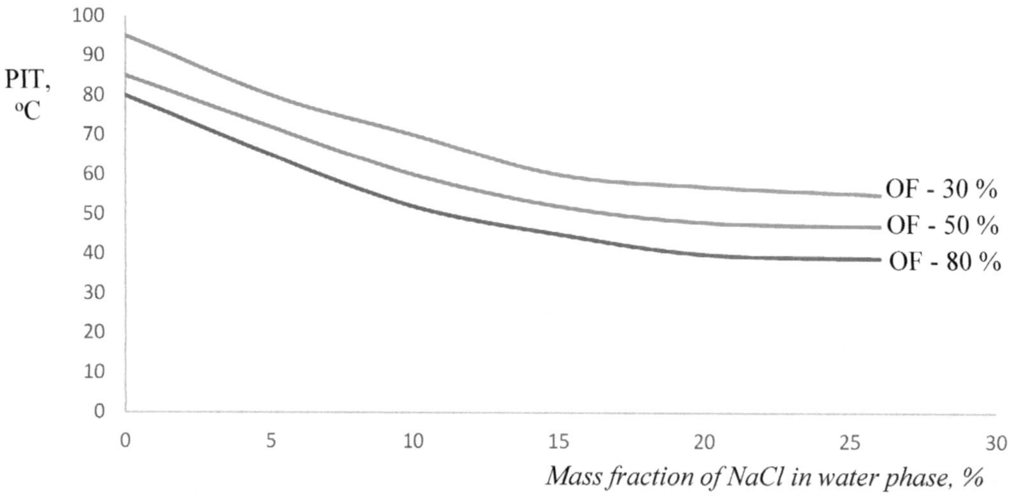

Figure 6. Dependence of PIT emulsions containing different ratios of OF and surfactant on the concentration of NaCl (total concentration of OF + surfactant = 15 % by weight).

Increasing the surfactant concentration increases the PIT of the emulsion in conditions of high electrolyte concentration, which is associated with an increased role of the electrostatic and structural-mechanical mechanisms for system stabilization.

The advantages of the electrostatic method of stabilization of direct emulsions are the minimum number of system components, a wide range of variations in rheological characteristics and PIT, and the absence of adsorption active reagents in the polymer composition. To increase the stability of direct emulsions and expand the temperature range of their use (increasing the PIT), it is necessary to increase the coefficients of steric and entropic stability of the system by increasing the concentration of oxyethylated surfactant, reducing the volume fraction of the oil phase, or using special additives that reduce the acid number of the oil. With an increase in the electrolyte concentration in the aqueous phase of the emulsion, the PIT decreases, remaining at a sufficient level (>40°C) in full salinity conditions for the use of emulsion compositions in low-temperature wells. Additional solubilization of polar oil-soluble components (fatty acids and esters of vegetable oils) leads to the development of coalescence and reduction of the PIT emulsion. On the contrary, the introduction of non-polar hydrocarbons (alkanes, polyolefins, mineral oil), as well as polyatomic alcohols (triethylene glycol, propylene glycol, glycerin) into the emulsion provides a moderate increase in PIT as a result of the hydrotropic effect.

References

[1] Belikov V T, Volkov V A and Turapin A N 2006 RU № 2279462
[2] Gaevoy E G, Zavorotny AV, Zaitsev K I and Magadov R S 2005 RU № 2246609
[3] Jamalinsky V K, Eremenko T A, Lysenkov E A, Sukhomlinov A P and Chetverikov A D 2003 RU № 2215868
[4] Ryabokon S A, Volters A A, Surkov A B and Glushchenko V N 1989 Fluids for well repair and their influence on reservoir properties of the reservoir *Overview information. Series "Oilfield business" Moscow VNIIOENG* pp 34-37
[5] Akhmetov A A, Galyautdinov A A, Dokichev V A, Dudov A N and Istomin N N 2002 RU № 2187532.
[6] Belyaeva A D, Dyabin A G, Zavolzhsky V B, Kan V A and Magadova L A 1997 RU № 2097547
[7] Gaykhman M G, Hodzyur I, Dmitruk, V V and Kleschenko I I 2003 RU № 2213762
[8] Rozeluk B V and Rusakov S Y 2002 RU № 2190657
[9] Belenko E V 2007 Creation and research of environmentally safe pseudoplastic surfactant emulsions *Environmental protection in the oil and gas industry* **№ 7** pp 28-32
[10] Belenko E V and Vakhrushev L P 2007 Physical and chemical properties of copolymers of ethylene and propylene oxides *Environmental protection in the oil and gas industry* **№ 8** pp 28-32
[11] Penkov A I, Vakhrushev L P, Belenko E V, Ostryagin A I and Dzhioev T E 2000 Influence of polyanion electrolytes on the state of adsorption layers of polyalkylene glycols in bentonite suspensions *News Of Universities.-North Caucasus region Natural science* pp 90-94

IOP Conf. Series: Materials Science and Engineering **921** (2020) 012003 doi:10.1088/1757-899X/921/1/012003

Natural zeolites: prospects for heavy metal polluted soil remediation

E A Bocharnikova[1], V.P.Shabayev[2], V.E.Ostroumov[2] , D.V. Demin [1]

[1]Institute Basic Biological Problems RAS, 142290, Pushchino, Russia
[2]Institute of Physicochemical and Biological Problems in Soil Science RAS, 142290, Pushchino, Russia

E-mail: nimedd@yandex.ru

Abstract The environmental concerns related to oil extraction, transportation and refining include the pollution with heavy metals (HMs). Currently, there is a growing interest in the study of an influence of silicon-rich materials, including such natural minerals as diatomite, zeolite, and others, on the HM behavior in the environment with regard to their exploitation in soil remediation strategy. In greenhouse pot experiment, the effects of natural zeolite on the biomass of barley plants grown in artificially cadmium (Cd)-contaminated soil and Cd uptake and translocation were studied. Zeolite applied to soil at a low rate (0.15%) mitigated the Cd phytotoxicity and provided the same plant biomass production as in uncontaminated soil without zeolite. Zeolite stimulated development of root system. The application of zeolite at a low rate enhanced the phytoextraction process–cleaning up the soil from HM at the early (booting) stage of plant development.

1.Introduction

Oil extraction and spillage result in the environment pollution by organic compounds and heavy metals [1]. The pollution by cadmium (Cd) is of severe concern due to its extreme persistence in the environment and high toxicity to plants, animals, and people. Most methods of purifying chemically polluted areas are expensive and don't make the soil suitable for plant growth [2]. At present, phytoremediation attracts high attention as a cost-effective technology for cleaning up polluted territories [3; 4]. The phytoremediation technology is based on using plants that are able to accumulate contaminants with subsequent removal of plant biomass from the ecosystem. One of the main problems that reduce the efficiency of this technology is long duration of a remediation process and toxic effect of pollutants on the plants used in phytoremediation resulting in small production of biomass [5].

Over the last decades, there has been a growing interest in the study of the effect of silicon (Si)-rich compounds, including natural zeolites, on HM-affected plants in regard to their use in soil remediation technologies [6; 7]. Silicon-rich compounds may adsorb HMs, thus reducing their mobility in the soil [8]. Silicon minerals dissolve, releasing silicic acid that can directly react with HM, thus impacting the HM bioavailability [9].

Zeolites are one of the most promising natural Si-rich minerals for use in the soil remediation technologies due to their abundance and low cost. Zeolites are crystalline porous aluminosilicates with a framework structure. Absorption properties of zeolites are identified by a unique crystal lattice, characterized by a developed inner surface with minimum pore diameters of 0.3 to 1.0 nm that provide the selective absorption of molecules depending on their size [10]. Today zeolites are widely studied for remediation of HM-contaminated soils. However, the results of these studies are quite contradictory. Many authors indicate a significant zeolite-induced reduction in the exchangeable forms of HMs in soil and in their translocation into plants [11; 12; 13]. Some data shows that for efficient HM immobilization application rates of zeolite should be about 40-50 t ha^{-1} in the case of low fertile soil. If soil has high clay content and high fertility level, the rates of zeolite should be even higher [14]. Study of the Iranian natural zeolite impact on the Cd leaching from soil showed that an optimal rate for reducing the HM mobility ranged between 9 to 15% depending on the soil texture [15].

Besides the remediation technologies, zeolites are used as soil amendment for increasing yield quality and quantity and for improving soil fertility [16; 17]. However, the recommended high application rates of zeolites often are not cost-effective. Balakhnina et al. [18] showed that low doses of zeolite have a positive impact on the plants under Cd toxicity. However, the effect of low zeolite rates on the migration and bioavailability of Cd in the soil–plant system has not been studied.

The aim of the research was to study the effect of natural zeolite on the growth and the uptake and translocation of Cd in barley plants grown in artificially Cd-contaminated Gray Forest soil.

2.Materials and Methods

Greenhouse pot experiment was conducted with barley (*H. vulgare L., cv. Suzdaletz*) grown in Gray Forest loamy soil (Phaeozems in FAO classification). Soil was collected from an abandoned field (from a depth of 0–20 cm) in the southern Moscow region, Russia. The soil had the following characteristics: pH (1 M KCl), 5.73; Corg, 0.87%; total N, 0.12%; Ca and Mg (1 M KCl), 5.2 and 1.0 cmol kg^{-1}, respectively; P_2O_5 and K_2O (0.2 M HCl) 129 and 136 mg kg^{-1}, respectively. Wet soil was thoroughly mixed and passed through a 1-cm sieve.

Soil was contaminated with Cd as chemically pure $Cd(NO_3)_2 \cdot 4H_2O$ at a rate of 10 mg kg^{-1} Cd. Ground (0.2-0.3 mm size) natural zeolite (clinoptilolite, $(Na,K,Ca)_2\text{-}3Al_3(Al,Si)_2Si_{13}O_{36} \cdot 12(H_2O)$, Khotynetz deposit, Orel region, Russia) was applied at a rate of 1.5 g kg^{-1} soil (0.15%). Zeolite had the following chemical composition: SiO_2, 69.0–74.0%; TiO, 0.08–0.16%; Al_2O_3, 11.4–14.0%; Fe_2O_3, 0.60–1.8%; MnO, 0.02–0.05%; CaO, 1.7–3.3%; MgO, 0.4–1.7%; K_2O, 0.5–5.0%; Na_2O, 0.4–0.9%; P_2O_5, 0.4%; and pH 6.5. The following treatments were conducted: control, Cd, Cd+zeolite; each in 4 replicates.

All treatments included application of N, P and K at a rate of 100 mg kg^{-1} of each element as NH_4NO_3, KH_2PO_4 and K_2SO_4. The treatments with $Cd(NO_3)_2$ were adjusted to the same level of N as that in the corresponding control variants of uncontaminated and N-fertilized soil by adding NH_4NO_3. Cadmium salt, zeolite and mineral fertilizers were thoroughly mixed with the soil 10 days before seeding. Twelve (12) germinated seeds were planted in a pot containing 800 g of soil and grown for 28 days until the booting stage. The air temperature in the growth area was kept at 20 ± 2° C during the day and 12 ± 2° C during the night. The light period was 12 h; the light intensity was 950 μmol photons m^{-2} s^{-1}. The relative air humidity was 45 ± 5% during the day and 70 ± 5% during the night. Distilled water was added to each pot daily to maintain soil moisture at 21%.

Plants were harvested and separated into leaves, stems and roots. Roots were separated from soil with tap water and washed with distilled water. Plant material was dried at 70° C, weighed and ground. A 0.4–0.5-g plant sample was digested with a 20-mL mixture of concentrated HNO_3 and $HClO_4$ (2:1 v/v) and analyzed for Cd with the Optima 5300 DV ICP optical emission spectrometer (Perkin Elmer, USA).

Values are means ± standard error of 4 replicates. Significant differences were calculated using Student's t test ($\alpha = 0.05$).

3. Results and Discussion

The results obtained evidenced the toxic effect of Cd on plant growth in the booting stage. The biomass of whole plants in Cd-contaminated soil was reduced by 16% and a decrease in the weight of shoots amounted to 18% compared to the control plants (Table 1).

Table 1. Weight of barley plants in the booting stage.

Treatment	Leaves	Stems	Leaves + stems	Roots	Whole plants
			Dry matter, g pot^{-1}		
Control	1.36 ± 0.15	0.63 ± 0.09	1.99 ± 0.30	0.26 ± 0.04	2.25 ± 0.25
Cd	1.18 ± 0.11	0.46 ± 0.07	1.64 ± 0.25	0.24 ± 0.02	1.88 ± 0.10
Cd + zeolite	1.42 ± 0.18	0.65 ± 0.12	2.07 ± 0.31	0.33 ± 0.03	2.40 ± 0.35

The zeolite applied to Cd-polluted soil has benefited the plant growth. The weight of shoots increased by 26% as compared to the Cd-exposed plants without zeolite application. Moreover, the weights of aboveground parts in the unpolluted soil without zeolite and in the Cd-polluted soil with zeolite were approximately the same. The Cd contamination had no adverse impact on the root biomass. Application of zeolite resulted in increases in the root biomass by 27 and 38% compared to the control and Cd-exposed plants, respectively.

Cadmium was not detected in the shoots of plants grown in uncontaminated soil, whereas the Cd content was significant in the Cd-exposed plants (Table 2). The maximum Cd content was observed in the roots. The Cd contents in stems and leaves were almost 4 and 8 times lesser, respectively, than those in roots. Zeolite resulted in an insignificant increase in the leaf Cd, a double increase in the stem Cd and a 6-fold increase in the root Cd compared to the Cd-exposed plants without zeolite.

Table 2. Cadmium content in barley plants in the booting stage.

Treatment	Leaves	Stems	Roots
		mg kg^{-1} of dry weight	
Control	n/d	n/d	n/d
Cd	7 ± 1	12 ± 2	45 ± 8
Cd + zeolite	8 ± 1	26 ± 6	270 ± 32

n/d - non-detectable.

The zeolite-mediated total Cd removal by shoots per pot was almost twice as high as that in the contaminated soil without zeolite (Fig. 1). Cadmium was mainly accumulated in the roots. The use of zeolite led to an almost 7-fold increase in the root Cd and to a 4-fold increase if calculated for whole plants in the booting stage.

Our results have shown that zeolite applied at a low rate eliminated the Cd toxic effect on barley growth. Zeolite at a rate of 0.15% promoted increased Cd in above-ground organs and its maximum accumulation in the roots at the early (booting) stage of plant development. An increase in the root Cd as influenced by Si-rich compounds was reported in several studies [19; 20].

Zeolite-mediated Cd increase in the roots, stems, and leaves indicates enhanced metal translocation from the soil to the vegetative organs. This finding could relate to the interaction between Cd and monosilicic acid, the product of zeolite dissolving. The effect of soluble Si on the HM plant-availability was found to depend strongly on the monosilicic acid concentration in the soil solution [9]. Monosilicic acid at lesser concentrations forms the water-soluble Si-HM complexes or it precipitates HM via the formation of slightly soluble silicates at higher concentrations, thus increasing or decreasing the HM mobility. A slight increase in monosilicic acid at a low rate of zeolite could result in the formation of soluble Cd-Si complexes.

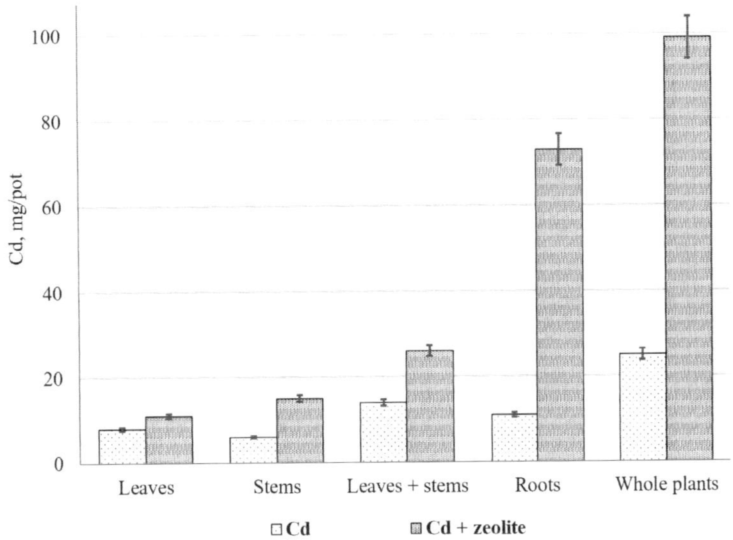

Figure 1. Cadmium uptake by barley plants in booting stage.

Zeolite-induced mitigation of adverse Cd impact on barley growth can be caused by the several processes in soil and plant. In soil, Cd may be adsorbed on the surface and in the pores of zeolite, resulting in a decrease in the metal availability, however, these processes are more pronounced at high application rates. Additional plant Si nutrition was shown to facilitate the root system development due to the fact that Si is a constituent of the root cap [21]. In our experiment, zeolite, despite the low rate, promoted the root growth even in comparison with the Cd free plants.

Some studies reported that the toxic effect of HMs, including Cd, on plants is associated with intensifying the production of reactive oxygen species (ROS) [22; 23]. Improved Si nutrition was found to reduce the ROS generation and stimulate enzymatic and non-enzymatic antioxidants [18]. It was supposed that mono- and polysilicic acids present inside or outside plant cell may precipitate or encapsulate and deactivate Cd [8; 20; 24]. As was determined, the concentrations of mono- and polysilicic acids in the plant sap exceed manifold those in the soil solution, this fact may induce intensifying Si-HM interactions inside plant [25].

4.Conclusions

The Cd-induced inhibition of barley growth was overcome by the soil-applied natural zeolite. The beneficial effect of zeolite on plant tolerance was attributed to increased root biomass. The use of natural zeolite at a low rate could be promising for increasing the efficacy of phytoremediation of Cd-contaminated areas.

5.Acknowledgement

This work was supported by Russian Ministry of Education and Science, themes # AAAA-A17-117030110139-9; AAAA-A18-118013190180-9; AAAA-A18-118013190181-6.

6.References

[1] Radulescu C, Stihi C, Popescu I, Toma L, Chelarescu E, Stirbescu R 2012 *J. Science and Arts* . **4(21)** 459
[2] Stupin D U 2009 *Soil pollution and newest restoration technologies*. St.Petersburg: Lan.
[3] Ali H, Khan E, Sajad M 2013 *Chemosphere* **91(7)** 869
[4] Badr N, Fawzy M, Al-Qahtani K M 2012 *World Appl. Sci. J.* **16** 1292

IOP Conf. Series: Materials Science and Engineering **921** (2020) 012003 doi:10.1088/1757-899X/921/1/012003

[5] Evangelou MW, Papazoglou E G, Robinson B H, Schulin R 2015 In: Ansari A A, Gill S S, Gill R, Lanza G R, Newman L eds. *Phytoremediation: management of environmental contaminants.* Switzerland: Springer 115
[6] Fu F, Wang Q 2011 *J. Environ Manage.* **92(3)** 407
[7] Wu J W, Shi Y., Zhu Y X, Wang Y C, Gong H J 2013 *Pedosphere* **23(6)** 815
[8] Matichenkov V V, Wei X, Liu D, Bocharnikova E A 2013 *Agric. Sci. Technol.* **14(3)** 498
[9] Ji X, Liu S, Huang J, Bocharnikova E, Matichenkov V 2016 *Chemosphere* **157** 132
[10] Shi W, Shao H, Li H, Shao M, Du S 2009 *J. Hazard. Mater.* **170** 1
[11] Chen Z S, Lee G J, Liu J C 2000 *Chemosphere* **41(1-2)** 235
[12] Hasanabadi T, Lack S, Ardakani M R, Ghafurian H, Modhej A 2015 *Biol Forum* **7(2)** 361
[13] Vrînceanu N O, Motelică D M, Calciu I, Tănase V, Preda M, Plopeanu G, Ivana I 2017 *AgroLife Scientific J.* **6(2)** 227
[14] Rehakova M, Cuvanova S, Dzivak M, Rimar J, Gavalova Z 2004 *Current Opinion in Solid State and Materials Science* **8** 397
[15] Mahabadi A A, Hajabbasi M A, Khademi H, Kazemian H 2007 *Geoderma* **137** 388
[16] Ahmed O H, Sumalatha G, Muhamad A N 2010 *Int. J. of Physical Science* **5(15)** 2393
[17] Campos Bernardi A C, Oliviera P P, de Melo Monte M B, Souza-Barros F 2013 *Microporous Mesoporous Mater.* **16** 16
[18] Balakhnina T, Bulak P, Matichenkov V, Kosobryukhov A, Wlodarczyk T 2015 *J Plant Growth Regul.* **75(2)** 557
[19] Ji X, Liu S, Juan H, Bocharnikova E A, Matichenkov V V 2017 *Environ Sci Pollut Res Int.* **24(11)** 10740
[20] Vaculík M, Lux A, Luxová M, Tanimoto E, Lichtscheidl I 2009 *Environ. Exp. Bot.* **67** 52
[21] Ma J F, Takahashi E 2002 *Soil, fertilizer, and plant silicon research in Japan* Amsterdam: Elsevier
[22] Balakhnina T I, Kosobryukhov A A, Ivanov A A, Kreslavskii V D 2005 *Russian J. Plant Physiology* **52(1)** 15
[23] Gratao P L, Polle A, Lea P J, Azevedo R A 2005 *Funct. Plant Biol.* **32(6)** 481
[24] Wei, X., Zhang, P., Zhao, D., Elena Bocharnikova, Vladimir Matichenkov, Dmitry Demin 2018 *Shengtai Xuebao/ Acta Ecologica Sinica* **38(5)**
[25] Matichenkov V V, Kosobryukhov A A, Biel K Y, Bocharnikova E A 2008 *Dokl. Biol. Sci.* **418(1)** 39

IOP Conf. Series: Materials Science and Engineering **921** (2020) 012004 doi:10.1088/1757-899X/921/1/012004

Technical solution for supplying a solvent of asphalt-resin-paraffin deposits (AFS) to a tubing string of oil wells equipped with electric centrifugal pumps.

Cheban S.E.[1], Valeev M.D.[2], Leontiev S.A.[3], Mayer A.V.[1], Samoilov A.R.[1]

[1]Yugra state university
[2]OOO NPP "VM system" Republic of Bashkortostan
[3]Tyumen state university

E-mail: cbanse@bk.ru

Abstract: For prevention and combating paraffin deposites different chemical regents are widely used (ihibitors and AFS solvents). In field conditions many ways of chemical regents dosing are applied. Injection of chemical reagents in oilwell tubing is the most effective for ARPD solvents; in this case consumption of reagents is largely decreasing by comparison with chemical reagents dosing through the well-casing annulus. The article describes design and operating processes of technology of ARPD-solvent dosing in electric centrifugal well pumps installation. This process opens fluid access to oilwell tubing at higher pressure at the wellhead. Reagent RT-1-3, is used as aromatic ARPD solvent. In RT-1-3there is butylbenzene fraction of Kazan Production Association Orgsintez containing a mixture of butylbenzene, isopropylbenzene and polyalkylbenzene.

1. Introduction

Every year the amount of new operational oil fields is being increased. The oils in these fields are marked with higher solidification point, viscosity and thixotropic properties. ARPD solvents appear inside oilwell tubings and decrease its throughput. At thickness 1 cm, pressure drop in oilwell tubing raises from 0.3 to 1 Mpa depending from its length [1 – 3]

To fight with paraffinization of well equipment different prevention and deposits termination methods are applied (also mechanical, thermal, chemical and combined methods). One of the ways of fighting with ARPD is using of solvents. This method is not universal and highly-effective because of necessity to match a solvent due to ARPD of the specific fields. [4 - 11].

2. Operating principle and description of technological automatic solvent feeder ARPD.

The conctruction of ECP is developed. The ECP opens access to water into oilwell tubing at increased pressure at the wellhead. Existing means [13-20] for pumping fluids into oilwell tubes from wellhead have some disadvantages, that do not allow to open check valve effectively.

Figure 1 illustrates the scheme of valve. The hull consists of an upper part and a lower part. In the body between lower (1) and higher (2) (not shown) there is thread connected stepped cylindrical

IOP Conf. Series: Materials Science and Engineering **921** (2020) 012004 doi:10.1088/1757-899X/921/1/012004

support (3) with gasket (4) in lower step and lock ring (5). In the lower part of upper support step (3) there are radial drains (6). The upper support end part (3) is covered by a bowl (7), above which there is a restrictor of its movement (8) with through-hole openings (9). The opening in upper end of support (3) makes a landing hole for the bowl (7). Inside of cylindrical support (3) there is a hermetically disposed water distributor (10) with through-hole radial drains (11), combined with radial drains (6) of support (3). Drain (11) is combined with vertical drain (12), the lower end of which is constructed in shape of saddle and covered by another bowl (13) , and supported by spring (14). Bowl 13 with spring 14 are situated in body 15 with through-hole drains 16. In the body 15, in lower part there is a regulating bolt 17 with movable plate 18.

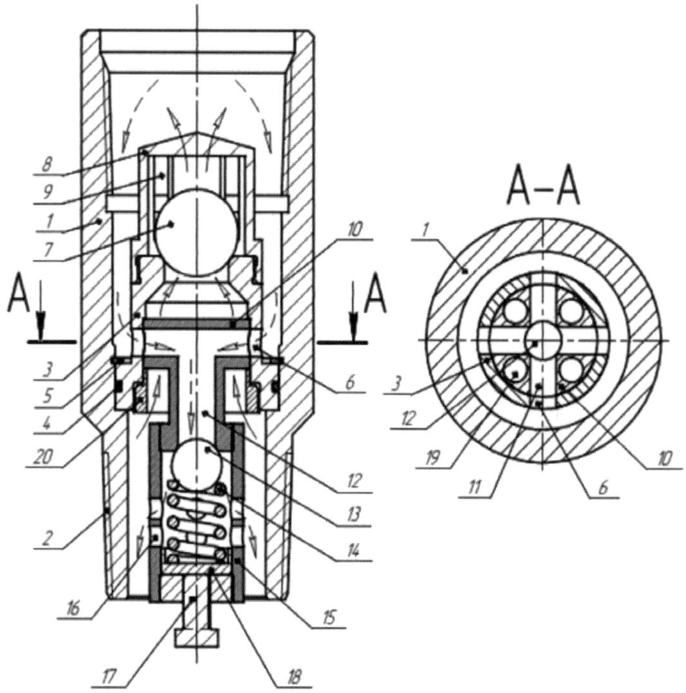

Figure 1 – valve scheme

In the body of water-distributing streams (10) there are through-hole openings (19). Water-distributing of streams (10) is holded in inner chamber (3) by checknut (20) with screw threaded connection.

The valve is installed in lower part of tubing string directly above submersible ESP (not shown in Figure 1)

The operating principle of valve.

At the ordinary operating of submersible ESP the extracted production of reservoirs after coming out of pump goes through the vertical drains (19) of water-distributor (10) and raising under the influence of pressure the bowl (7), goes into tubing through the openings (9) (solid arrow on the figure).

When deposites are generated in tubing or in operating details of submersible ESP, for ex. inorganic salts or mechanic dashes wirh ARPD, there appears significant decreasing of its delivery. It will be affected in information on reservoir gauging.

For sludge disposal well is being stopped by shutdown of ESP. After stopping the bowl (7) is placed in saddle under its own weight and blocks draining the liquid from tubing strings into the well through operating elements of the pump.

Then the calculated amount of solvent is injected into tubing string under overpressure. The solvent through radial drains (6,11) flows into valve and fixes a bowl (13) and creates a confined space in the pump without liquid circulation. In that way the dissolution of sediments with its following extraction happens.

Then the pump is launching. Well product lifts a bowl (7), goes into tubing string and flows on the surface into a system for gathering oil.

For securing the trouble-free operation of pump and effective disposal of draught the string 14 is fixed by screw 17 so that the bowl 13 overlaps its saddle at pump stopping. For solvent injecting into tubing string at the wellhead it is necessary to increase pressure up to estimated value, at which bowl 13 releases string 14 and starts flowing the liquid.

3. Conclusion

1. Solvent injection into tubing string effectively decreases consumption of reagents in comparison with solvent injection through the annular space of well.
2. Accumulating of impurities and deposites on filter and receiving part of the pump require ordinary pump washing with incoming fluid without chemical reagents.
3. Feasibility benefits of the valve are simplicity and reliability of valve opening, also an opportunity of sludge disposal from submersible equipment using less amounts of chemical reagents.

References

[1] N.G. Ibragimov Complications in oil production / A.R. Khafizov, V.V. Shaidakov and others; Ed. N.G. Ibragimova, E.I. Ishemguzhina.-Ufa: LLC "Publishing house of scientific and technical literature" Monograph "", 2003.-302p

[2] Galikeev R.M. Vsledovanie regularity of structure formation of paraffin-containing oils in the system of production and oil gathering / R.M. Galikeev: dis. ... tech. sciences. - Tyumen: TyuGNGU, 2011.- 113 p.

[3] Leontiev S. A. Resource-saving technologies in systems for collecting borehole products of oil fields (scientific generalization, research and implementation results) / S. A. Leontiev: dis. ... doctor. tech. sciences. - Tyumen: Tsogu, 2012. - 299 p.

[4] R.M. Galikeev Methodology for the study of chemical reagents for the prevention and dissolution of paraffin oil deposits of JSC Gazpromneft - Noyabrskneftegaz / S.A. Leontiev, V.V. Misnik // Oilfield business. - 2010- No. 9. - S. 36-39.,

[5] Glushchenko V.N. Prevention and elimination of asphaltene-tar-paraffin deposits / Glushchenko V.N., Silin M.A., Gerin Yu.G. // Oilfield chemistry. - M .: Intercontact Nauka, 2009 .-- V. V. - 475 p.

[6] Bortnikov EA On the results of laboratory modeling of the processes of interaction of formation fluid with injected water under conditions that simulate intensive fluid withdrawal from the formation / A. E. Bortnikov, K. E. Kordik, V. N. Moroz, S. A. Leont'ev, M.D. Valeev // Geology, geophysics and development of oil and gas fields. - 2015. - No. 2. –66-69 s.

[7] Galikeev R.M. Laboratory study of the rheological properties of oil and graphical determination of the temperature of the transition point / R.M. Galikeev, S.A. Leont'ev, M. Yu. Tarasov, E.V. Portnyagina // Territory of NEFTEGAZ. - 2010. - No. 8. –14-17 s.

[8] Mastobaev B.N. Chemical means and technologies in oil pipeline transportation / B.N. Mastobaev, A.M. Shammazov, E.M. Movsumzade. - M .: Chemistry, 2002 .-- 296s.

[9] Mazepa B.A. The fight against paraffin deposits during oil production abroad / B.A. Mazepa. - M .: Gostoptekhizdat, 1961 - 89 p.

IOP Conf. Series: Materials Science and Engineering **921** (2020) 012004 doi:10.1088/1757-899X/921/1/012004

[10] Kuchumov R.Ya. Analysis and modeling of the efficiency of operation of wells complicated by paraffin deposits / R.Ya. Kuchumov, M.F. Pustovalov, R.R. Kuchumov. - M.: VNIIONG, 2005 - 186 p.
[11] Mursalova M.A. Development and application of methods for controlling paraffin deposits in oil and gas fields / M.A. Mursalova, N.G. Efendiev, N.N. Kyazimova. - M.: VNIIEgazprom, 1986. - Issue. 12. - 48 p.
[12] Pump production of high viscosity oils from deviated and flooded wells. / K.P. Urazakov, E.I. Praying Mantis, J.C. Seitpagambetov. et al. Edited by M.D. Valeev. - M .: Nedra, 2003.-302 p.
[13] Minnivaleeva A.N. Improving the cleaning of tubing from asphalt-resin-paraffin deposits (ASPO) / A.N. Minnivaleeva, L.M. Zaripova, M.S. Gabdrakhimov, R.R. Ismagilov // Electronic Scientific Journal. Oil and gas business. 2013. -№2.-C.218-226.
[14] Valeev M.D. The statistical relationship between the nonequilibrium characteristics of oils and the content of paraffin hydrocarbons in them / M.D. Valeev, M.S. Gabdrakhimov, L.M. Zaripova, A.K. Zaripov // News of higher educational institutions. Oil and gas. -2018. -№ 3. -C. 108-113.
[15] Bogdanov A.A. Submersible centrifugal electric pumps for oil production. - M.: Nedra. 1968. 54 sec
[16] Patent RU No. 56940 U1, Non-return valve / Ulyanov S.S., Sagyndykov R.I., Mishchenko M.A., and others. // decl. 04/07/2006, publ. 09/27/2006.
[17] Patent RU 72268 U1 Valve for a borehole centrifugal pump / Gabsalyamov R.R., Smotryk D.V., Yakhin I.N //, decl. 12/26/2007, publ. 04/10/2008
[18] Patent RU 2441139 C1 Purified borehole filter / Kuneevsky V.V //. declared 07/12/2010. publ. 01/27/2012

IOP Conf. Series: Materials Science and Engineering **921** (2020) 012005 doi:10.1088/1757-899X/921/1/012005

Centrosymmetric and acentric whitlockite-type phosphates and vanadates among $Ca_3(VO_4)_2$-$Ca_3(PO_4)_2$-Y_2O_3 compositions

D V Deyneko[1], V V Titkov[1], S M Aksenov[2], S Yu Stefanovich[1] and B I Lazoryak[1]

[1] Department of Chemistry, Moscow State University, 119991, Moscow, Russia
[2] Laboratory of Nature-Inspired Technologies and Environmental Safety of the Arctic, Kola Science Centre, Russian Academy of Sciences, Apatity, 184209, Russia

E-mail: ecofond-oka@yandex.ru

Abstract. Identifications and studies of whitlockite-type phases for the first time are fulfilled in the cation-anion mixed phosphate-vanadate ferroelectric systems $Ca_{10.5}(V_{1-x}P_xO_4)_7$, $Ca_9Y(V_{1-x}O_4)_7$, $Ca_{10.5-1.5x}Y_x(VO_4)_7$ and $Ca_{10.5-1.5x}Y_x(PO_4)_7$. These systems are the sections of a more extensive multi-component system $Ca_3(VO_4)_2$ - $Ca_3(PO_4)_2$ - Y_2O_3, but are all built on the basis of well-known individual compounds of the type of mineral whitlockite. The solid solutions formed between them are also of ferroelectric whitlockite-type, but some compositions are not ferroelectric, and according to the nonlinear optical phase analysis possess centrosymmetric structure. Ferroelectric phase transitions were detected in the region of 1000–1300 K by the method of dielectric spectroscopy. A precision X-ray study of some compositions of solid solutions using synchrotron radiation revealed a significant deviation in the cationic-anionic arrangement from being completely disordered. Partial ordering of structural tetrahedral groups of PO_4^{3-} and VO_4^{3-} determines a significant increase in nonlinear optical activity in the mixed anions whitlockite-type compounds.

1. Introduction

Tricalcium vanadate $Ca_3(VO_4)_2$ [1] is a forefather of the extensive family of high-temperature ferroelectric (FE) and nonlinear optical phosphates and vanadates with the structure of the mineral whitlockite [2,3]. Like natural mineral whitlockite, all other representatives of this family regularly have a noncentrosymmetric structure described by a polar space group (sp. gr.) *R*3*c* [4]. The ferroelectric and nonlinear optical properties of members of this family are most pronounced in compounds $Ca_9M^{III}(XO_4)_7$, where X = P or V, and M^{III} – trivalent cation [2–7]. Moreover, vanadates are characterized by a higher optical nonlinearity and temperature of the FE phase transformations.

Compounds of the $Ca_9M^{III}(XO_4)_7$ group are capable of forming solid solutions both between themselves [8], and with tricalcium vanadate. It is more convenient to present its formula as $Ca_{10.5}(XO_4)_7$ [9]. This formula embraces all structural positions *M*1-*M*6 occupied by divalent and trivalent cations which can be clearly identified. These positions are divided into the closest octahedral site (*M*5), three more extensive three-fold sites (*M*1, *M*2 and *M*3) and two additional sites (*M*4 and *M*6) [10]. The *M*4 and *M*6 sites can be either both empty, as in $Ca_9M^{III}(XO_4)_7$, or the *M*4 site is partially occupied, in the case when $M^{III} \rightarrow Ca$, so $Ca_{10.5}(VO_4)_7 \equiv Ca_3(VO_4)_2$. All whitlockite-type compounds with the latter formula are certainly noncentrosymmetric, sp. gr. *R*3*c*, and the compounds themselves are ferroelectrics [11]. It should be noted that, above the FE phase transition, they transform into a centrosymmetric ion-conducting state, corresponding to the through motion of Ca^{2+} through the *M*4 and *M*6 sites, which are equally occupied by Ca^{2+} and become structurally identical [3]. From a structural point of view, whitlockite-type compounds $Ca_9M^{III}(XO_4)_7$ have more prerequisites for a centrosymmetric structure, since *M*4 and *M*6 positions are equally empty. The temperatures of transitions FE - centrosymmetric paraelectric phase in this group are lower than in $Ca_3(XO_4)_2$ (X = P,V) [12]. In addition, among the compounds $Ca_9RE(PO_4)_7$, where M^{III} are represented by rare earth elements,, there are a significant number of antiferroelectrics with centrosymmetric structures even under normal conditions [2]. Such centrosymmetric whitlockite-type phosphates have no analogues among vanadates.

In this contribution we report on synthesis and characterization of whitlockite-type phases in the phosphate and vanadate systems, including the mixed in the anionic and cationic parts: $Ca_{10.5}(V_{1-x}P_xO_4)_7$, $Ca_9Y(V_{1-x}P_xO_4)_7$, $Ca_{10.5-1.5x}Y_x(VO_4)_7$ and $Ca_{10.5-1.5x}Y_x(PO_4)_7$. Special attention is paid to determining the presence or absence of an inversion center in their crystal structure. The latter, as is known, determines the possibility of the formation of nonlinear optical and ferroelectric properties in substances.

2. Experimental section

2.1. Materials and general methods

The solid solutions $Ca_{10.5}(V_{1-x}P_xO_4)_7$, $Ca_9Y(V_{1-x}P_xO_4)_7$, $Ca_{10.5-1.5x}Y_x(VO_4)_7$ and $Ca_{10.5-1.5x}Y_x(PO_4)_7$ were synthesized by solid state route from stoichiometric amounts $CaCO_3$ (99.99%), $Ca(HPO_4)_2{\bullet}2H_2O$ (99.99%), NH_4VO_3 (99.98%) and Y_2O_3 (99.9%) in air at 1273 K for 150 hours with intermediate grinds every 50 hours.

Powder X-ray diffraction (PXRD) patterns were collected on a Thermo Termo ARL powder diffractometer (Cu $K\alpha_1+\alpha_2$), Bragg–Brentano geometry, Peltier-cooled CCD detector at RT in a range of $2\theta = 5°–65°$ with a step of 0.02°. The second harmonic generation (SHG) signal was measured with a Q-switched YAG:Nd laser at λ_ω = 1064 nm in the reflection mode for powder series. The experimental set-up and arrangement have been described in [13]. The powders were ground in ethanol for 30 minutes before the formation of a suspension and then dried in air. The similar procedures were used to prepare quartz powders, which were used as references. To separate the powders into fractions according to crystallite size, time-controlled deposition of powders in an ethanol solution was used. Certification of particle size distribution was carried out by constructing histograms of light scattering of powders in solutions, measured using a laser analyzer CILAS 1180.

For electrical properties, examinations the ceramic pellets with 8 mm in diameter and 2–3 mm thickness were prepared. Electrodes were applied to the flat surfaces of the disks by burning colloidal platinum containing paste SPI-Chem. The upper heat treatment temperature was 1273 K, temperature exposure was 1 hour. The dielectric constant and dielectric loss measurements were carried out by a two-contact method using Novocontrol Beta-N Impedance Analyser with Probostar A-cell at frequencies from 0.3 Hz to 1 MHz in a 293–1273 K temperature range.

Synchrotron PXRD data for $Ca_9Y(P_{1-x}V_xO_4)_7$, x = 0 - 1 and $Ca_{10.5}(P_{0.5}V_{0.5}O_4)_7$ were measured with a large Debye-Scherrer camera at the BL15XU beamline of SPring-8. The intensity data were collected in a 2θ range from 1° to 62° with a step of 0.003°. The incident beam was monochromatized at λ = 0.65298 Å. The sample was packed into a Lindemann glass capillary with the inner diameter of 0.1 mm, that was rotated during the measurement. The absorption coefficient was also measured. Rietveld analysis [14] was performed using JANA2006 programme package [15].

3. Results and Discussion

3.1. System $Ca_{10.5-1.5x}Y_x(PO_4)_7$

Obtained XRD patterns for $Ca_{10.5-1.5x}Y_x(PO_4)_7$ compositions with x = 0, 0.167, 0.333, 0.500, 0.667, 0.833, 1 are similar to more simple whitlockite-type compounds [2-7], all reflections in the XRD were in accordance with JCPDSPDF-2 №46-402 data. The unit cell parameters a, c and V are given in table 1. The volume V of the unit cell shows non-monotonic change during $Ca^{2+} \rightarrow Y^{3+}$ substitution.

Although substitution with smaller cations usually leads to a decrease in unit cell parameters [16,17], in this case the replacement adduct to formation of vacancy in the *M*4 site according to the scheme $3Ca^{2+} = 2Y^{3+} + \square$ [18]. So, a decreasing of the unit cell parameters is observed, as a result.

Table 1. Unit cell parameters, SHG signal and phase transition temperatures for $Ca_{10.5-1.5x}Y_x(PO_4)_7$

x	a, Å	c, Å	V, Å^3	$I_{2\omega}/I_{2\omega}(SiO_2)$	T, K
0.000	10.4024(1)	37.380(1)	3503.0(1)	1.85	1203
0.167	10.4182(3)	37.384(2)	3514.0(2)	1.75	1159
0.333	10.4323(1)	37.376(1)	3522.8(1)	1.35	1123
0.500	10.4297(2)	37.386(7)	3522.0(3)	1.20	1073
0.667	10.4306(1)	37.376(1)	3519.0(2)	0.70	1043
0.833	10.4346(1)	37.390(2)	3525.6(2)	0.35	-
1.000	10.4309(1)	37.360(2)	3520.3(1)	0.3	863

Taking into account smaller ionic radii of the Y^{3+} (r_{VI} = 0.89 Å) in comparison with Ca^{2+} (r_{VI} = 1.00 Å), the yttrium ions occupy smaller sites, ranking from the sequence *M*5 > *M*1 > *M*2 > *M*3. The range of substitution in the *M*3 site is the smallest: *M*3 < 0.01 [19]. The occupation of *M*5 site by Y^{3+} ions increases the lattice cell, while the location in the *M*1, *M*2 sites reduces them [20]. Thus, cell volume changes are associated with the redistribution of substitutional cations in crystallographic positions.

The SHG study on powders with a grain size of 5 μm showed low-intensity SHG signals characteristic to phosphates compounds (table 1). The signal intensity for $Ca_9Y(PO_4)_7$ coincides with the references data, and monotonically decreases in the solid solution with declining yttrium concentration.

For compound with x = 0.667 temperature dependence of SHG signal was measured (figure 1). The SHG signal intensity gradually decreases with rising temperature until it disappears. The declining of the response occurs at temperature T_c, corresponding to the transition from a noncentrosymmetric to a centrosymmetric state. Subsequent cooling will restore the signal. A smooth decay of the signal near T_c shows that this transition is close to a second-order type. This fact complicates the study of phase transition by scanning calorimetry method.

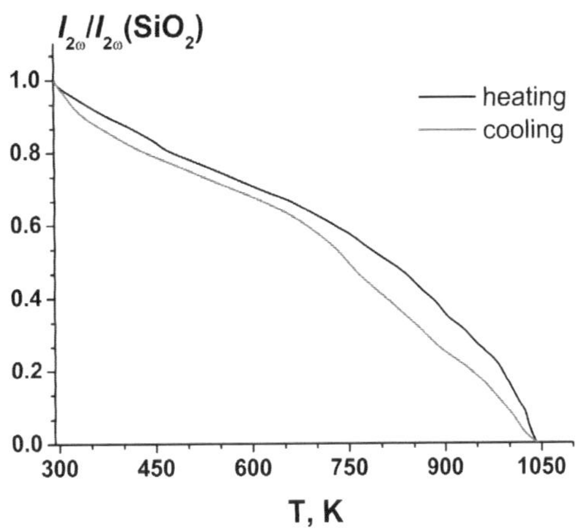

Figure 1. Temperature dependences of SHG signal for $Ca_{10.5-1.5x}Y_x(PO_4)_7$, x = 0.667 in the heating-cooling circle.

The temperatures of phase transitions in the $Ca_{10.5-1.5x}Y_x(PO_4)_7$ solid solution decreasing with enrichment of yttrium content, which corresponds to the previously obtained results [2]. The temperatures of phase transitions obtained by the SHG method are given in the table 1.

The temperature of phase transitions was also measured by dielectric spectroscopy, and the ferroelectric nature of solid solution was confirmed. The temperature dependences of the dielectric constant and the dielectric loss tangent of $Ca_{10.5-1.5x}Y_x(PO_4)_7$ solid solution are shown in the figure 2. The dielectric constant ε has a maximum in the range 1163-1229 K. Though temperature positions of the maximum do not depend on the frequency of the electric field, the peaks are significantly broadened, similar to a diffuse phase transition in $Ca_3(PO_4)$ [21]. As a whole, the forms of the ε(T) and the tanδ(T) curves correspond to the behavior of ferroelectrics.

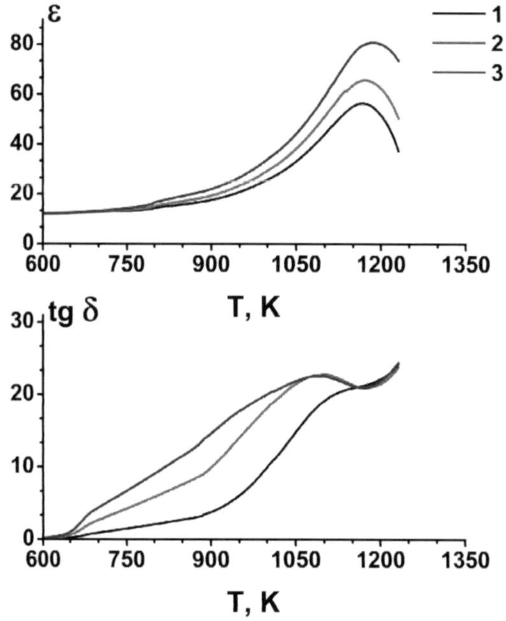

Figure 2. Dielectric constant ε and the loss tangent for $Ca_{10.5-1.5x}Y_x(PO_4)_7$ ($x = 0.167$) at the frequency of the measuring field: 750 kHz (1), 1.5 MHz (2) and 3 MHz (3).

3.2. System $Ca_{10.5-1.5x}Y_x(VO_4)_7$

All the compounds obtained in this system have unit cell parameters (*a*, *c*, and *V*) close to those of whitlockite vanadates $Ca_9RE(VO_4)_7$ and the characteristic spatial symmetry of the *R*3*c* group, confirmed by the presence of the second harmonic generation. Unit cell parameters increase with increasing content of yttrium (table 2). The SHG study was carried out on samples of the same dispersion (20 μm). The data are given in table 2. The signal intensity for $Ca_3(VO_4)_2$ was 8 units of quartz standard, nonlinearly increasing with increasing of Y^{3+} content to 60 units.

Table 2. The unit cell parameters and SHG signal of $Ca_{10.5-1.5x}Y_x(VO_4)_7$

x	a, Å	c, Å	V, Å^3	$I_{2\omega}/I_{2\omega}(SiO_2)$
0	10.7929(2)	37.922(2)	3825.7(1)	8
0.167	10.8048(8)	37.934(3)	3835.3(5)	12
0.333	10.8098(6)	37.917(2)	3837.1(4)	13
0.500	10.8105(2)	37.895(2)	3835.3(1)	18
0.667	10.8197(2)	37.908(1)	3843.2(1)	40
0.833	10.8308(2)	37.907(5)	3850.9(3)	50
1.000	10.8369(9)	37.916(3)	3856.2(6)	60

The dielectric constant for $Ca_{9.5}Y_{0.667}(VO_4)_7$ has a sharp temperature maximum at 1223-1273 K, which is characteristic of the FE phase transition. The position of the maximum does not depend on the frequency of the electric field. The temperature minimum on the tanδ(T) curve precedes the ε maximum, which also corresponds to the ferroelectric behavior.

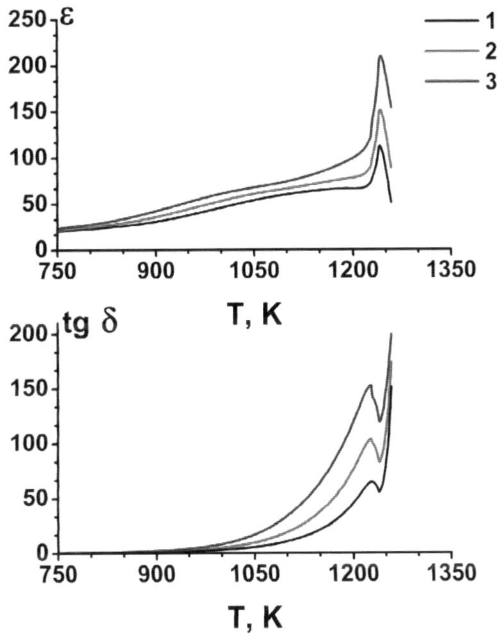

Figure 3. Dielectric constant ε and the loss tangent for $Ca_{9.5}Y_{0.667}(VO_4)_7$ at the frequency of the measuring field: 93 kHz (1), 187 kHz (2) and 375 kHz (3)

The phase transition temperature of $Ca_{9.5}Y_{0.667}(VO_4)_7$ is just between T_c = 1383 K for $Ca_{10.5}(VO_4)_7$ and $Ca_9Y(VO_4)_7$ (1213 K). Taking into account the structural mechanism of ferroelectric-to-paraelectric phase transition in the whitlockites [6,20], one may conclude that an additional population of *M*4 position by calcium atom prevents from free reorientation of the $X1O_4$ tetrahedron which is adjacent to *M*4 positions. In such neighborhood reorienation of $X1O_4$ tetrahedron to cetrosymmetry requires more thermal energy, and therefore the phase transition into centrosymmetric paraelectric phase in $Ca_{10.5}(VO_4)_7$ is higher in temperature.

3.3. *System* $Ca_{10.5}(P_xV_{1-x}O_4)_7$

Fragments of XRD patterns of $Ca_{10.5}(P_xV_{1-x}O_4)_7$ (x = 0, 0.2, 0.4, 0.5, 0.6, 0.8, 1) presented in the figure 4. All samples obtained are single phase and do not contain additional phases besides similar to $Ca_3(PO_4)_2$ and $Ca_3(VO_4)_2$ (JCPDS №48-247). Dependences of unit cell parameters *a, c* and volume *V* from vanadium concentration are shown in figure 4. The growth of all three parameters with *x* is a sequence of the incorporation larger in compared with phosphorus vanadium atom into the structure.

Table 3. Unit cell parameters and SHG values for powders of different dispersion in solid solution $Ca_{10.5}(P_xV_{1-x}O_4)_7$.

x	*a*, Å	*c*, Å	*V*, Å^3	SHG (17 μm)	SHG (3 μm)
0.0	10.7922(9)	37.928(5)	3825.8(9)	8.0	4
0.2	10.7124(3)	37.760(1)	3752.7(2)	14	5
0.4	10.6259(3)	37.607(1)	3677.4(2)	16	7
0.5	10.5738(8)	37.519(3)	3632.9(6)	15	6
0.6	10.5251(2)	37.449(1)	3592.8(2)	7.0	5
0.8	10.4439(3)	37.379(1)	3530.9(2)	2.5	1.3
1.0	10.3677(7)	37.201(2)	3463.1(4)	1.2	0.6

All obtained compounds have the sp. gr. *R*3c usual for most of whitlockite-type compounds. The indication of powder patterns led to a hexagonal whitlockite cell (table 3). The nonmonotonic nature of the change in the parameters $Ca_{10.5}(P_xV_{1-x}O_4)_7$ near x = 0.5–0.6 is noteworthy. Apparently, in this concentration range, the distribution of phosphorus and vanadium over possible anionic positions E1-E3 is mostly ordered. This also corresponds to a marked increase in the nonlinear optical activity of the solid solution, which is clearly manifested on powders with different crystallite sizes (table 3). It is significant also that the luminescence of guest europium atoms in this system [22] has a maximum intensity near x = 0.7. The destruction of the V – V relaxation mechanisms proposed in [22] is probably due to an increase in the distance between the vanadium – oxygen tetrahedra as a result of their alternation with the phosphorus – oxygen.

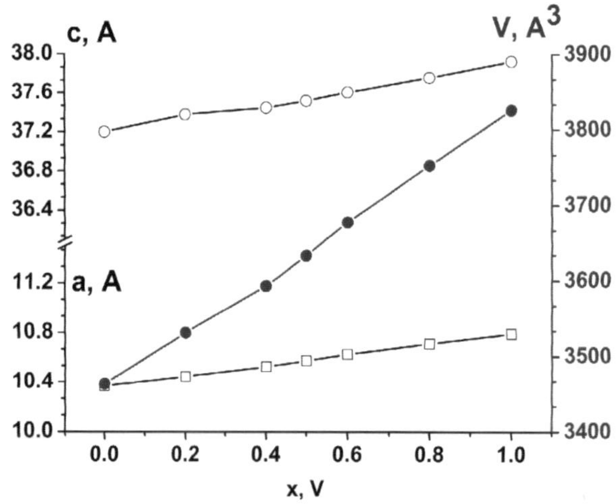

Figure 4. Concentration dependences of unit cell parameters *a*, *c*, *V* in $Ca_{10.5}(P_xV_{1-x}O_4)_7$.

IOP Conf. Series: Materials Science and Engineering **921** (2020) 012005 doi:10.1088/1757-899X/921/1/012005

To refine the crystal structure of $Ca_{10.5}(P_{0.5}V_{0.5}O_4)_7$, the structure of calcium phosphate was used as an initial model [23]. At the first stage, f-curves for phosphorus were calculated and the coordinates of calcium were refined. Next, vanadium atoms were introduced and bound by coordinate and atomic displacement parameters. The coordinates of all atoms were clarified. The refinement R-factors show good agreement between the calculated and experimental curves. The adjusted unit cell parameters, taking into account the differences in the chemical composition of the samples, are in satisfactory agreement with the data in figure 4. The calculated composition is close to theoretical.

The cations of phosphorus and vanadium occupy all three positions E1, E2, E3. However, the position distribution of the uneven positions E1, E2 is predominantly occupied by vanadium cations, while phosphorus cations preferably occupy the positions E3, E2. In calcium phosphate $Ca_3(PO_4)_2$ and vanadate $Ca_3(VO_4)_2$, the average distances E1-O, E2-O, E3-O are approximately the same (1.540 Å, 1.538 Å, 1.536 Å – for phosphate and 1.691 Å, 1.700 Å, 1.694 Å – for vanadate). However, for $Ca_{10.5}P_{3.451}V_{3.549}O_{28}$ significant asymmetry is observed: predominantly occupied by vanadium E1, E2 are larger than E3 (E3-O = 1.602 Å). The scatter of bond lengths in one tetrahedron is 0.06-0.10 Å, instead of 0.02-0.03 Å for extreme compounds. The asymmetric distribution over the positions, as well as a significant distortion of the bond lengths, leads to an increase in optical nonlinearity.

3.4. System $Ca_9Y(P_{1-x}V_xO_4)_7$

A series of solid solutions in the $Ca_9Y(P_{1-x}V_xO_4)_7$ system begins and ends with double-phosphate and vanadate calcium-yttrium having the polar whitlockite-type structure. It was previously assumed that between them there is a continuous series of solid solutions. However, large single crystals of compositions close to $x = 0.5$ grown by the Czochralski method [8] showed no SHG effect. This indicated that the crystals are centrosymmetric (CS). It was of interest to study in detail the structural features of the samples in this system, provided they were obtained by standard solid-state synthesis. For a more reliable refinement of their structure of powders in the system $Ca_9Y(P_{1-x}V_xO_4)_7$, x = 0. 0.1, 0.2, 0.3, 0.4, 0.5, 0.6, 0.7, 0.8, 0.9 and 1, their X-ray diffraction patterns were obtained using synchrotron radiation. A detailed X-ray phase analysis showed that all the obtained samples contained a small amount of YVO_4 impurity. The position of the peaks in the diffraction patterns is close or coincides with their location for $Ca_9Y(PO_4)_7$ and $Ca_9Y(VO_4)_7$. Table 4 shows the unit cell parameters a, c and volume V from the vanadium concentration. All three parameters naturally rising with increasing x, which is due to the larger size of the vanadium anions in comparison with phosphorus.

Table 4. Unit cell parameters, SHG and T_c values for $Ca_9Y(P_{1-x}V_xO_4)_7$ system ($0 \leq x \leq 1$).

y	SG	a, Å	c, Å	V, Å^3	SHG	T_c, K
0	$R3c$	10.43116(3)	37.3335(1)	3517.98(1)	0.30	823
0.1	$R3c$	10.47607(4)	37.3354(1)	3548.53(2)	0.15	-
0.2	$R\bar{3}c$	10.52250(9)	37.3736(3)	3583.72(5)	0	CS-phase
0.3	$R\bar{3}c$	10.56593(6)	37.4111(2)	3616.99(3)	0	CS-phase
0.4	$R\bar{3}c$	10.60703(5)	37.4648(2)	3650.41(2)	0	CS-phase
0.5	$R\bar{3}c$	10.64756(5)	37.5203(2)	3683.81(3)	0.15	CS-phase
0.6	$R3c$	10.68738(2)	37.5978(2)	3719.07(1)	0.35	-
0.7	$R3c$	10.72038(4)	37.6555(1)	3747.83(2)	0.8	-
0.8	$R3c$	10.74430(3)	37.7014(1)	3769.15(8)	1.8	1123
0.9	$R3c$	10.78648(7)	37.7998(3)	3808.72(4)	4.0	1193
1.0	$R3c$	10.83206(1)	37.8877(5)	3849.92(1)	6.0	1207

Weakening to null of the SHG signal intensity for mixed yttrium calcium phosphate vanadates dramatically contrasts in comparison with its augmentation with described above $Ca_{10.5}(P_xV_{1-x}O_4)_7$ system, and makes it possible to reliably establish the existence of a centrosymmetric whitlockite-type phase at $0.2 \leq x \leq 0.5$. This centrosymmetric phase exists only in yttrium-rich compositions.

According to SHG data even 15-20 at.% shortage of yttrium leads to disappearance of centrosymmetry at the P^{5+}-V^{5+} substitution.

SHG intensities for the non-centrosymmetric compounds $x = 0, 0.1, 0.2, 1$ demonstrate temperature dependences typical for ferroelectrics (figure 5). With increasing temperature, the SHG gradually decreases until disappearance at the Curie point (T_c) corresponding to the FE transition into the centrosymmetric paraphase. Upon subsequent cooling, the signal is reversibly restored. The smooth decay of the signal near T_c for $Ca_9Y(VO_4)_7$ ($x = 0$) shows that this transition is close to the second order FE transition found on the single crystal in [20]. As phosphorus enrichment occurs, the T_c this decreases. As evidenced by a stepwise change in the SHG near the T_c, the phase transition acquires the features of a first-order transformation. A slight change in SHG with temperature up to a sharp disappearance of the effect at $T = T_c$ in the case of $Ca_9Y(PO_4)_7$ ($x = 1$) indicates a first-order transition. The temperatures of phase transitions obtained by the SHG method are given in the table 4.

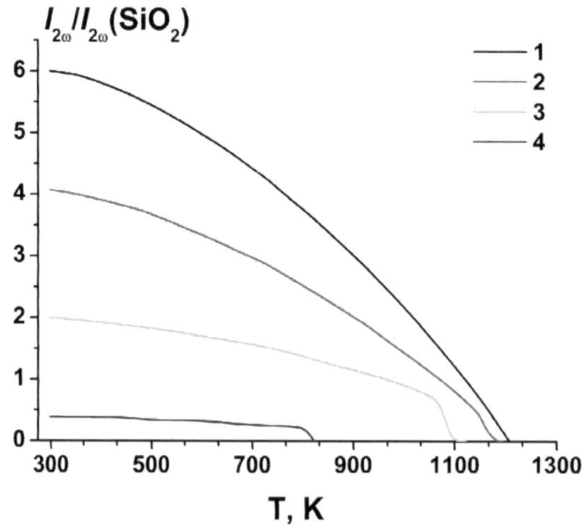

Figure 5. SHG temperature dependences for noncentrosymmetric compositions in the system $Ca_9Y(P_{1-x}V_xO_4)_7$, $x = 1$ (1), 0.9 (2), 0.8 (3), 0 (4).

The crystal structure of the whitlockite-type phase was refined for two compositions according to the SHG data located in the centrosymmetric region $0.2 \leq x \leq 0.5$.

The structure of $Ca_9Y(PO_4)_{4.511}(VO_4)_{2.489}$ was refined in two models: centrosymmetric (CS) and noncentrosymmetric (NCS). The initial NCS model was $Ca_9Y(VO_4)_7$, for which the coordinates of all large cations were refined based on the measured array. Then, f-curves for Ca and V were constructed and yttrium and phosphorus atoms were introduced, linking them with the corresponding atoms. The final refinement of all atoms to the R-factor value of 3.16 was carried out. As the initial CS model, we used the data for $Ca_9Y(P_{4/7}V_{3/7})O_7$ single crystals from [8]. The final refinement of all atoms to the R-factor value of 3.03 was carried out.

A similar model calculation was performed for the composition $Ca_9Y(PO_4)_{3.5}(VO_4)_{3.5}$, the structure of which was refined in the $Ca_9Y(P_{4/7}V_{3/7})O_7$ model [8]. For both compositions, the best agreement between the calculated and experimental data was obtained for the CS model. The choice of the CS model also relies on the absence of a noticeable SHG effect in powders of the corresponding compounds.

4. Conclusion

Structural features, corresponding to the transformation from NCS to CS type structure are of great scientific and practical interest. Previous study shown that mixed vanadate-phosphate is a suitable matrix for synthesis of new CS-type whitlockites. Present investigation confirms that the problem of a stabilization of CS-type structure may be solved using only cation ordering in the whitlockite-type host. It is demonstrated that among whitlockite-type compounds with NCS structure in the system $Ca_3(VO_4)_2$-$Ca_3(PO_4)_2$-Y_2O_3 there is a specific concentration region where CS phase is stabilized at room temperature. This fact greatly relieves future examination of crystal structure characteristics of paraelectric phases for a number of whitlockite-type ferroelectrics with ferroelectric-to-paraelectric phase transitions in the region of 1000 K or above.

The sections studied in this paper $Ca_{10.5}(V_{1-x}P_xO_4)_7$, $Ca_9Y(V_{1-x}P_xO_4)_7$, $Ca_{10.5-1.5x}Y_x(VO_4)_7$ and $Ca_{10.5-1.5x}Y_x(PO_4)_7$ are characterized by homovalent substitutions in the anionic part (P - V) and heterovalent in the cationic part (Ca-Y). Heterovalent substitutions are accompanied by a change in the filling of the *M*4 cation position near the XO_4-tetrahedron, whose polar orientation determines the non-centrosymmetric structure in the ferroelectric phase. The temperature of transition from the ferroelectric to the paraelectric phase (T_c) depends on the stereochemical possibility of reorientation of half of the XO_4 tetrahedra. This possibility is limited if a calcium cation is located in the nearby *M*4 position. This results in higher T_c for $Ca_{10.5}(XO_4)_7$ (X = P, V) against a plurality of $Ca_9M^{III}(XO_4)_7$ compounds with significantly lower ferroelectric transition temperatures. Note also that filling the *M*4 position due to electrostatic compression of the corresponding oxygen polyhedron reduces the volume of the entire unit cell, which also prevents its symmetrization and shifts the existence of the CS phase to the high temperature region.

The systems with isovalent anionic substitution of P^{5+}-V^{5+} studied in this work make significant refinements of the "composition-structure-properties" correlations described above. Systems $Ca_9Y(P_{1-x}V_xO_4)_7$ and $Ca_{10.5}(P_{1-x}V_xO_4)$ demonstrate completely different behavior of the unit cell parameters, intensity of the SHG, and phase transitions temperatures. A continuous solid solution of a whitlockite-like structure with a polar space group *R*3*c* is formed in the $Ca_{10.5}(P_{1-x}V_xO_4)$. The P^{5+}/V^{5+}- mixing leads in this system to leads to an increase in the second harmonic effect with a pronounced maximum at x=0.6. Regular increase in optical nonlinearity is not here unexpected, since it is achieved due to the greater electron polarizability of vanadium ions with respect to phosphorus. The presence of the maximum SHG in the system $Ca_{10.5}(P_{1-x}V_xO_4)_7$ at $x = 0.6$ can be associated with the initial localization of vanadium chiefly in the T1 position, primarily responsible for the nonlinear optical susceptibility of the whitlockites [13,24].

It may be resumed that the CS-type state in whitlockites in the $Ca_3(VO_4)_2$-$Ca_3(PO_4)_2$-Y_2O_3 system is controlled by a partial ordering of V^{5+} and P^{5+} cations among tetrahedral sites with further orientational disorder of T1-tetrahedron around center of symmetry. That becomes possible because of the absences of any cations in *M*(4)-site and increasing of the unit-cell parameters by the insertion of V cations. However, the field of stability of CS state decreases with the Y→Ca substitution. The further structural investigations of compounds belonging to $Ca_3(VO_4)_2$-$Ca_3(PO_4)_2$-Y_2O_3 system with the structural solutions are necessary to understand the crystal chemical features of CS-type whitlockite.

Acknowledgments
The study was supported by RFBR (Project № 20-03-00929).

References

[1] Glass A M, Abrahams S C, Ballman A A and Loiacono G 1977 *Ferroelectrics* **17** 579–82
[2] Teterskii A V, Morozov V A, Stefanovich S Y and Lazoryak B I 2005 *Russ. J. Inorg. Chem* **50** 986–9
[3] Lazoryak B I, Belik A A, Stefanovich S Y, Malakho Morozov V A, A P, Shel'menkova, O V, Leonidov, I A, Leonidova and O N 2002 *Dokl. Akad. Nauk SSSR* **384** 780–5
[4] Kim M S, Lah M S and Kim H K 2002 *Bull. Korean Chem. Soc.* **23** 98–102
[5] Lazoryak B I, Morozov V A, Belik A A, Stefanovich S Y, Grebenev V V., Leonidov I A,

Mitberg E B, Davydov S A, Lebedev O I and Van Tendeloo G 2004 *Solid State Sci.* **6** 185–95
[6] Lazoryak B I, Aksenov S M, Stefanovich S Y, Dorbakov N G, Belov D A, Baryshnikova O V., Morozov V A, Manylov M S and Lin Z 2017 *J. Mater. Chem. C.* **5** 2301
[7] Loiko P A, Yasukevich A S, Gulevich A E, Demesh M P, Kosmyna M B, Nazarenko B P, Puzikov V M, Shekhovtsov A N, Kornienko A A, Dunina E B, Kuleshov N V. and Yumashev K V. 2013 *J. Lumin.* **137** 252–8
[8] Kosmyna M B, Mateychenko P V., Nazarenko B P, Shekhovtsov A N, Aksenov S M, Spassky D A, Mosunov A V. and Stefanovich S Y 2017 *J. Alloys Compd.* **708** 285–93
[9] Belik A A, Izumi F, Ikeda T, Lazoryak B I, Morozov V A, Malakho A P, Stefanovich S Y, Grebenev V V, Shelmenkova O V, Kamiyama T, Oikawa K, Leonidov I A, Leonidova O N and Davydov S A 2002 *Phosphorus. Sulfur. Silicon Relat. Elem.* **177** 1899–902
[10] Lazoryak B I 1996 *Usp. Khim.* **65** 307–25
[11] Zhukovskaya E S, Deyneko D V., Baryshnikova O V., Belik A A, Leonidov I I, Ishchenko A V., Stefanovich S Y, Morozov V A and Lazoryak B I 2019 *J. Alloys Compd.* **787** 1301–9
[12] Dorbakov N G, Grebenev V V, Titkov V V, Zhukovskaya E S, Stefanovich S Y, Baryshnikova O V, Deyneko D V, Morozov V A, Belik A A and Lazoryak B I 2018 *J. Am. Ceram. Soc.* **101**
[13] Beskorovaynaya D A, Deyneko D V., Baryshnikova O V., Stefanovich S Y and Lazoryak B I 2016 *J. Alloys Compd.* **674** 323–30
[14] Rietveld H M 1969 *J. Appl. Crystallogr.* **2** 65–71
[15] Petricek V, Dusek M and Palatinus L 2014 *Z. Krist.* **229** 345–352
[16] Teterskii A V., Stefanovich S Y, Lazoryak B I and Rusakov D A 2007 *Russ. J. Inorg. Chem.* **52** 308–14
[17] Deineko D V., Stefanovich S Y, Mosunov A V, Baryshnikova O V and Lazoryak B I 2013 *Inorg. Mater.* **49** 507–12
[18] Brixner L H and Flournoy P A 1965 *J. Electrochem. Soc.* **112** 303
[19] Golubev V N and Lazoryak B I 1991 *Russ. J. Inorg. Mater.* **27** 480–3
[20] Lazoryak B I, Deyneko D V., Aksenov S M, Stefanovich S Y, Fortalnova E A, Petrova D A, Baryshnikova O V., Kosmyna M B and Shekhovtsov A N 2018 *Zeitschrift fur Krist. - Cryst. Mater.* **233** 453–62
[21] Deyneko D V., Morozov V A, Zhukovskaya E S, Nikiforov I V., Spassky D A, Belik A A and Lazoryak B I 2020 *J. Alloys Compd.* **815** 152352
[22] Zhou X and Wang X 2013 *Russ. J. Phys. Chem. A* **87** 1859–63
[23] Yashima M, Sakai A, Kamiyama T and Hoshikawa A 2003 *J. Solid State Chem.*
[24] Deyneko D.V., Nikiforov I.V., Spassk y A.D., Dikhtyar Y.Y, Aksenov S.M., Stefanovich S.Y, Lazoryak B.I. CrystEngCom 2019, **21**.5235-42

IOP Conf. Series: Materials Science and Engineering **921** (2020) 012006 doi:10.1088/1757-899X/921/1/012006

The influence of rheological properties of diesel fuel in the pumping process for collapsible pipelines in low temperature conditions

D A Drozdov, K M Plotnikova, L V Seoev, A V Elkin

Federal Autonomous Establishment «The 25-th State Research Institute of Himmotology, Ministry of Defence of Russian Federation», 10 Molodogvargeyskaya Street, Moscow, Russia

E-mail: nio21-25gosnii@yandex.ru

Abstract. The results of studies of the influence of rheological properties of diesel fuels on the pumping process for collapsible pipelines in low temperature conditions. It is established that diesel fuel when they are cooled below the solidification point represent a viscoplastic medium with curves of current, which describes a universal equation of Buckley-Herschel. A refined method for hydraulic calculation of collapsible tubing to determine the head losses to reflect changes in the rheological properties of diesel fuels at low temperatures. The simulation of the processes of transfer of winter diesel fuel for collapsible pipelines in low temperature conditions and offers recommendations for their application in low-temperature conditions.

Introduction

Policy of realization of strategy of development of the Arctic zone of the Russian Federation and ensuring national security in the Arctic involves the development of transport infrastructure, which entails the organization of the system of fuel supply that can operate at low temperatures down to - 60 °C. The Solution of the problem of fuel supply can be realized with the use of collapsible pipelines.

Experience in the application of collapsible pipelines shows that the transportation of diesel fuels at temperatures below their temperature of filterability possible. While there is a significant deviation of the actual parameters of the pipeline design due to the high viscosity of the pumped product.

The existing method of calculation of parameters of the work collapsible pipelines does not influence the rheological characteristics of the fuels on the process of their transportation in conditions of low temperatures, which leads to inaccurate results of the hydraulic characteristics. The current method of determination of hydraulic losses, based on the use of kinematic viscosity of fuels passport data, has sufficient accuracy, which also leads to inaccurate assessment of the hydraulic characteristics of the pumping process at low temperatures. The influence of the viscosity of the fuel in the collapsible pipelines mode of operation in low temperatures causes the actuality of researches of rheological properties of diesel fuels, determining the rheological model of viscous fluid needed for the hydraulic calculation of collapsible pipelines and reliable operation of the pipeline equipment, and development of refined methods for the assessment of the hydraulic characteristics of the process of transportation of diesel fuels on collapsible pipelines in the conditions of low temperatures.

As objects of research were selected 13 samples of diesel fuel (DF) of different plants. In tables 1-3 lists the actual values of the main physico-chemical and operational parameters of the test specimens of diesel fuel.

Table 1. Actual values of the main physico-chemical and operational performance of diesel fuel EURO Class 3 and Class C in accordance with the requirements of GOST R 52368-2005

Name of the indicator	Brand diesel fuel				
	Sample No. 1	Sample No. 3	Sample No. 5	Sample No. 6	Sample No. 8
	Class 3, type III	Grade C, type III	Grade C, type III	Grade C, type III	Class 3, type III
Density at 15 °C, kg/m^3	806	827	839	835	806
Kinematic viscosity while 40 (20) °C, mm^2/s (cSt)	1.6 (2.3)	2.99 (4.76)	2.96 (4.65)	2.79 (4.31)	1.69 (2.41)
Fractional composition:					
to a temperature of 180°C, % (by volume)	6.3	-	-	-	7.0
at a temperature of 250°C, % (by volume)	-	23	28	31	-
at 350°C, % (by volume)	98.9	98	97	98	98.0
95 % (by volume) distills at a temperature of °C, %	277	332	343	330	290
Cloud point,°C	no higher – 38	not less than 9	not exceeding 6	not exceeding 6	not less than 40
Cold filter plugging point, °C	not less than 40	not less than 14	not less than 15	no higher – 13	no higher – 51
Pour point, °C	no higher – 44	not less than 21	not less than 22	not less than 22	no higher – 53

Table 2. Actual values of the main physico-chemical and operational performance of diesel fuel EURO Class 2 and of Class C in accordance with the requirements of GOST R 52368-2005

Name of the indicator	Brand diesel fuel			
	Sample No. 10	Sample No. 11	Sample No. 12	Sample No. 13
	Grade C, type III	Class 2, type III	Grade C, type III	Grade C, type III
Density at 15 °C, kg/m^3	830	824	828	838
Kinematic viscosity while 40 (20) °C, mm^2/s (cSt)	3.16 (5.10)	1.784	2.86 (4.61)	2.44
Fractional composition:				
to a temperature of 180°C, % (by volume)	-	7.7	-	-
at a temperature of 250°C, % (by volume)	23	-	32	42
at 350°C, % (by volume)	97	99.0	96	94
95 % (by volume) distills at a temperature of °C, %	340	293	344	356
Cloud point, °C	-	Not exceeding 33	not less than 9	not exceeding 6
Cold filter plugging point, °C	not above – 11	no higher – 34	not above – 11	not above – 8
Pour point, °C	-	no higher – 37	not less than 23	not less than 16

Table 3. Actual values of the main physico-chemical and operational characteristics of summer diesel fuel EURO Class C environmental class 5 in accordance with the requirements of GOST 32511-2013

Name of the indicator	Brand diesel fuel			
	Sample No. 2	Sample No. 4	Sample No. 7	Sample No. 9
Density at 15 °C, kg/m^3	829	843	839	837
Kinematic viscosity while 40 (20) °C, mm^2/s (cSt)	2.88	2.84 (4.5)	3.17	3.16 (5.06)
Fractional composition:				
to a temperature of 180°C, % (by volume)	-	-	-	-
at a temperature of 250°C, % (by volume)	33	33	26	26
at 350°C, % (by volume)	94	98	95	94
95 % (by volume) distils at a temperature of °C, %	352	334	345	353
Cloud point, °C	not less than 5	not less than 14	-	not less than 5
Cold filter plugging point, °C	no higher – 13	not less than 15	not above – 12	not less than 5
Pour point, °C	not above – 18	not less than 24	-	not less than 17

Studies have been conducted and analysis of the viscosity-temperature characteristics the following samples, EURO diesel, type III (sulfur content less than 10 mg/kg) according to GOST R 52368-2005 (EN 590:2009) Rev. 1:

– sample No. 1. Class 3 – manufacturer: JSC "NNK - Khabarovsk refinery"; component composition: diesel hydro-treating fraction 286-304 °C – 99.9% by mass, anti-wear additive "of Coltec DS 7739" – 0.035% by mass., Cetane Improver additive "of Coltec DS 1406" – 0.01% by mass;

– sample No. 3. Grade C – manufacturer: JSC "NNK - Khabarovsk refinery"; component composition: Hydro-treated diesel fraction 340-371 °C – 99.947% by mass anti-wear additive "of Coltec DS 7739" – 0.05 % by mass, Cetane Improver additive "of Coltec DS 1406" – 0.05% by mass;

– sample No. 5. Grade C – manufacturer: JSC "Gazpromneft - ONPZ"; component composition: hydro-treated diesel fraction 375 °C – 100% by mass., anti-wear additive DC-32 Total – 0.12% by mass;

– sample No. 6. Grade C – manufacturer: JSC "ANHK"; component composition: Hydro-treated diesel fraction 301-317 °C – Of 99.02 by mass%, anti-wear additive "Baikal" – 0.02% by mass, Cetane Improver additive "Ecosetol" – 0.06% by mass;

– sample No. 8. Class 3 – manufacturer: JSC "NNK - Khabarovsk refinery"; component composition: Hydrotreated diesel fraction 378-399 °C – 99.945% by mass, anti-wear additive "of Coltec DS 7739" – 0.05% by mass, Cetane Improver additive "of Coltec DS 1406" – 0.005% by mass;

– sample No. 10. Grade C – manufacturer: JSC "NNK - Khabarovsk refinery"; component composition: hydro-treated diesel fraction 286-304 °C – 99.955% by mass, anti-wear additive "of Coltec DS 7739" – 0.035% by mass., Cetane Improver additive "of Coltec DS 1406" – 0.01% by mass;

– sample No. 11. Class 2 – manufacturer: JSC "ANHK"; component composition: hydro-treated diesel fraction 349-362 °C – 99.889% by mass, anti-wear additive "Baikal" – 0.03% by mass, Cetane Improver additive "Ecosetol" – 0.08% by mass; anti-oxidation additive "Agidol-1" – 0.001% by mass;

– sample No. 12. Grade C manufacturer: JSC "Slavneft - YANOS"; component composition: hydro-treated diesel fraction L-24-6 325-330 °C – 23.81% by mass, diesel fraction (hydrocracking) 395-391 °C – of 29.84% by mass, hydro-treating diesel fraction L-24-7 325-330 °C – 21.73% by mass, hydro-treating diesel fraction 325-330 °C DT – 24.60% by mass, anti-wear additive "Dodilube"-4940 "Clariant" – 0.0225% by weight., anti-static additive Stadis firm "Innospect" – 0.002% by mass;

IOP Conf. Series: Materials Science and Engineering **921** (2020) 012006 doi:10.1088/1757-899X/921/1/012006

– sample No. 13. Grade C – manufacturer: OOO "Gazprom Neftekhim Salavat"; component composition: hydro-treated diesel fraction 348-366 °C – A 99.97% by mass, anti-wear additive "ADDY TOP CM"/AddiTOP-1 – 0.03% by mass

Studies have been conducted and analysis of the viscosity-temperature characteristics of samples the following summer EURO diesel, environmental class K5 for GOST 32511-2013:

– sample No. 2. Grade C – manufacturer: JSC "TANECO"; component composition: hydro-treated diesel fraction from counter 355 to 358 °C – 89.77% by mass, hydro-treated kerosene fraction 295-296 °C – 2.18% by mass, diesel fraction (hydrocracking) 346-424 °C – 6.43% by mass, diesel fraction of oil production 362-363 °C and 1.57% by mass, anti-wear additive "GT-16" – 0.03% by mass, Cetane Improver additive "Kerobrisol EHN" – 0.02% by mass;

– sample No. 4. Grade C – manufacturer: JSC "ANPZ VNK"; component composition: hydro-treated diesel fraction 331-338 °C – Of 99.96% by mass, anti-wear additive "Baikal" – 0.022% by mass, Cetane Improver additive "Ecosetol" – 0.016% by mass;

– sample No. 7. Grade C – manufacturer: OOO "RN - Komsomolsk refinery"; component composition: hydro-treated diesel fraction 334-365 °C – Efficiency of 99.78% by mass, anti-wear additive "Baikal" – 0.029% by mass, Cetane Improver additive "Ecosetol" – 0.016% by mass;

– sample No. 9. Grade C – manufacturer: JSC "Republic popular"; component composition: hydro-treated diesel fraction 361-366 °C – Of 99.96% by mass, anti-wear additive "Baikal" – 0.022% by mass, Cetane Improver additive "Ecosetol" – 0.016% by mass.

Results

Laboratory investigations of the physico-chemical properties of the samples DF, three winter and ten summer showed that at temperatures above the cloud point temperature of the investigated the samples do not form the structure and behave as a Newtonian fluid (figure 1, 2), which allows to Express the viscosity-temperature characteristics of diesel fuel coefficient of kinematic viscosity on temperature analytical dependencies.

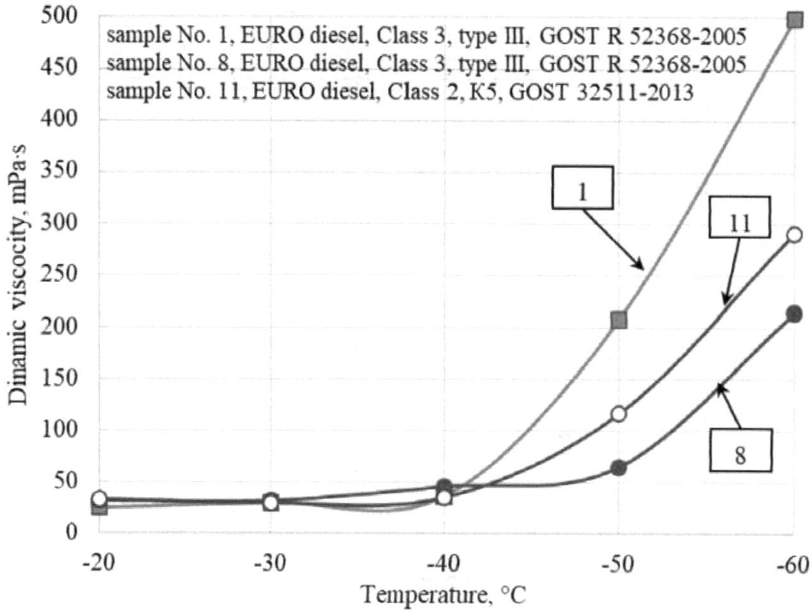

Figure 1. Viscosity-temperature dependence winter diesel fuels at low temperatures

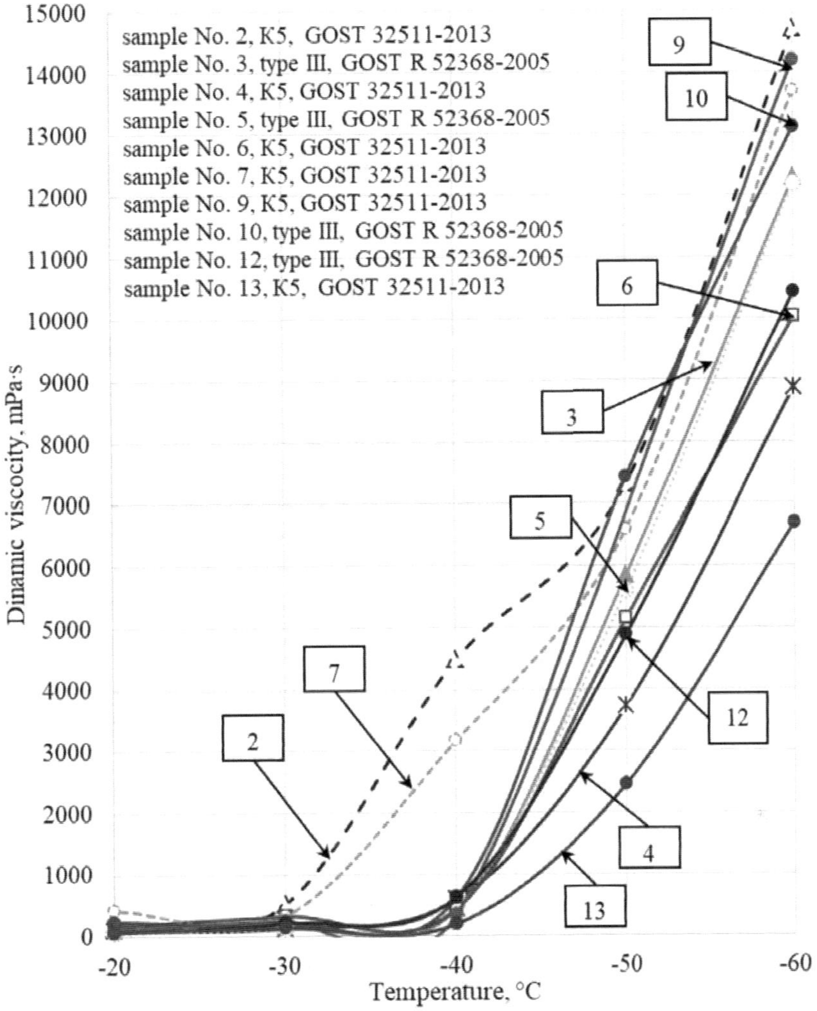

Figure 2. Viscosity-temperature dependence of summer diesel fuel (diesel fuel EURO, grade C) to low temperatures

Figure 1 shows that at temperatures below - 40 degrees, winter diesel fuel starts to show anomalies of viscosity, then the expression of their viscosity-temperature properties in this manner is incorrect. By reducing the fuel temperature below its limiting temperature of filterability can be formed microcrystals of n-paraffin hydrocarbons, there is a sharp, abrupt increase in the values of the parameter viscosity [1]. For winter fuels, these changes are less pronounced. Therefore, to conduct hydraulic calculations necessary to investigate the rheological properties of diesel fuels in the temperature range from - 20 to - 60 degrees.

The nature of fluid responsiveness to external stimuli are divided into Newtonian and non-Newtonian. Newtonian fluids are called viscous fluids, reporting in its flow of viscous friction law of Newton, i.e. shear stress and velocity gradient in a liquid is linearly dependent. Rheological model of the fluid flow Newton [2] according to equation:

$$\tau = \mu D \,, \tag{1}$$

where: μ – dynamic viscosity, Pa·s; D – shear rate, s^{-1}.

The nature of the reaction fluid on the external influences are divided into Newtonian and non-Newtonian. To describe the behavior of non-Newtonian fluids there are more than 30 rheological models that pass within a certain range of shear stresses individual features of real currents. The greatest practical distribution to describe the viscoplastic behavior of structured disperse systems received a rheological model of the flow of non-Newtonian fluids Shvedov-Bingham [3] according to equation:

$$\tau = \tau_0 + \mu_{pl} D \,, \tag{2}$$

where: τ – shear stress, Pa; τ_0 – ultimate dynamic shear stress (yield stress), the excess of which leads to viscous flow, Pa; μ_{pl} – the ratio of plastic (structural) viscosity, Pa·s; D – shear rate, s^{-1}.

During the rheological studies of samples of DT on the rotary viscometer Rheotest RN 4.1 using the cylindrical measuring system according to GOST 1929 the obtained flow curves at low temperatures, which determine the rheological model of the fluid flow.

Values of dynamic viscosity were recorded with the gradual cooling of DF from - 20 to - 60 degrees. Additionally, a study was conducted of the rheological behavior of the winter DF in terms of: pour point, cloud point and cold filter plugging point. The build results based on changing the magnitude of shear stress from the gradient of shear rate, the curves of the currents for several samples of DF are presented in figures 3-5.

It is established that DF when they are cooled below the solidification point represent a viscoplastic medium with curves of current, which describes a universal equation of Buckley-Herschel for nonlinear viscoplastic liquid [4] according to equation:

$$\tau = \tau_{st} + kD^n \,, \tag{3}$$

where: τ_{st} – critical shear stress, Pa; k – the coefficient of consistency; n – the rate of flow (index flow).

The transition from Newtonian to non-Newtonian fluids (viscoplastic and nonlinear viscoplastic) is carried out smoothly as the temperature is decreasing. Comparison of results of calculations of currents of DF curves of the sample No. 1 obtained by experimental means with rheological models of fluid flow Newton equation (1), Shvedov-Bingham equation (2) and Buckley-Herschel equation (3) in terms of low temperatures presented in figure 6.

To quantify the accuracy of determining rheological model for yield curves of the sample No. 1 of DF were used, the relative deviation, the smaller the error the higher the accuracy.

It is seen that the results of the calculations of the curves of the currents to the rheological models of Newton coincide with experimental data at temperatures above the temperature of cloud point of diesel fuel (figure 6a, b). At lower temperatures there is a deviation of the curve of the flow of diesel fuel to the rheological models of Newton (figure 6b, d). The appearance of the ultimate dynamic shear stress occurs at a temperature close to the melting point (figure 6b). Such a non-Newtonian fluid obeys the equation of a Shvedov-Bingham. This is confirmed by the calculation results of flow curve, which good agreement with experimental data.

At temperatures below - 50°C in the range low velocity of the flow curve is non-linear with decreasing shear rate. Such a move is contrary to a valid curve model Shvedov-Bingham (figure 6d). This contradiction is eliminated in the mathematical model Buckley-Herschel for nonlinear viscoplastic environment.

The analysis of the curves of the currents revealed that when the critical shear rate the dynamic viscosity of the fuel becomes constant for each particular temperature.

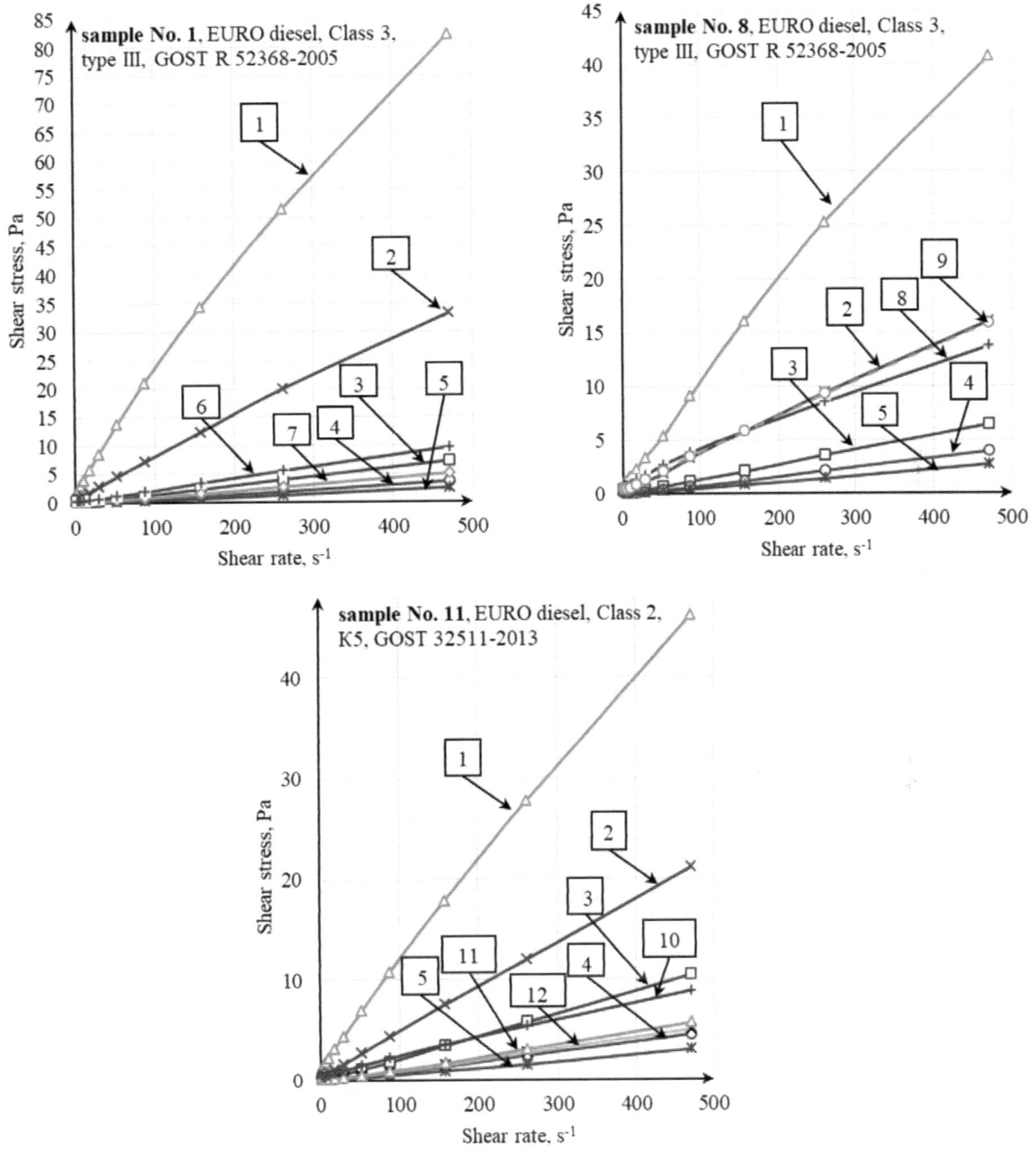

Figure 3. Flow Curves of winter diesel fuel samples No. 1, 8 and 11 (1 – at -60 °C; 2 – at -50 °C; 3 – at -40 °C; 4 – at -30 °C; 5 – at -20 °C; 6 – at -44 °C; 7 – at -38 °C; 8 – at -51 °C; 9 – at -53 °C;10 – at -37 °C; 11 – at -34 °C; 12 – at -33 °C)

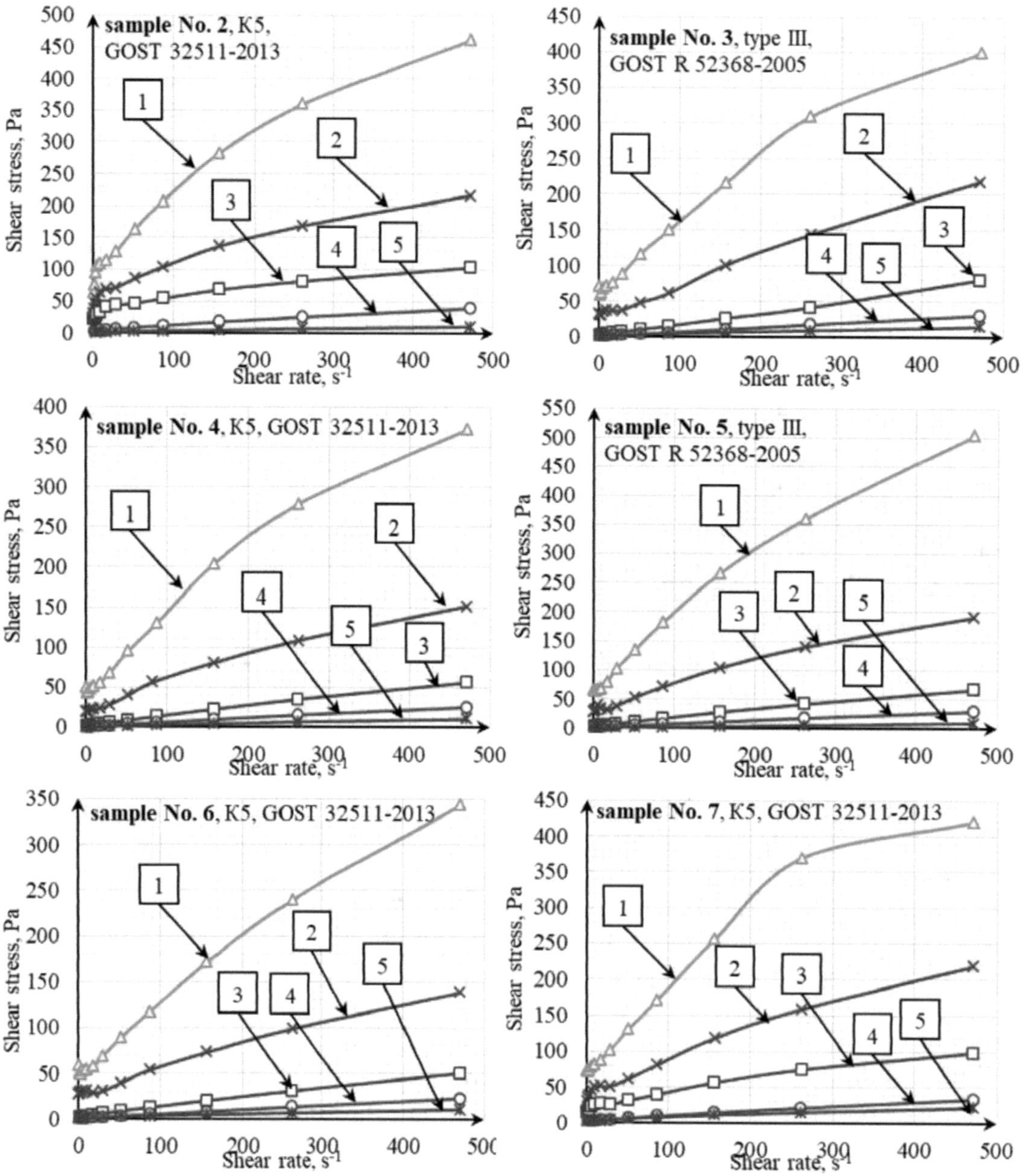

Figure 4. Flow Curves of diesel fuel samples № 2, 3, 4, 5, 6, 7 (1 – at -60 °C; 2 – at -50 °C; 3 – at -40 °C; 4 – at -30 °C; 5 – at -20 °C)

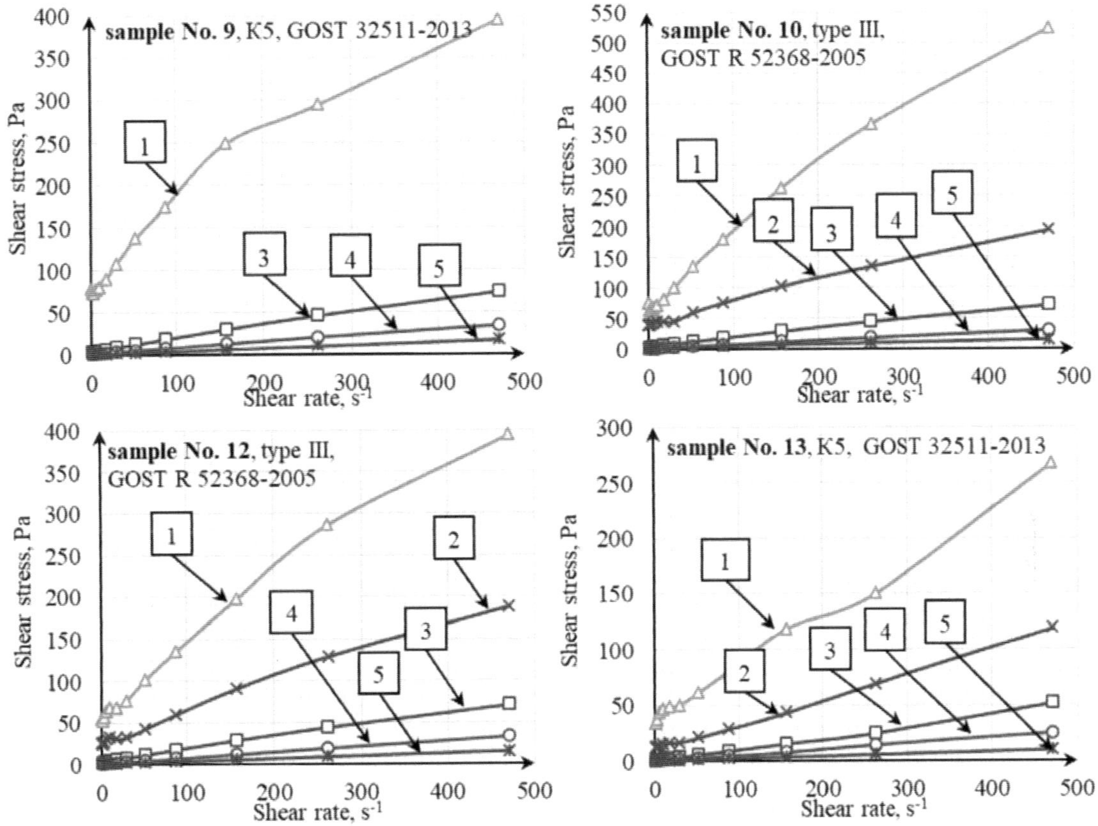

Figure 5. Flow Curves of diesel fuel samples No. 9, 10, 12, 13 (1 – at -60 °C; 2 – at -50 °C; 3 – at -40 °C; 4 – at -30 °C; 5 – at -20 °C)

A refined method of hydraulic calculation of collapsible pipelines, allows to determine the pressure loss, taking into account changes of the rheological properties of diesel fuels at low temperatures. One of the main objectives when performing hydraulic calculations of pipelines is the determination of the magnitude of the head losses due to friction h_c calculated according to Darcy-Weisbach equation:

$$h_{\mathrm{f}} = \frac{\lambda l w^2}{d 2g}, \tag{4}$$

where: h_{f} – friction head loss, m; λ – Darcy friction factor; l – estimated pipeline length, m; d – pipeline inner diameter, m; w – average fluid velocity, m/s; g – acceleration of gravity, m/s^2.

The Darcy friction factor λ is determined depending on the flow regime. To determine if there are many different empirical formulas. The boundaries of the flow regimes and areas of hydraulic friction set depending on the Reynolds number *Re* and the coefficient of relative roughness of the internal surface of the pipe ε.

There are three modes of fluid flow: laminar $\mathrm{Re} < 2320$, transition zone $2320 < \mathrm{Re} < 10000$ and turbulent $\mathrm{Re} > 10000$. Turbulent regime for Newtonian liquids conventionally divided into the three areas of the hydraulic friction: smooth $10000 < \mathrm{Re} < 10/\varepsilon$ mixed $10/\varepsilon < \mathrm{Re} < 500/\varepsilon$ quite rough (quadratic) $\mathrm{Re} > 500/\varepsilon$.

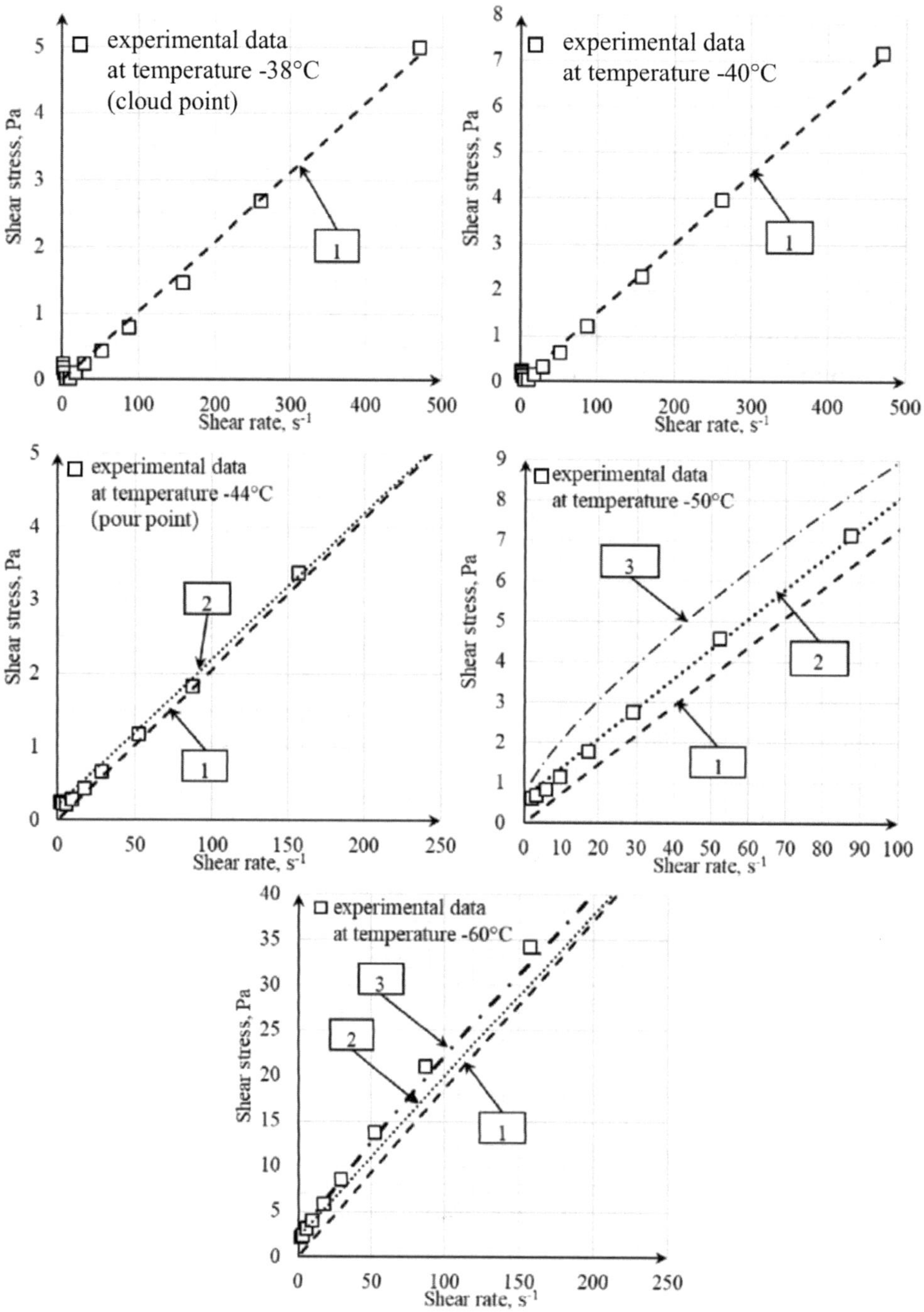

Figure 6. Comparison of rheological models of flow and experimental the curves of the flows of diesel fuel sample No. 1 at different temperatures: (1 – model of the Newton; 2 – model of Shvedov-Bingham; 3 – model of the Buckley-Herschel)

The coefficient of relative roughness of the pipe is determined from equation:

$$\varepsilon = \frac{k_e}{d}, \tag{5}$$

where: ε – coefficient of relative roughness of the inner surface of the pipe; k_e – the coefficient of equivalent roughness of the inner surface of the pipe, m.

The Reynolds number for Newtonian and non-Newtonian fluids is defined as follows.

The Reynolds number for a Newtonian fluid is determined from equation:

$$Re = \frac{wd}{\nu} = \frac{wd\rho}{\mu}, \tag{6}$$

where: ν – coefficient of kinematic viscosity, mm^2/s (cSt); ρ – fluid density, kg/m^3.

The Reynolds number for non-Newtonian viscoelastic (rheological model Shvedov-Bingham) fluids is determined from equation [5]:

$$Re_0 = \frac{Re}{\left(1 + \frac{Il}{6}\right)}, \tag{7}$$

where: Re_0 – the Reynolds number for non-Newtonian viscoplastic (rheological model Shvedov-Bingham) fluids; Re – the Reynolds number for a Newtonian fluid; Il – the number of Ilyushin for non-Newtonian viscoplastic (rheological model Shvedov-Bingham) fluids.

The number of Ilyushin is the ratio of initial shear stress to the stress of viscous friction and is determined from equation [6]:

$$Il = \frac{\tau_0 d}{\mu_{pl} w}. \tag{8}$$

The Reynolds number for non-Newtonian nonlinear viscoelastic (rheological model Buckley-Herschel) liquid is determined from equation:

$$Re_1 = \frac{d^n w^{2-n} \rho}{\frac{k}{8}\left(\frac{6n+2}{n}\right)^n} \cdot \frac{8}{Il_1 + \frac{8}{3n+1} \cdot \left\{ n + \left[(2n+1)^2 + \frac{Il_1 n(3n+1)}{4} \right]^{1/2} \right\}} \cdot \frac{3}{2} \cdot \frac{(3n+1)^2}{(2n+1)(5n+3)}, \tag{9}$$

where: Re_1 – the Reynolds number for non-Newtonian nonlinear viscoplastic (rheological model Buckley-Herschel) fluids; Il_1 – the number of Ilyushin for non-Newtonian nonlinear viscoplastic (rheological model Buckley-Herschel) fluids.

Option Ilyushin is determined from equation:

$$Il_1 = \frac{\tau_{st} \cdot d^n}{\frac{k}{8}\left(\frac{6n+2}{n}\right)^n w^n}. \tag{10}$$

With respect to the hydraulic calculations of pipeline transport of oil products to determine the Darcy friction factor is recommended to use the formulas presented in table 4 [5].

Based on the results of research of rheological properties and characteristics of the current diesel fuel at low temperature conditions by using the mathematical model of hydrodynamic processes of pumping diesel fuel at low temperatures, it is possible to expeditiously determine the losses in the hydraulic calculation.

The study was conducted modeling of the pumping of diesel fuel in the collapsible pipelines different diameters 100 and 150 mm in low temperatures.

The graphs (figures 7-9) the resulting show a loss of pressure from performance of the pump at the collapsible pipelines of different diameters in conditions of low temperatures.

Table 4. The formula for determining the Darcy friction factor

Flow regime	Determination formula for Darcy friction factor
1. Laminar	
1.1 Newtonian fluid, $Re < 2320$ (Stokes formula)	$\lambda = \frac{64}{Re}$
1.2 Non-Newtonian viscoplastic fluid with an ultimate dynamic shear stress τ_0, $Re_0 \leq 2320 \cdot (1+Il/6)$	$\lambda = \frac{64}{Re_0}$
2. Transition	
2.1 Newtonian fluid, $2320 < Re < 10000$ (Wullis-Ginzburg formula)	$\lambda = \frac{64}{Re} \cdot (1-\gamma) + \frac{0.3164}{Re^{1/4}} \cdot \gamma$, $\gamma = 1 - exp[-0.002 \cdot (Re - 2320)]$
2.2 non-Newtonian viscoplastic fluid with an ultimate dynamic shear stress τ_0	$\lambda = \frac{64}{Re_0} \cdot (1-\gamma_0) + \frac{0.3164}{Re_0^{1/4}} \cdot \gamma_0$, $\gamma_0 = 1 - exp[-0.002 \cdot (Re_0 - 2320)]$
3. Turbulent	
3.1 Newtonian fluid	
− smooth friction area, $10000 < Re < 10/\varepsilon$ (Blasius formula)	$\lambda = \frac{0.3164}{Re^{1/4}}$
− mixed friction area, $10/\varepsilon < Re < 500/\varepsilon$ (Altshul formula)	$\lambda = 0.11\left(\frac{68}{Re} + \varepsilon\right)^{1/4}$
− complitely rough (quadratic) friction area, $Re > 500/\varepsilon$ (Shifrisson formula)	$\lambda = 0.11(\varepsilon)^{1/4}$
3.2 non-Newtonian viscoplastic fluid with critical shear stress τ_{st}, $10000 < Re_0 < 100000$	$\lambda = \frac{0.3164}{Re_0^{1/4}}$

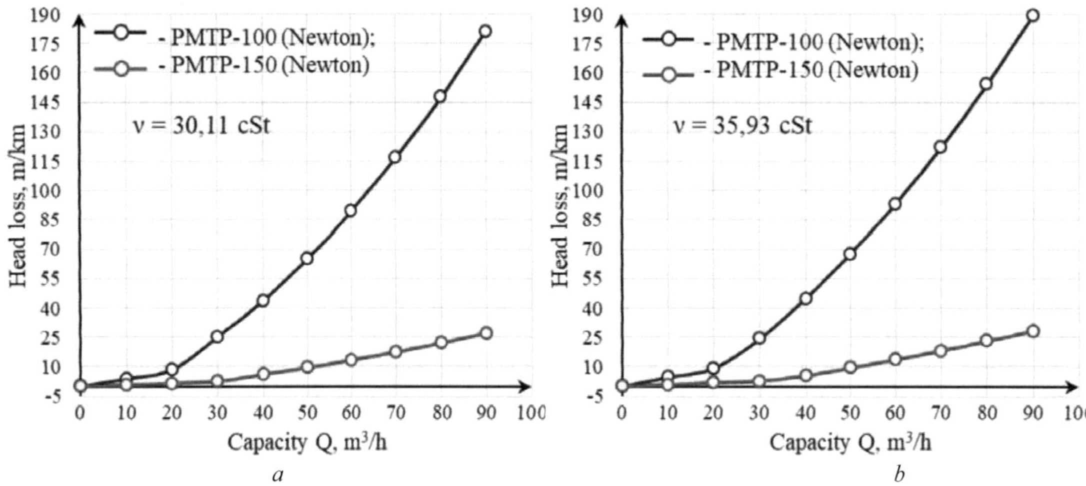

Figure 7. The Dependence of the head loss and noise transfer of different diameters collapsible pipelines winter diesel fuel at temperatures of 20°C (*a*) and 30°C (*b*)

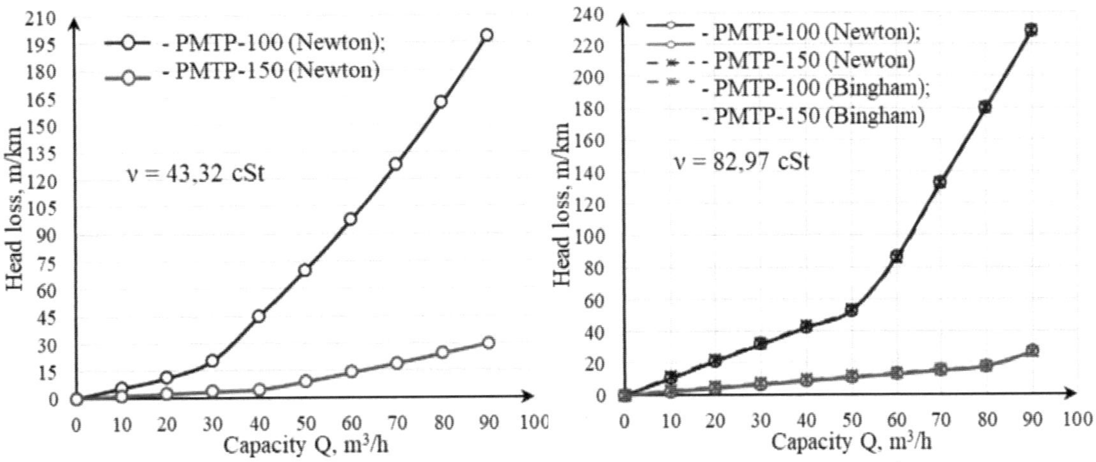

a *b*

Figure 8. Dependence of the head loss and noise transfer of different diameters collapsible pipelines winter diesel fuel at temperatures of 40°C (a) and 44°C (b)

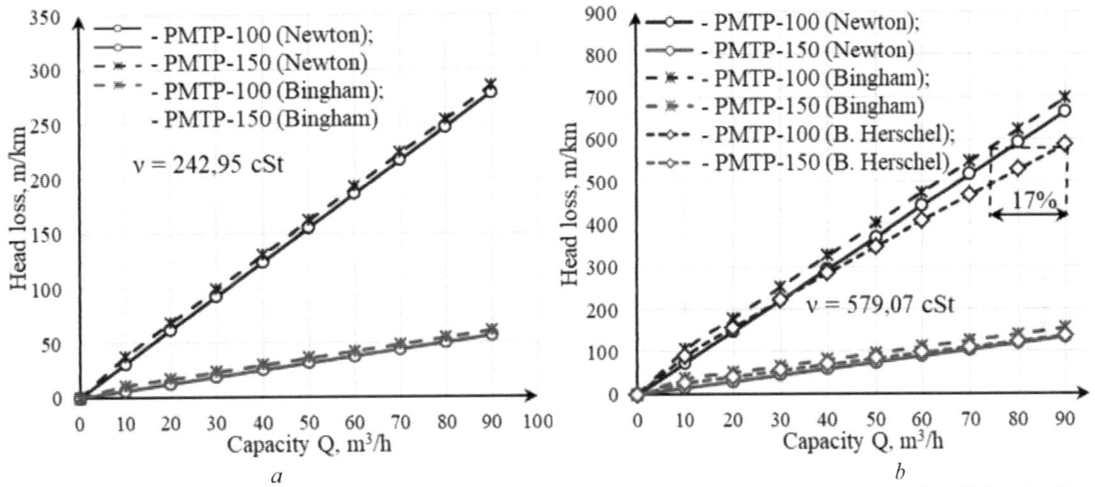

a *b*

Figure 9. The Dependence of the head loss and noise transfer of different diameters collapsible pipelines winter diesel fuel at temperatures 50°C (a) and 60°C (b)

Determined that the choice of a wrong rheological model of the flow can lead to errors in hydraulic calculations and reduce the expected productivity of transfer in the collapsible pipelines by up to 17%. The study also determined that depending on the flow, head losses for collapsible pipelines with a nominal diameter of 100 mm can be up to 700 m/km, 7 times higher than with a diameter of 150 mm. the foregoing allows to conclude that the use of collapsible pipelines with a diameter of 100 mm is impractical under these conditions, as before the mobile pumping unit PNU-75 will not be able to provide pumping for no more than 1 km. Use collapsible pipelines with a diameter of 150 mm allows to move the one PNU -75 for a distance of 9 km.

In the technical documentation of the hydraulic characteristics of the pump means of pumping are presented at nominal frequency of rotation of the shaft of the pump, taken on water or on light oil products

IOP Conf. Series: Materials Science and Engineering **921** (2020) 012006 doi:10.1088/1757-899X/921/1/012006

at positive temperature and small values of the kinematic viscosity. For conversion of the hydraulic characteristics of centrifugal pumps with regard to the influence of the viscosity of the fluid became widespread based on the theory of similarity and dimensions the method proposed by Sukhanov and Eisenstein [7].

This resampling method accounts for the effects of kinematic viscosity on the hydraulic characteristics of the pumps are obtained by test on water, by introducing correction factors to the performance K_Q, the pressure K_H and efficiency K_η. Consequently, for liquids with a viscosity greater than water, the design characteristics of get the appearance, according to equation:

$$Q_v = K_Q Q_w;\ H_v = K_H H_w;\ \eta_v = K_\eta \eta_w, \tag{11}$$

where: Q – flow rate; H – pump head; η – pump efficiency; in equation (11) the indices (v, w) refer to the work of the pumps, respectively, at the viscous liquid and water.

The values of the coefficients K_Q, K_H and K_η depend on the Reynolds number (Re_p), according to equation:

$$Re_p = \frac{Q}{D_{eq}\nu}, \tag{12}$$

where: D_{eq} – equivalent diameter of the impeller of the pump.

The equivalent diameter of the impeller of the pump D_{eq} is determined from the equation [7]:

$$\frac{\pi D_{eq}^2}{4} = \pi D_2 b_2 k_r, \tag{13}$$

where: D_2 – outer diameter of the impeller of the pump; b_2 – the width of the impeller blades of the pump on the outside diameter; k_r – coefficient of restriction of the impeller blades at the outlet, the calculations shall be equal to 0,92-0,95.

From the expression (13) follows equation:

$$D_{eq} = 4D_2 b_2 k_r, \tag{14}$$

where: D_2 – outer diameter of the impeller of the pump; b_2 – the width of the impeller blades of the pump on the outside diameter.

In engineering practice, the coefficients K_Q, K_H and K_η it is recommended to determine if the rated capacity of the pump and make permanent for the working area of the pump. The value of the nominal flow rate of pump utilized performance value flow rate of pump at maximum efficiency. The boundaries of the working area are taken in the capacity range from 0.8 to 1.2 of nominal pump flow rate [8].

In the application of funds transfer in the collapsible pipelines is likely to work in a wider range of performance values, it is therefore advisable factors K_Q, K_H and K_η determined from the calculated, not the nominal performance values.

The coefficients K_Q and K_H defined on graphs [7] or by approximating the dependencies [9], according to equations:

$$K_Q = -0.774 + 0.580 \lg Re_p;\ 100 < Re_p \le 600, \tag{15}$$

$$K_Q = 0.412 + 0.153 \lg Re_p;\ 600 < Re_p < 7000, \tag{16}$$

$$K_Q = 1;\ Re_p \le 7000, \tag{17}$$

$$K_\eta = -0.852 + 0.483 \lg Re_p;\ 100 < Re_p \le 2300, \tag{18}$$

$$K_\eta = 0.201 + 0.170 \lg Re_p;\ 2300 < Re_p < 50000, \tag{19}$$

$$K_\eta = 1;\ Re_p \le 50000. \tag{20}$$

Ratio K_H is determined from the graphs or from equation:

$$K_Q = K_H^{2/3}. \tag{21}$$

IOP Conf. Series: Materials Science and Engineering **921** (2020) 012006 doi:10.1088/1757-899X/921/1/012006

At values of Reynolds number $Re_P > 7000$, the coefficients K_Q and K_H differ little from 1, while the ratio K_η in these limits Re_P changing to a much greater extent.

Methodological apparatus (11-21) allows the recalculation of characteristics for centrifugal pump with the values of the coefficient of specific speed n_S in the range from 50 to 130. In terms of the hydraulic characteristics of this method in a specified range of values of the coefficient of speed n_S the value of the relative deviation of the obtained values is less than 5%. For pumps with greater or lower rate of speed n_S the method is extended conditionally and with a certain approximation.

The coefficient of rapidity n_S – universal characteristics used to establish the fact of their similarity, calculation of geometrical parameters and characteristics and for classification of pumps. Calculation of specific speed is carried out according to the equation [7]:

$$n_S = 3{,}65n\frac{Q^{1/2}}{H^{3/4}}, \tag{22}$$

where: n_S – pump specific speed; pump speed.

The pump speed taken to determine if the rated pump flow rate and corresponding the pump head value.

To assess the possibility of applying this methodology at different pump operation modes the value of the pump speed must be qualified taking into account changes in the characteristics of the head pump and the pump efficiency.

To justify the use of mobile pumping units in the conditions of low temperatures, for example PNU-75, you have converted the basic characteristics of pump operation 80-60 when pumping diesel fuel. Data for conversion of the hydraulic characteristics of pump mobile pumping unit PNU-75 (CN 80-60): outer diameter of the impeller of the pump (D_2) 197 mm; width of the impeller blades of the pump on the outside diameter (b_2) 9.5 mm; flow and head corresponding to maximum efficiency on the water 80 m^3/h and 580 m, respectively. Taking the coefficient of restriction of the impeller blades at the outlet (k_r) is equal to 0.94, according to the above formulas is defined by the equivalent diameter of the impeller of the pump D_{eq} and the Reynolds number Re_P for values of the volumetric pump flow rate. The equivalent diameter of the impeller of the pump D_{eq} equal 83.9 mm. On the basis of the obtained values was determined as the magnitude of the correction factors K_Q, K_H and K_η. and the calculated values of the hydraulic characteristics of the pump head, pump flow rate and pump efficiency.

Data to build the hydraulic characteristics of pump operation CN 80-60 obtained based on the calculation results for winter diesel fuel in low temperatures presented in tables 5.

Table 5. Results of the recalculation of characteristics of the pumping unit PNU-75 (CN 80-60) when working at the winter diesel fuel

The pumping temperature, °C	-20	-30	-40	-44	-50	-60
The product viscosity, ν, cSt	30.11	35.93	43.32	82.97	242.95	579.07
The Reynolds Number, Re_P	8798	7373	6115	3193	1090	457
Correction factors:						
to pump flow rate, K_Q	1	1	0.991	0.948	0.877	0.769
to pump efficiency, K_η	0.87	0.86	0.84	0.8	0.62	0.43
to pump head, K_H	1	1	0.994	0.965	0.916	0.839

The graphs (figures 10-12) presents a recalculation of the hydraulic characteristics of mobile pumping unit PNU-75 (CN 80-60), with the maximum values being taken from the engines of the means of transfer of power, from water to diesel fuel at different low temperatures.

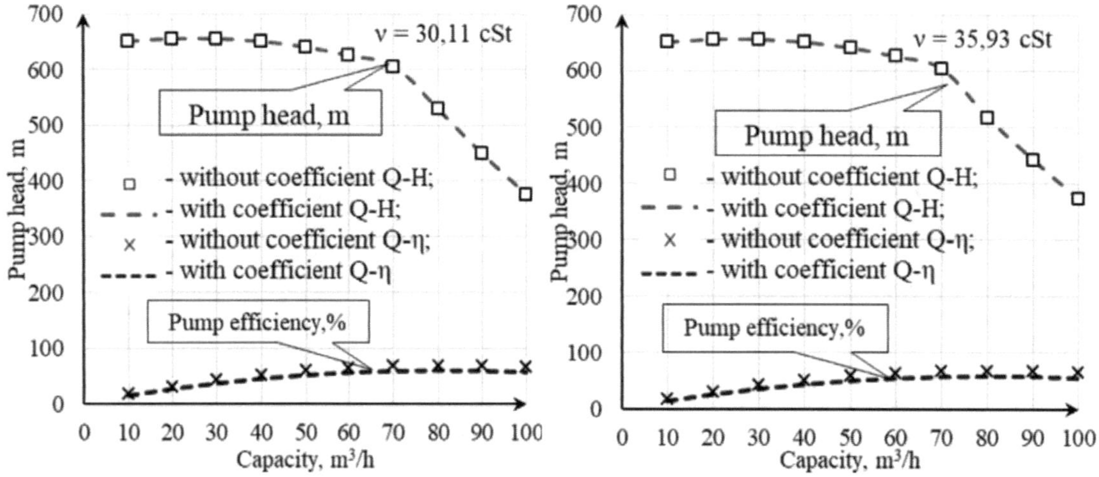

a *b*

Figure 10 – The estimated performance of PNU-75 (CN 80-60) at winter diesel fuel at temperatures of 20°C (*a*) and 30°C (*b*)

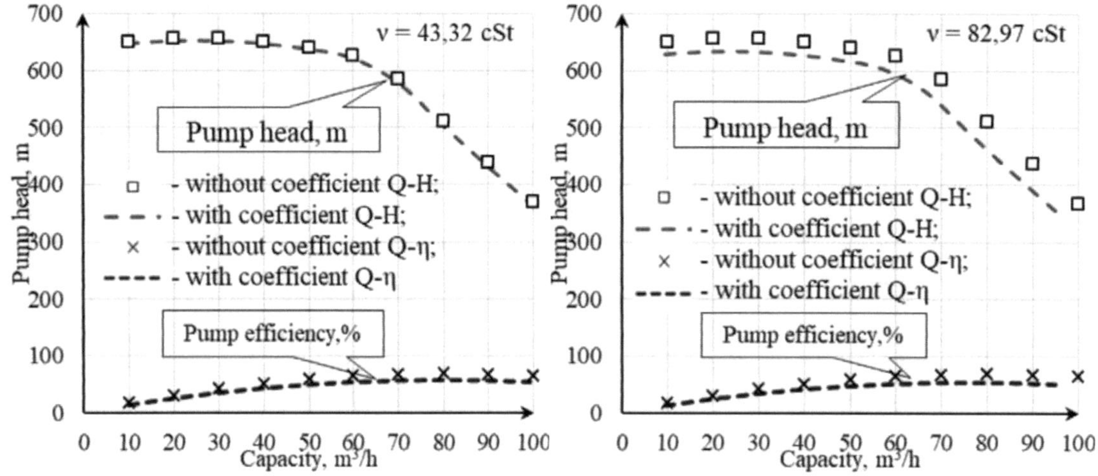

a *b*

Figure 11. The estimated performance of PNU-75 (CN 80-60) for winter diesel fuel at temperatures of 40°C (a) and 44°C (b)

According to the results of the simulation of centrifugal pumps, mobile pump units established that increase the viscosity of the fluid hydraulic characteristics of the pump means pumping deteriorate, there is a decrease in pressure, flow and efficiency, increased power consumption, so you need to clarify on the technique [7] and considered in the hydraulic calculation.

Studies to assess the impact of rheological properties on the process of transportation of diesel fuels on collapsible pipelines in the low temperatures of the obtained flow curves with which it is possible to determine how the rheological models must be given environment at various temperatures. It is established that DF upon cooling below the freezing temperature represent a viscoplastic medium with curves of current, which describes a universal equation of Buckley-Herschel. It is revealed that when the critical shear rate dynamic viscosity becomes constant at a specific temperature.

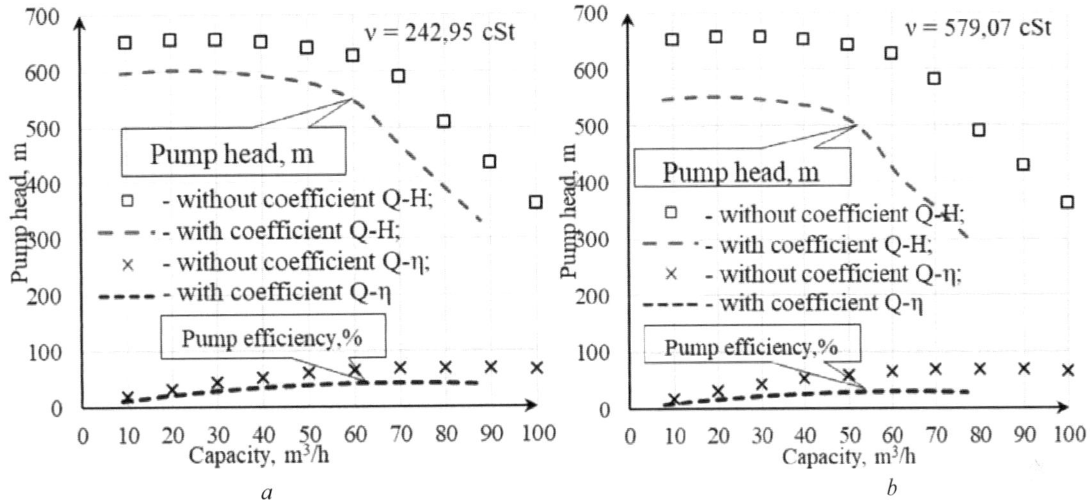

Figure 12. The estimated performance of PNU-75 (CN 80-60) for winter diesel fuel at temperatures of 50°C (a) and 60°C (b)

This has helped to clarify the methodology of hydraulic calculations, allowing to determine the pressure loss, taking into account changes of the rheological properties of diesel fuel in low temperatures.

Conclusion

The result of the simulation of the DF pumping at the collapsible pipelines established that:

– the wrong choice of the rheological model of the flow can lead to errors in hydraulic calculations and reduce the expected performance of pumping through the pipeline by up to 17%.

– use tubing with a diameter of 100 mm is impractical in low temperature conditions, for practical application it is recommended to use tubing with a diameter of 150 mm

As a result of simulation of operation of centrifugal pumps mobile pumping unit it is established that with increasing viscosity worsen the hydraulic characteristics of the pump that must be considered in hydraulic calculations

Referense

[1] Grisnin N N and Sereda V V 2016 *Encyclopedia of Himmotology* (Moscow: Pero Press) p 960
[2] Antipyev V N 1983 *Utilization of Associated Petroleum Gas* (Moscow: Nedra Press) p 160
[3] Wilkinson W L 1964 *Non-Newtonian Fluids. Fluid Mechanics, Mixing and Heat Transfer* (Moscow: MIR Press) p 216
[4] Shulman Z P 1975 *Convective Heat and Mass Transfer Reologicheskie Complex Fluids* (Moscow: Energy Press) p 352
[5] Maron V I 2009 *Hydrodynamics Single-Phase and Multiphase Flows Through Pipeline* (Moscow: MAKS Press) p 344
[6] Lurie M V 2012 *Mathematical Modeling of Pipeline Transportation of Oil and Gas* (Moscow: Gubkin Russian State University of Oil and Gas Press) p 456
[7] Aizenstein M D 1957 *Centrifugal Pumps for the Oil Industry* (Moscow:

IOP Conf. Series: Materials Science and Engineering **921** (2020) 012006 doi:10.1088/1757-899X/921/1/012006

Gostoptekhizdat Press) p 364
[8] Kolpakov L G 1985 *Centrifugal Pumps of Oil Pipelines* (Moscow: Nedra Press) p 185
[9] Tugunov P I, Novoselov V F, Korshak A A, Shammazov A M 2008 *Model Calculations in the Design and Operation of Tank Farms And Oil Pipelines* (Ufa: Designprogram Service Press) p 658

IOP Conf. Series: Materials Science and Engineering **921** (2020) 012007 doi:10.1088/1757-899X/921/1/012007

About use of supercapacitors in autonomous power supplies

M I Ganiev, A I Dyomko and O Yu Semenov

Surgut State University, Surgut, Russia

E-mail: ous.tutor.phinma@mail.ru

Abstract. The article describes the use of supercapacitors in power supplies designed for a long time of unattended operation in difficult meteorological conditions. A comparison is made of the energy efficiency factor of energy storage devices - batteries and supercapacitors when using solar panels. The optimization of the power supply circuit is considered, taking into account the provision of the maximum battery life with a limited number of charge / discharge cycles.
Key words: power source, storage battery, supercapacitor, solar battery, energy efficiency factor.

1. Introduction

Wireless technologies confidently take positions in many areas of our lives. In most cases this involves self-powered with different scenarios ensure efficiency: using network chargers stationary power sources or alternative power sources (wind generators, solar cells, tidal power, etc.). A necessary element of autonomous devices is an autonomous power supply (APS), which can use non-renewable energy sources (disposable chemical power sources - batteries) and renewable (ionistors, batteries, etc.) [1].

The weight, size and duration of autonomous operation of APSs are determined not only by their structure, but also by the requirements for power parameters and climatic conditions of work. In addition, the weight and size parameters of the APS depend on the fraction of the energy of the primary sources, which can be used during operation. The energy efficiency of consumers is characterized by a coefficient of performance, however, this parameter is not directly related to the energy parameters of the source.

2. Theory

For primary energy sources can introduce energy efficiency coefficient K_e characterizing the fraction (in relative units or percentage) of energy, which energy is used when data conditions (time, temperature, load, etc.). [2]:

$$K_e = \frac{E_1}{E}, \tag{1}$$

where E_1 - available for use the energy of one primary source; E is the total energy that the source can give to the consumer under optimal conditions (for a battery, this value is related to the electric capacity, measured in A*h or W*h); N is the number of primary energy sources used.

Consider the energy efficiency ratio for lithium cells are widely used to supply stand-alone devices. They are characterized by optimal current consumption of only 2 - 6 mA [3], and with a larger current consumption, the electric capacity of the battery is significantly reduced. As an example in Fig. 1 shows the values of the energy efficiency coefficient calculated on the basis of [3] for the lithium cell SL2880 with a nominal capacity of 19 A*h.

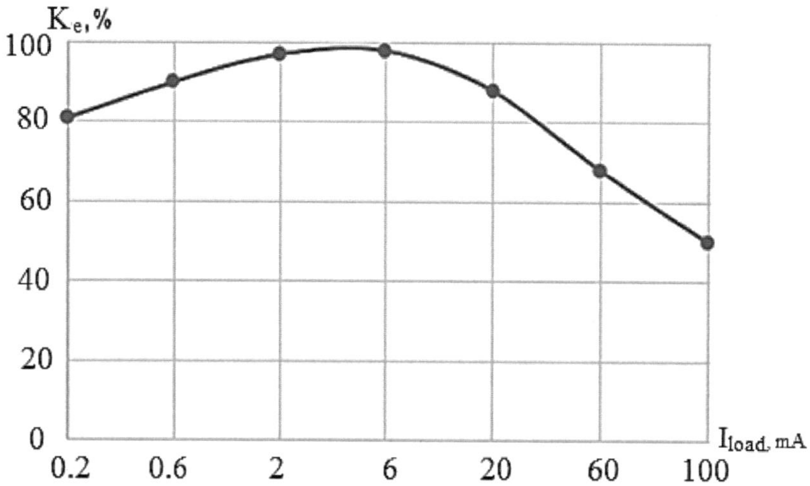

Figure 1. The dependence of the energy efficiency coefficient of a lithium cell SL2880 on the current consumption

As follows from calculation, only 50 will be used in non-optimal load - 80% of the battery power (9 - 15 A*h).

There is a method of increasing the energy efficiency coefficient of a battery with a pulsed nature of the load by using an intermediate controlled DC / DC converter, a ballast resistor and a buffer storage capacitor [3]. Controlled DC / DC converter boosts the output voltage of 100 mV increments to an external command. Thus the amount of current consumed by the battery does not exceed the set value, while the buffer capacitor current discharge may be substantially greater magnitude. The block diagram of the device is shown in Fig. 2, and the diagrams of the current consumption from the battery and the capacitor voltage in Fig. 3.

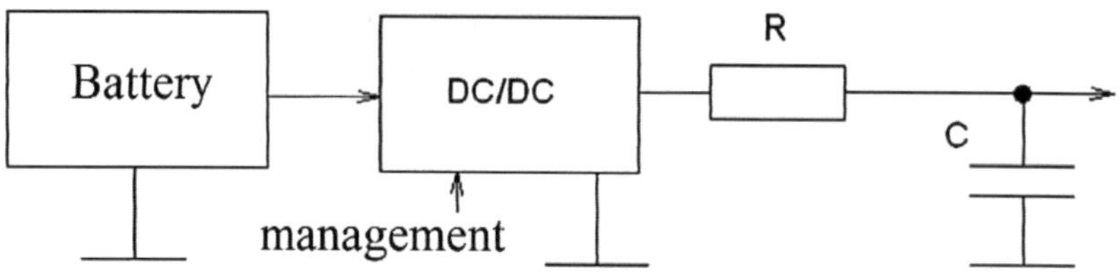

Figure 2. The structural diagram of the phased charge buffer storage capacitor

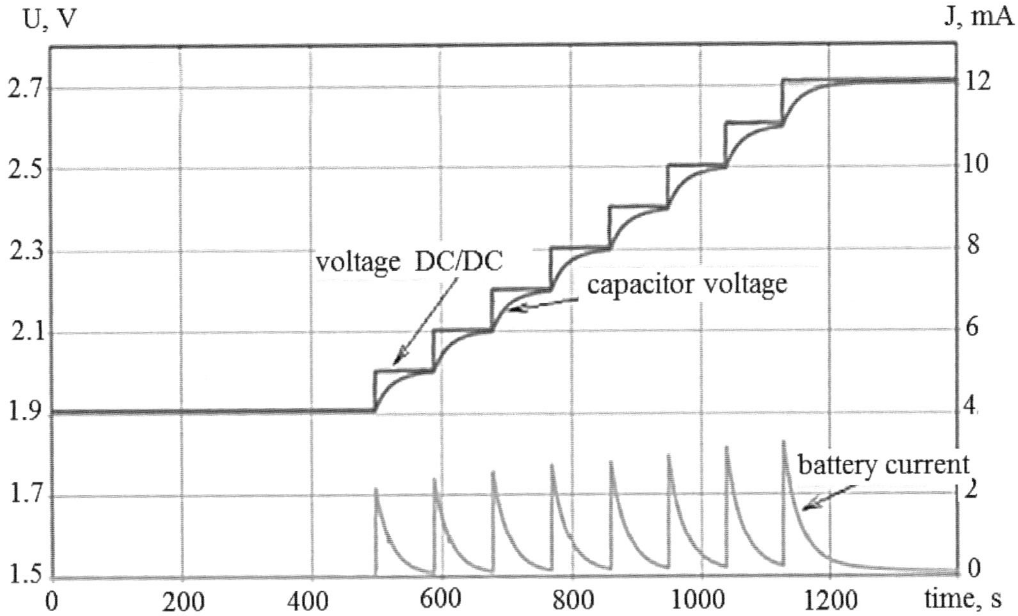

Figure 3. Phase charge diagrams of the buffer storage capacitor

Another example of improving power efficiency coefficient battery refers to autonomous devices, for a long time at low temperatures at which capacitance battery, its output voltage and the maximum load current are reduced [4].

The increase of the load current (say when the radio auxiliary device) can result in an unacceptable reduction in the output voltage, causing a standalone device stops working. To prevent such an emergency, it is necessary to use either the above method of dosed current extraction from the battery or its heating [5-8].

For autonomous APSs operating in a wide temperature range, it is advisable to use capacitors of large capacity (ionistors) that are charged from solar panels. Features of supercapacitors are low internal resistance, the possibility of working at low negative temperatures and a large number of charge / discharge cycles, which advantageously distinguishes them from batteries [9-12].

We calculate the energy efficiency coefficient of the ionistor. The ionistor is discharged from the initial voltage of 2.7 V to the final E_{fin}, determined by the capabilities of the energy consumer.

There are three options for taking energy from the ionistor:

1) with a constant load resistance R_{load},
2) with constant current consumption,
3) with a varying load current resistance (load in the form of a DC / DC converter).

3. Experimental Results

For example, we determine the energy efficiency coefficient of the ionistor for constant load resistance R_{load}. The voltage on the supercapacitor is determined by solving the problem on the transition process and has the form shown in Fig. 4. In this case, the current through the resistor will change in proportion to the voltage on the ionistor according to the exponential law.

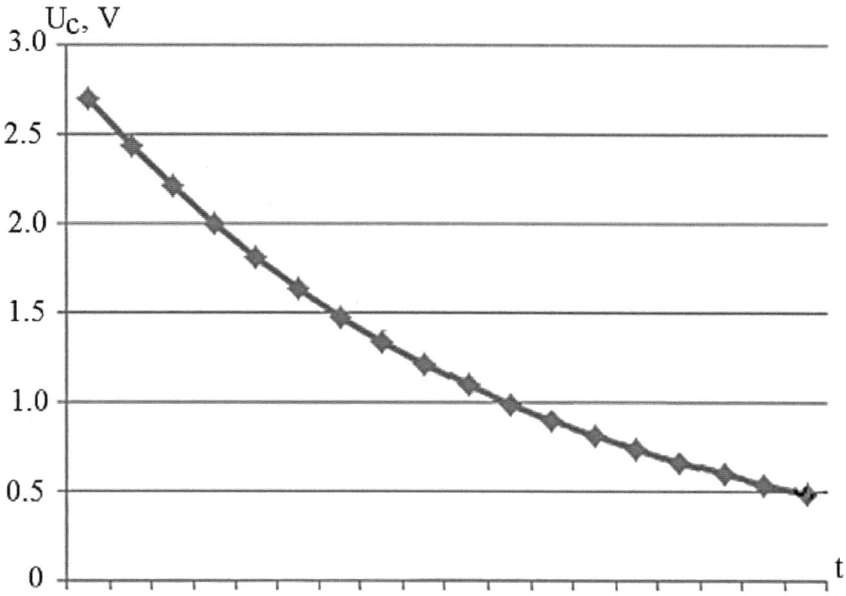

Figure 4. The dependence of the voltage of the ionistor on time with a resistive load

Full energy stored in supercapacitors:

$$E=\int_0^\infty p(t)dt=\int_0^\infty u_c^2(t)dt/R_н=\frac{1}{R_н}\int_0^\infty U_{c0}^2 e^{-\frac{2t}{R_н C}}dt=$$

$$=-\frac{R_н C}{2R_н}U_{c0}^2 e^{-\frac{2t}{R_н C}}\Big|_0^\infty=\frac{CU_{c0}^2}{2}=C\frac{2{,}7^2}{2}=3{,}645C \quad (2)$$

$$K_{e1}=\frac{Э_1}{Э}=\frac{2{,}7^2-E_{fin}^2}{2{,}7^2}=1-\left[E_{fin}/2{,}7\right]^2. \quad (3)$$

When series connection supercapacitors energy efficiency ratio increases:

$$K_{eN}=\frac{Э_N}{Э}=\frac{(N\cdot 2{,}7)^2-E_{fin}^2}{(N\cdot 2{,}7)^2}=1-\left[E_{fin}/(N\cdot 2{,}7)\right]^2=1-\frac{1}{N^2}\left[E_{fin}/2{,}7\right]^2. \quad (4)$$

4. Results Discussion

Figure 5 shows the dependences of the energy efficiency coefficient of ionistors during its discharge on a voltage of 2.7 V to the final E_{fin} for one and two series-connected ionistors.

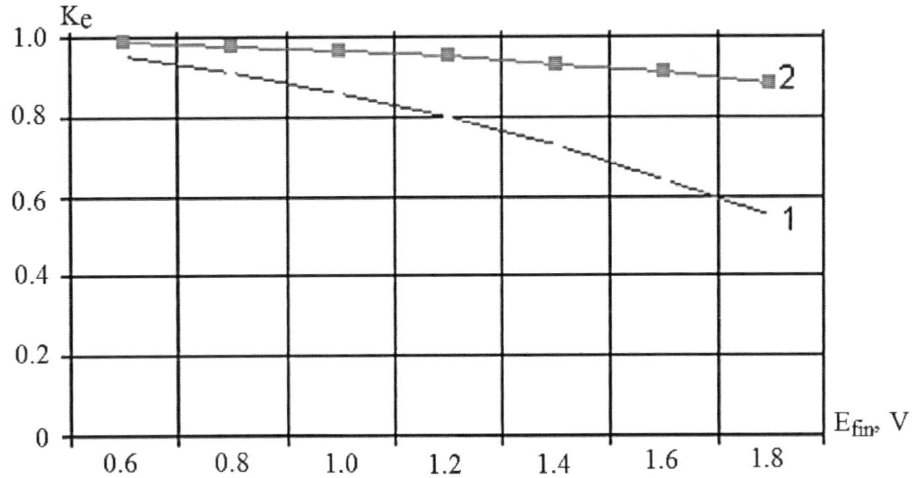

Figure 5. The energy efficiency coefficient of one (1) and two (2) series-connected supercapacitors

In most cases, a stable voltage is required to power the equipment, while the power is supplied not directly from the ionistor, but through a stabilizing DC / DC pulse converter.

As an example, consider a DC / DC boost converter LTC3525-3.3 with an output voltage of 3.3 V when the input voltage changes from 0.8 to 2.7 V. In Fig. 6 and Fig. 7 shows the experimentally measured parameters of the converter at a constant load (current 3.14 mA, load resistance 1050 Ohms).

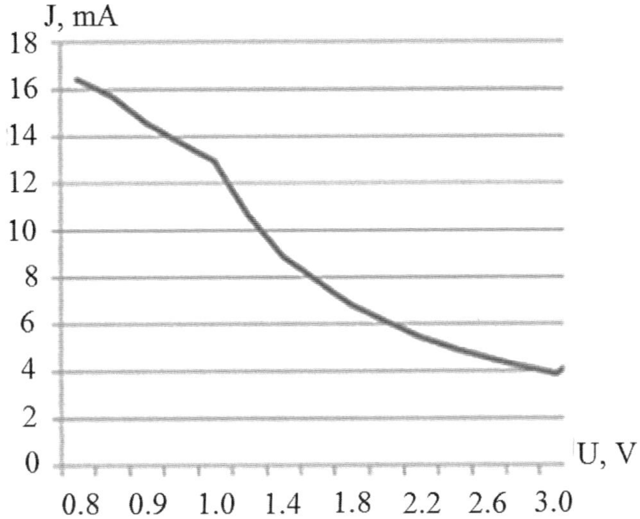

Figure 6. The dependence of the current consumption of the Converter LTC3525-3.3 from the input voltage

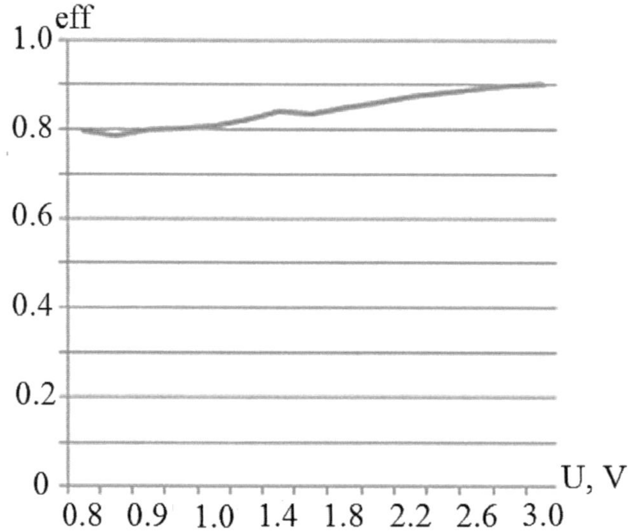

Figure 7. The dependence of the efficiency of the LTC3525-3.3 Converter on the input voltage

As follows from fig. 6, the current consumed from the ionistor does not decrease, but increases with decreasing voltage on the ionistor. To determine the energy efficiency coefficient (by analogy with (2)) in this case, approximations of the experimental current dependences and efficiency were preliminarily performed:

$$i = (4{,}015u^2 - 20{,}7u + 29{,}991), \text{ мА}, \qquad (5)$$

$$\eta = 0{,}76 + 0{,}045u. \qquad (6)$$

Results of calculation of the energy efficiency coefficient supercapacitors during its connection to the stabilizing DC / DC converter is pulsed with 3.3 LTC3525-3.3 The output voltage shown in Fig. 8.

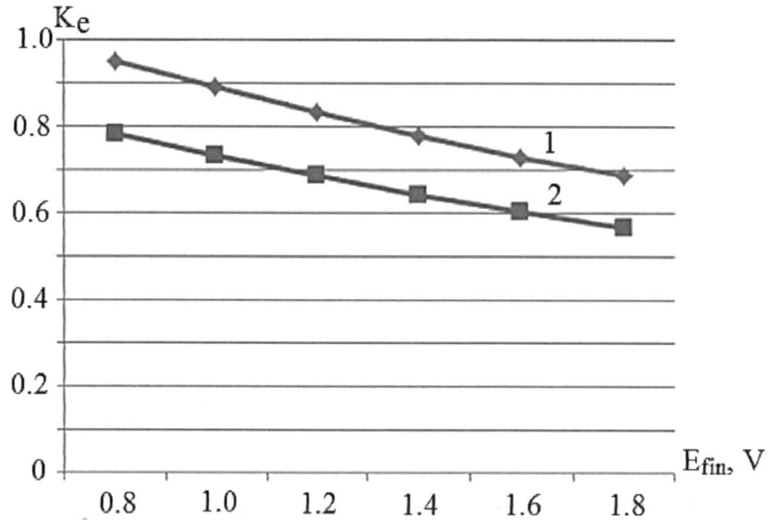

Figure 8. The energy efficiency coefficient of the ionistor (1) with a changing load, (2) taking into account the efficiency of the LTC3525-3.3 converter

The final discharge voltage supercapacitors E_{fin} ranged from 0.8 V (corresponds to the lower boundary of input voltages for which the operating DC-DC converters) to 1.8 V (corresponds to the lower boundary of the supply voltages, at which employ microcontrollers and wireless modems) [13-fifteen]. Similar calculations can be made for the battery.

5. Conclusions

1. The use of supercapacitors in conjunction with solar panels it is advisable to stand-alone power supply with long-time operation in a wide temperature range.
2. The electric double layer capacitors have a high value (0.95) Energy efficiency factor that helps to minimize weight and size independent power supply.
3. The use of supercapacitors together with a stabilizing DC / DC switching converter slightly (1.2-fold) reduces the ratio of energy efficiency supercapacitors.
4. The above results may be useful in the design of autonomous power supplies.

Acknowledgments
The research was conducted with the financial support according to the state task of the Education and Youth Policy Department grant of the Khanty-Mansiysk Autonomous Region – Yugra (order No 1281 of 25 August 2017)

References

[1] Dyomko I, Petrov B A and Radomsky S A 2016 *Autonomous intelligent power supply Patent No 2615985 for an invention.*
[2] Dyomko I and Semenov O Yu 2019 ["Power supply of autonomous devices,"] in *Oil and Gas Engineering-2019 AIP Conference Proceedings*, edited by Myshljavcev A V *et al.* (OSTU Omsk) pp 193. Russian.
[3] Lithium batteries "Tadiran". Catalog. http://www.platan.ru/docs/pdf/ tadiran_ 2014.pdf.
[4] Pushkarev O 2015 How to achieve maximum operating time of a wireless node with autonomous power "*Electronics News*" vol 2. Russian.
[5] Alferov Zh I, Andreev V M and Rumjancev V D 2004 *Semiconductors* vol **38 (8)** pp 937–948. Russian.
[6] Nikitin M 2012 [Electronics: Science Technology Business] vol **4** pp 50–56. Russian.
[7] Odeim F, Roes J and Heinzel A 2015 *Energies* vol **8** pp 6302–6327.
[8] Ganiev M I, Dyomko A I and Semenov O Yu 2019 ["The use of supercapacitors in the autonomous power supplies"] in *Problems of Improving Energy Efficiency in the Northern Regions of the Russian Oil and Gas Industry* International Scientific and Practical Seminar, edited by Salnikov V G *et al. (NSU, Nizhnevartovsk)* pp 59–64. Russian.
[9] Jaysree K S, Ezhilarasan G, Monish K, Arundhilipan V and Sumalatha G 2014 *Int. J. Sci. Eng. Res.* vol **5** pp 214–218.
[10] Kyriakarakos G, Piromalis D, Dounis A I, Arvanitis K and Papadakis G 2013 *Appl. Energ* vol **103** pp 39–51.
[11] Dyomko A I., Radomskij S A and Petrov B A 2017 *RF Patent No 20161028/01* (17 January 2017). Russian.
[12] Kollimalla S K, Mishra M K and Narasamma N L 2014 *IEEE Trans Sustain Energy* vol **5** pp 1137–1144.
[13] Ye Y and Cheng K W E 2015 *IEEE J EM SEL TOP P* vol **3** pp 977–983.
[14] Dyomko A I, Dyomko I A and Semenov O Yu 2019 [*Embedded program of power source microprocessor intelligent device operation*] Software Registration Certificate No 2019662143/69 (02 December 2019). Russian.
[15] Rana M M *et al* 2017 *Int. J Elec. Power* vol **84** pp 225–231.

IOP Conf. Series: Materials Science and Engineering **921** (2020) 012008 doi:10.1088/1757-899X/921/1/012008

Determination of the Geological Age of Oil Using Diamondoids

M Giruts, G Gadzhiev and G Gordadze
National University of Oil and Gas «Gubkin University», Moscow, Russia

E-mail: moxixh@yahoo.com

Abstract. Based on the study of a large number of Western and Eastern Siberia oils, it was shown that adamantanes can be used as a means of correlation in oils of different genotypes and ages. Based on the relative distribution of mono-, di-, and trialkyl-substituted C_{11}–C_{13} adamantanes marine and teristerial oils and Cenozoic and Proterozoic oils were distinguished. This approach could be used when there are no generally recognized biomarker hydrocarbons or indicators based on them are uninformative. We found that Jurassic and Cretaceous oils of Kalmykia can be distinguished by the distribution of methyl- and dimethyldiamantanes. It was concluded that the distributions of alkyl substituted diamantanes could also be applied for the differentiation of oils of different ages.

1. Introduction

Taxonometric indication or, in other words, the determination of the geological age of initial bioorganic molecules (the age of oil) is one of the most interesting tasks of petroleum geochemistry. It is much more complicated to determine the age than the geochemical (genetic) types of oils. Nevertheless it is very important as this parameter is associated with the distribution of organic compounds in the living organisms of the modern era and the distant past.

The age of oil is related to the age of the oil source strata that generated it. Since oil is migratory, it can be generated in ancient sediments and, as a result of migration, end up in younger reservoirs. Therefore, identifying the source of generation of hydrocarbons and their migration paths is of great applied importance in the search and exploration of oil fields. It is needed to build correlations in oil-oil, oil-organic matter of rocks (EOM) systems.

The most commonly used indicator of the age of oils or organic matter of rocks is the ratio of regular C_{28}/C_{29} steranes [1]. It was shown that with the decrease of geological age of organic matter there is an increase in the relative concentration of C_{28} steranes and an increase in C_{28}/C_{29} ratio. Hence this parameter has limitaion as it cannot be used in case of oils of terrestrial genesis. These oils are initially characterized by a high relative concentration of C_{29} sterane. The use of sterane and terpane indicators is impossible in the case of light oils due to the absence of high molecular weight fractions and highly mature oils due to cracking of these hydrocarbons.

Indicators of the age may also be some biomarkers specific to certain geological era. The example is 12- and 13-methylalkanes which mark Precambrian oils [2,3]. On the basis of detailed studies of oils from the Timan-Pechora oil and gas province, it was found that the relative content of biomarker hydrocarbons such as pregnanes and cheilanthanes also changes with geological age of oil [4,5].

Diamondoids are not biomarkers, but they can also be useful in geochemical studies [6–15]. They are applied in the geochemical correlations of biodegraded oils and gas condensates, when bacterial activity changes the relative content of biomarkers [16,17]. We have shown that diamondoids are omnipresent and are found in oils and organic matter of both clay and carbonate strata, in organic matter of various maturities. In this paper we propose to correlate the age and the genotype of organic matter of rocks and oils using adamantanes.

2. Adamantanes as a mean of correlation of oils of different age and genotype

Figure 1a shows typical mass chromatograms with characteristic ions of steranes and terpanes of highly mature oil. It can be seen that steranes and terpans are practically absent, which is associated with their cracking in the occurrence conditions. At the same time from mass chromatograms on figure 1b one can see adamantanes of the same oil. These hydrocarbons were not cracked due to thermal stability and can be used for geochemical studies.

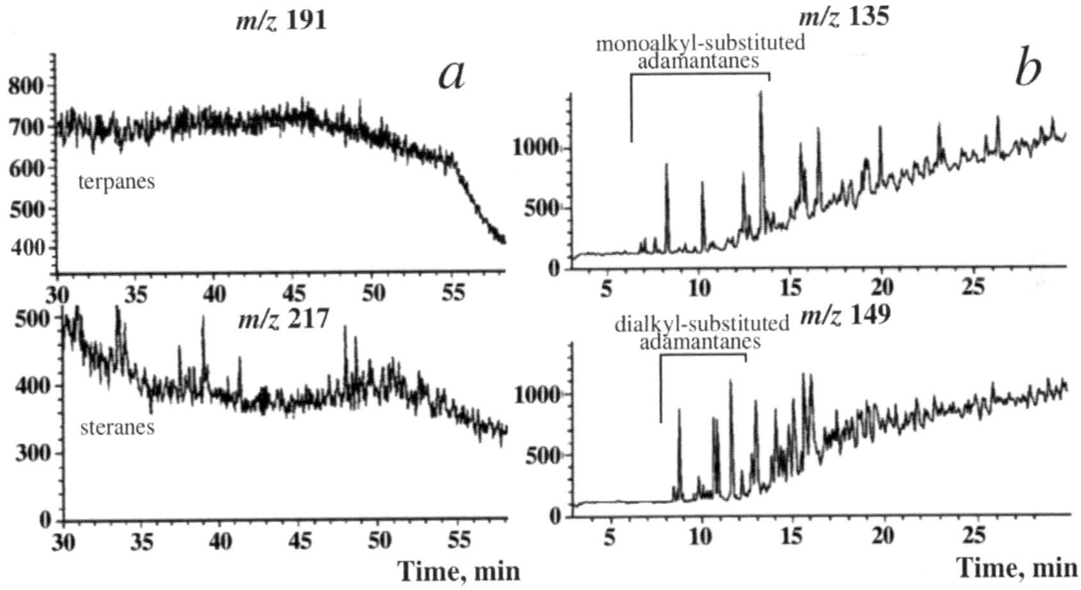

Figure 1. Mass chromatograms of steranes, terpanes and adamantanes of highly mature oil. Steranes and terpanes are practically absent and uninformative, while adamantanes can be used in geochemical studies.

In order to prove that adamantanes can be used to differentiate the age of oil, we studied the correlation of sterane and adamantane indicators in a large number of oils of different age and genotypes in Western and Eastern Siberia.

Figure 2 demonstrates the steranes and adamantanes ratios of oils of various genotypes – terrestrial and marine. Based on the distribution of regular C27–C29 steranes, it is easy to distinguish between marine and terrestrial oils (Figure 2a). Terrestrial oils have a significantly higher relative content of stigmastane C_{29} (more than 80%) compare to marine ones. The same oils differ well in the ratios of mono-, di-, and trialkyl-substituted adamantanes: C_{11}/C_{13} (on average 0.3 for terrestrial oils and 1.0 for marine oils) and C_{12}/C_{13} (0.8 for terrestrial and ~ 2.8 for marine oils) (figure 2b). Thus, we can talk about the good accordance of adamantane indicators with common sterane parameters for genetic correlations in oil–oil and oil–EOM.

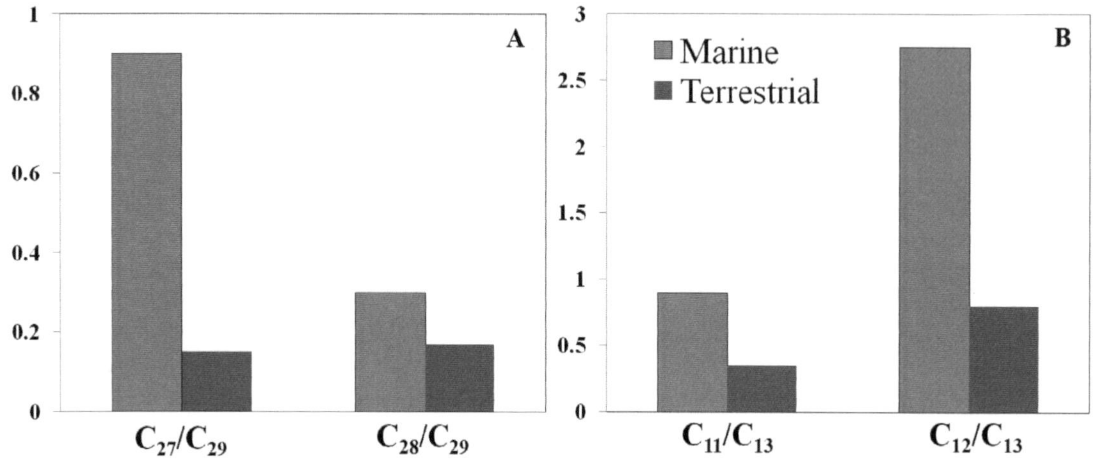

Figure 2. Steranes (a) and adamantanes (b) in oils of various genotypes (an average data).

We propose that the distribution of C_{11}–C_{13} mono-, di-, and trialkyl-substituted adamantanes can also be applied for differentiating young and ancient oils when the taxonometric Grantham coefficient is not aplicable like in light oils, terrestrial oils and highly mature oils. On figure 3a it can be seen that according to the ratios of the regular steranes C_{27}/C_{29} and C_{28}/C_{29}, the Cenozoic oils sharply differ from the oldest Proterozoic oils. If we consider the ratios of the adamantanes C_{11}/C_{13} and C_{12}/C_{13} used above (Figure 2b), we can conclude that they can also be used to differentiate these oils. Thus, it is seen that the C_{11}/C_{13} ratio of the adamantanes is on average 1.5 for the Cenozoic oils and 0.8 for the Proterozoic oils, while the C_{12}/C_{13} ratio is 3.2 for the Cenozoic and 2.5 for the Proterozoic ones.

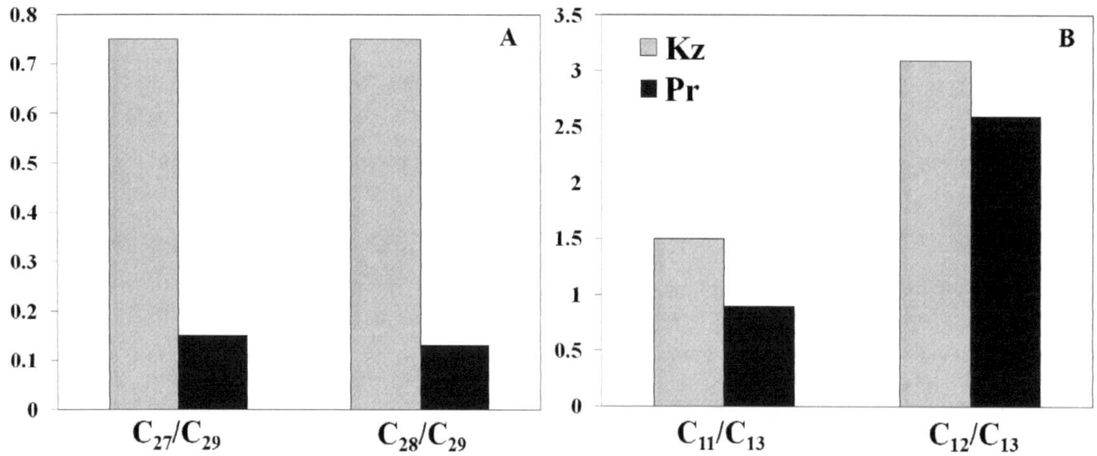

Figure 3. Steranes (a) and adamantanes (b) in oils of different ages (an average data).

3. Diamantanes as a mean of correlation of oils of different age

Another example of using adamantanoids distributions in oil-oil correlations is differentiation of oils of different age (the Jurassic and the Cretaceous) on the example of Kalmykia oils.

We studied the patterns of distribution of biomarkers and methyl- and dimethyldiamantanes in seven Jurassic (Bajocian) and fifteen Cretaceous (Aptian) oils of Kalmykia. It should be noted that according to the relative distribution of biomarkers – steranes – these oils are not possible to distinguish. Figure 4 presents a stellar diagram that shows the main sterane parameters, which are generally used in organic geochemistry. It can be seen that in both the Jurassic and the Cretaceous oils the values of these parameters are very close.

IOP Conf. Series: Materials Science and Engineering **921** (2020) 012008 doi:10.1088/1757-899X/921/1/012008

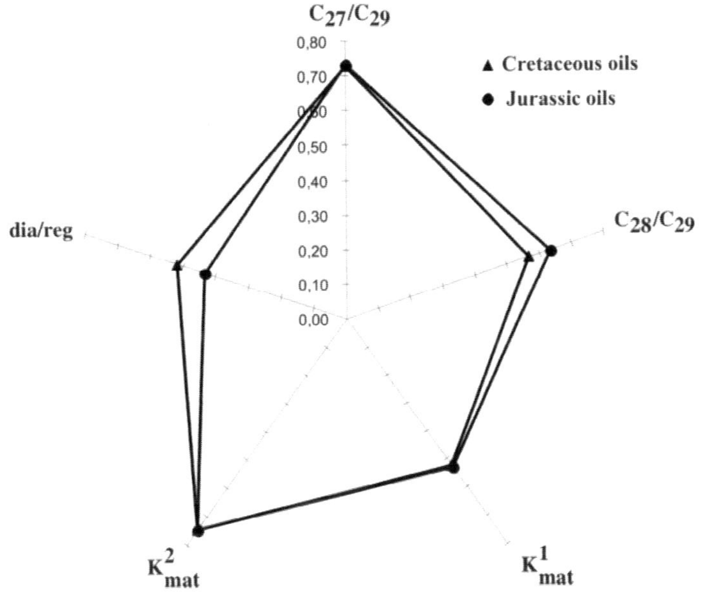

Figure 4. Geochemical characteristics of Jurassic and Cretaceous oils of Kalmykia based on steranes distribution.

At the same time, we found that Jurassic and Cretaceous oils can be distinguished by the distribution of methyl- and dimethyldiamantanes. Figure 5a shows a triangular diagram with the distribution of 1-, 3- and 4-methyldiamantanes. It is clearly seen that oils of different ages fall into different areas of the diagram. A similar conclusion can be made based on the distribution of 3,4-, 4,8- and 4,9-dimethyldiamantanes (Figure 5b), where oils of different ages also fall into different areas of the triangular diagram. It is interesting to note that oil from the Vostochno-Kamyshanskoye field, well 95, despite being selected from the Jurassic sediments corresponds to Cretaceous oils according to diamantanes distribution. It is possible that in this case there was an error in determining the geological age of the deposits.

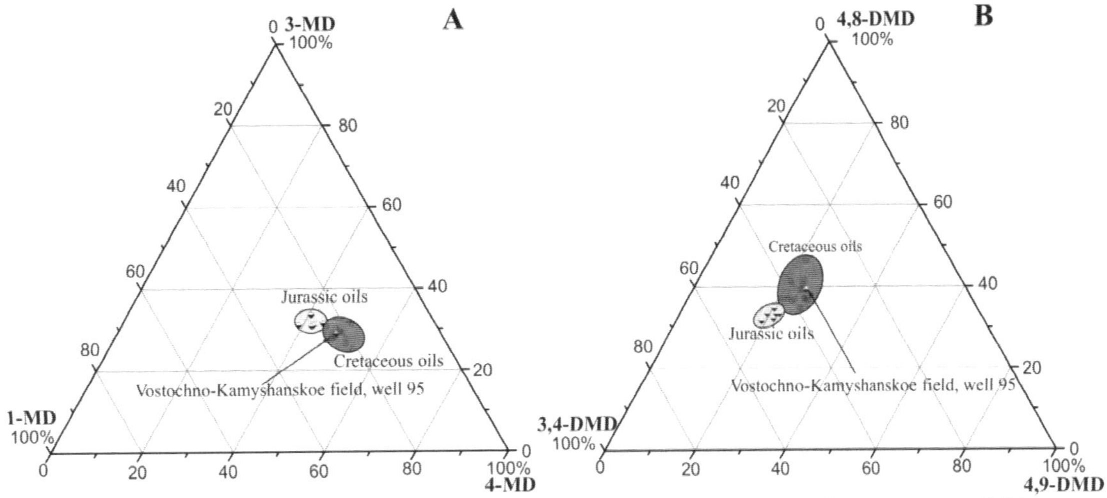

Figure 5. Triangular diagrams for the distribution of (a) monomethyldiamantanes and (b) dimethyldiamantanes in Jurassic and Cretaceous oils of Kalmykia. MD – methyldiamantane, DMD – dimethyldiamantane.

IOP Conf. Series: Materials Science and Engineering **921** (2020) 012008 doi:10.1088/1757-899X/921/1/012008

Conclusion

On the example of oils of different genotypes (marine and terrestrial) and of different ages (Cenozoic and Proterozoic), we showed that characterization of oils can be carried out on the basis of the distribution of C_{11}–C_{13} adamantanes. Conclusion based on the distribution of adamantanes coincides with conclusion based on commonly used biomarkers – steranes. We propose that the distribution of C_{11}–C_{13} mono-, di-, and trialkyl-substituted adamantanes can be used for differentiating young and ancient oils when the taxonometric Grantham coefficient is not applicable. Mono- and dialkyl substituted diamantanes can also serve as a means of differentiating oils of different ages. As it was shown on the example of Jurassic and Cretaceous oils of Kalmykia. For this, triangular diagrams in the coordinates of 1- / 3- / 4-methyldiamantanes and 3,4- / 4,8- / 4,9-dimethyldiamantanes can be used.

References

[1] Grantham P and Wakefield L 1998 *Org. Geochem.* **12** 61–73

[2] Makushina V, Aref'ev O, Zabrodina M and Petrov A 1978 *Petr. Chem.* **18** 847–54

[3] Klomp U *Org. Geochem.* 1986 **10** 807-14

[4] Matveeva I and Gordadze G 2001 *Geochem. Int.* 455-60

[5] Gordadze G, Kerimov V, Giruts M, Poshibaeva A and Koshelev V 2018 *Fuel* **216** 835–42

[6] Dahl J, Moldowan J, Peters K, Claypool G, Rooney M, Michael G, Mello M and Kohnen M 1999 *Nature* **399** 54–7

[7] Schulz L, Wilhelms A, Rein E and Steen A 2001 *Org. Geochem.* **32** 365–75

[8] Peters K, Walters C and Moldowan J 2005 *The Biomarker Guide* (Cambridge: Cambridge University Press)

[9] Chen J, Fu J, Sheng G, Liu D and Zhang J 2006 *Org. Geochem.* **25** 170–90

[10] Gordadze G 2008 *Petr. Chem.* **48** 241–53

[11] Azevedo D, Tamanqueira J, Dias J C, Carmo A P, Landau L and Goncalves F T 2010 *Fuel* **87** 2122–30

[12] Springer M, Garcia D, Gonçalves F, Landau L and Azevedo D 2010 *Org. Geochem.* **41** 1013–18

[13] Mansoori G 2012 *Diamondoid Molecules With Application in Biomedicine, Material Science, Nanotechnology & Petroleum Science* (Singapore: World Scientific Publishing Co. Pte. Ltd)

[14] Gordadze G 2015 *Hydrocarbons in Petroleum Geochemistry: Theory and Practice* (Moscow: National University of Oil and Gas «Gubkin University»)

[15] Giruts M and Gordadze G 2017 *Chemistry and geochemistry of diamond-like hydrocarbons* (Moscow: LLC Publishing House Nedra)

[16] Grice K, Alexander R and Kagi R I 2000 *Org. Geochem.* **31** 67–73

[17] Giruts M, Stroeva A, Gadzhiev G, Stokolos O, Koshelev V and Gordadze G 2014 *Petr. Chem.* **54** 10–15

IOP Conf. Series: Materials Science and Engineering **921** (2020) 012009 doi:10.1088/1757-899X/921/1/012009

Permeability Alteration of Carbonate Reservoir Rock Under Cyclic Geomechanical Treatment

I M Indrupskiy[1,2], I I Ibragimov[2], R A Zakiryanov[2], I F Idiyatullin[2], I I Girfanov[3], M M Remeyev[3] and A Mansurov[3]

[1]Oil and Gas Research Institute of the Russian Academy of Sciences (OGRI RAS), Gubkina str. 3, Moscow, 119333, Russia
[2]Almetyevsk State Oil Institute (ASOI), Lenina str. 2, Almetyevsk, 423450, Tatarstan, Russia
[3]TatNIPIneft Institute of PJSC "Tatneft", M. Djalilia str. 32, Bugulma, 423230, Tatarstan, Russia

E-mail: i-ind@ipng.ru

Abstract. The paper presents integral analysis of data obtained on carbonate reservoir rocks of Bashkir and Tournai formations (Republic of Tatarstan, Russia) under specially designed experimental studies for measuring rock permeability alteration under changes of pore pressure with constant triaxial outer (geostatic) stresses. Characteristic hysteretic effects of permeability alteration under cyclic changes of pore pressure are revealed and related to the reservoir and mechanical properties of rock samples, as well as different fluid saturation. It is shown that cyclic geomechanical treatment can promote significant improvements of flow characteristics of a carbonate rock material.

1. Introduction

It is well known that flow (transport) and mechanical properties of reservoir rocks vary significantly under changes of geostatic stresses and fluid pressure in pores (pore pressure). In most applications, it is assumed that the conception of effective stresses can satisfactorily describe these variations [1]. In petroleum applications, the simple Terzaghi relation for the effective stresses is usually adopted [2]:

$$\sigma_{eff} = \sigma - p, \tag{1}$$

with σ and σ_{eff} being the geostatic and the effective normal stress along a certain direction, and p being the pore (fluid) pressure. As proposed by relation (1), the pressure depletion typical for oil or gas production from petroleum reservoirs is assumed equivalent to the increase of geostatic stresses. Thus, most laboratory studies of reservoir rock behavior under changing loading (stress-pressure) conditions are performed on rock samples with constant pore pressure and changing confining pressure in the experiment [3, 1], and even with pore pressures much lower than at actual in-situ conditions. Though this approach has been shown to be appropriate in most situations for conventional sandstones [4], it is questionable for tighter and brittle rocks like carbonates.

An extended theory describing the effect of pore pressure on effective stresses was developed by Biot [5] with

$$\sigma_{eff} = \sigma - \alpha p, \tag{1}$$

with α being the Biot coefficient, which is close to 1 for highly-porous rocks (like weakly unconsolidated sandstones) and tends to 0 for nearly solid rocks with very low values of porosity. However, evaluation of α for tight rocks is quite tricky, and it turns out to depend on loading conditions [1].

Typically petroleum reservoir rocks demonstrate decreasing permeability (and porosity) values under pore pressure depletion [6]. Hysteresis is also typical, with only partial restoration of initial permeability when pressure is increased to its initial level and under further cyclic changes of pressure [7]. However, for carbonate rocks and some sandstones it was shown that if pore pressure is decreased (effective stress is increased) below a certain level, shear failure can occur in the rock sample with formation of micro fissures, which leads to a significant increase of permeability as pore pressure is restored [8-11]. This process is opposed to hydraulic fracturing which happens when pore pressure is increased above the tensile failure limit [12], but it can also be used as a method for increasing productivity of oil and gas wells [8-10]. Large-amplitude cyclic changes of pore pressure make the process more efficient, especially in tight and brittle carbonate rocks [13]. In this case, it is called the cyclic geomechanical treatment.

The possibility to increase rock permeability and thus productivity of wells by large-amplitude decrease in pore pressure was clearly shown for deep reservoirs (~2.5 km and deeper) with quite high initial fluid pressures and overburden stresses [8-10, 14]. For carbonate rocks at depths of about 1 km with initial fluid pressures ~100 bar it is questionable whether same effects can be obtained. The goal of this study is to investigate possible permeability variations under cyclic changes of pore pressure on rock samples with different reservoir and mechanical properties and fluid saturation from two carbonate oil-bearing formations of Republic of Tatarstan, Russia.

2. Experimental program

In this study, 12 laboratory experiments were performed on 6 rock samples of the Bashkir formation and 6 samples of the Tournai formation. In addition to different flow properties, effect of saturating fluid was also studied. Key paarameters of the samples are summarized in Tables 1 and 2.

The experiments were conducted on a certified laboratory setup under triaxial loading conditions corresponding to the actual in situ state of the reservoirs. Axial load, corresponding to vertical overburden stress, was planned to be 21.18 MPa for the Bashkir formation and 26.28 MPa for the Tournai formation. However, due to some technical issues, it was over overstated in most experiments which should be taken into account during interpretation. The confining pressure, corresponding to minimum lateral stress, was estimated by the Eaton's formula [15] and set to 14.16 MPa for the Bashkir formation and 17.45 MPa for the Tournai formation. The initial pore (fluid) pressure was P_0=8.9 MPa for the Bashkir samples and 10.83 MPa for the Tournai samples.

During the experiments on each sample, axial load and confining pressure were maintained, while pore pressure was changed stepwise in the following cycle: decrease – increase – decrease. Each experiment consisted of 14 pressure stages. Fig. 1 shows the typical program of the experiments which was slightly changed for some samples during the study.

Table 1. Parameters of the Baskir samples

Sample	Rock type	Porosity, %	Gas permeability, 10^{-3} μm^2	Saturating fluid
26	Limestone	12.03	5.74	water
32	Limestone	14.11	46.80	water
33	Limestone	15.60	32.07	water
34	Limestone	13.76	39.92	kerosene
35	Limestone	12.93	12.28	kerosene
36	Limestone	14.21	35.71	kerosene

IOP Conf. Series: Materials Science and Engineering **921** (2020) 012009 doi:10.1088/1757-899X/921/1/012009

Table 2. Parameters of the Tournai samples

Sample	Rock type	Porosity. %	Gas permeability, 10^{-3} μm^2	Saturating fluid
1	Limestone	15.14	179.76	kerosene
2	Limestone	14.49	129.65	water
3	Limestone	13.97	75.48	water
4	Limestone	12.00	35.53	kerosene
5	Limestone	13.97	31.59	water
6	Limestone	16.35	80.50	kerosene

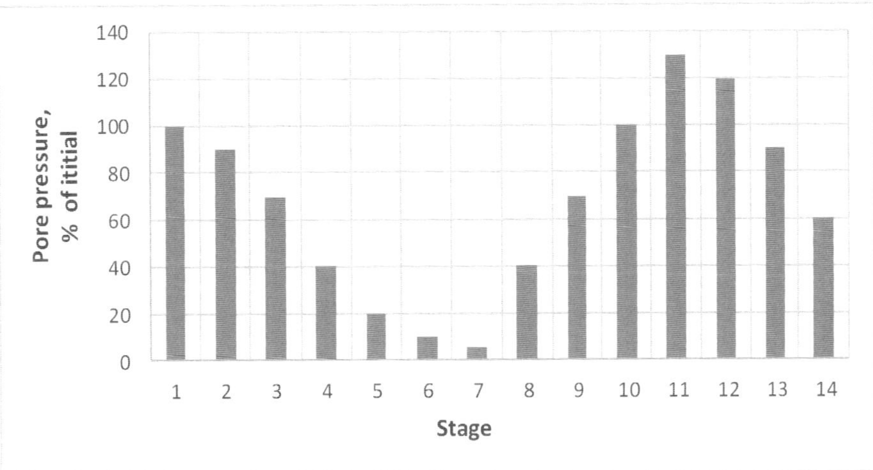

Figure 1. Typical program of the experiments

At each pressure stage, permeability to saturating fluid was measured after stabilization. Also, dynamic Young modulus and Poisson's ratio were measured by the acoustic wave method. It was controlled that changes to elastic properties were in compliance with measured changes in permeability.

3. Results and discussion

Fig. 2 shows the graphs of permeability (for saturating fluid) vs. pressure obtained in the experiments on the Bashkir samples. The following observations can be made.

• Saturation of the sample significantly affects the permeability vs. pressure dependence.

For samples 33 and 32 saturated with water, the primary decrease in permeability is many-fold and irreversible, i.e. permeability is not restored with the subsequent increase in pressure. For sample 26 also saturated with water, permeability practically did not change with decreasing and increasing pressure, and it increased only when pressure exceeded the tensile failure limit of the sample.

For sample 34 saturated with kerosene, the permeability changed nonmonotonically in the pressure decrease-increase cycle. But when the pressure was restored to the initial level, an increase in permeability of about 20% was recorded compared to the initial value. For samples 35 and 36 saturated with kerosene, nonmonotonic changes in permeability were also observed during the decrease and reverse increase in pressure. Nevertheless, a general increase in permeability of about 20% is also observed for these samples after restoration and / or reverse decrease in pressure to the initial values. (Note that in all experiments with Bashkir samples the axial load was overstated if compared to in-situ conditions.)

IOP Conf. Series: Materials Science and Engineering **921** (2020) 012009 doi:10.1088/1757-899X/921/1/012009

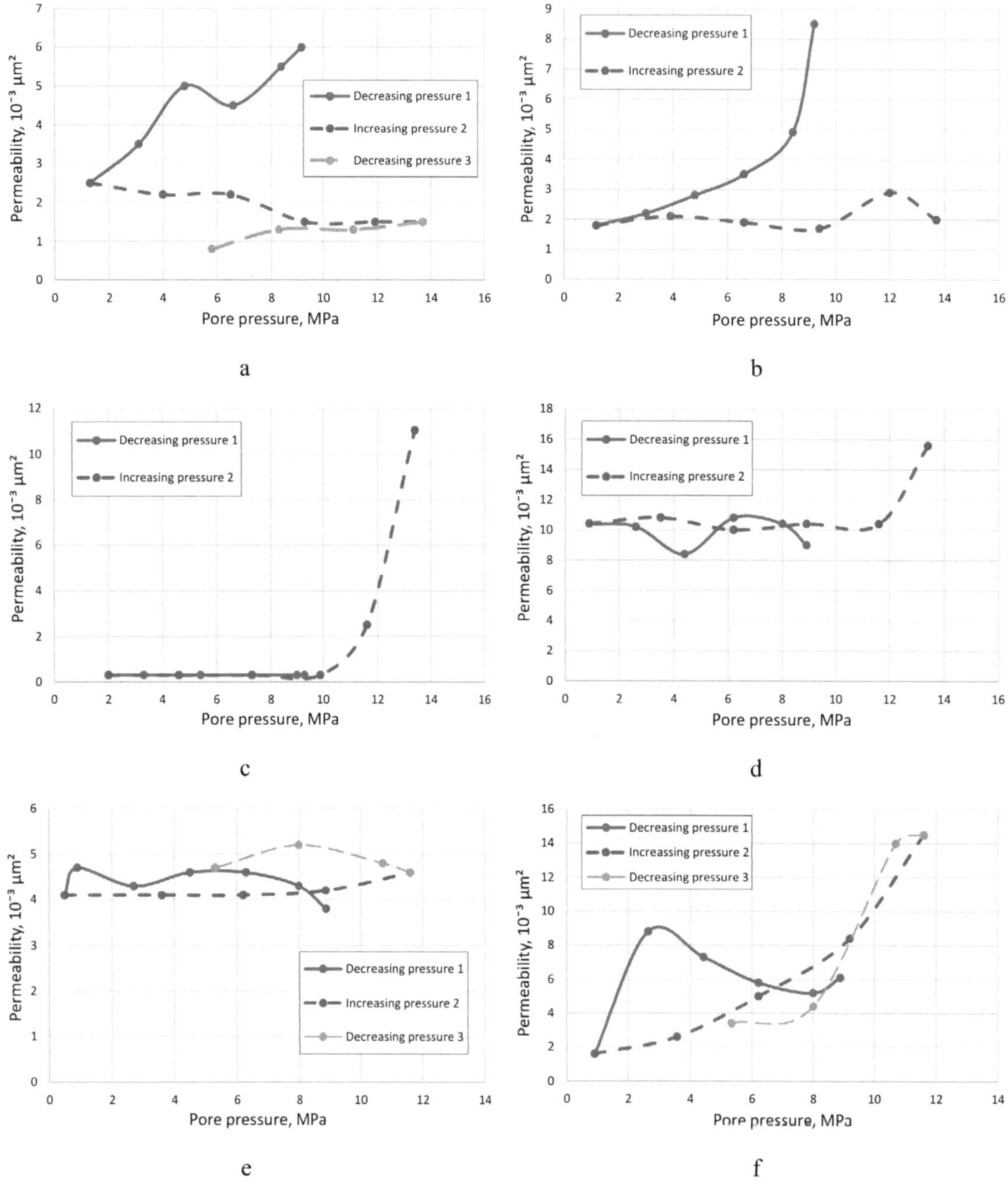

Figure 2. Experimental graphs of permeability vs. pressure for Bashkir samples. Samples: a – 33, b – 32, c – 26, d – 34, e – 35, f – 36

Fig. 2 also shows that for all samples saturated with kerosene, in some or other range of pressure values during reduction, an increase of permeability was observed. While for water-saturated samples only an intensive decrease (samples 33 and 32) or preservation of permeability (sample 26) was recorded.

• The nature of permeability vs. pressure dependence is also significantly affected by the internal structure (porosity, presence of microcracks) and associated mechanical characteristics of the samples.

IOP Conf. Series: Materials Science and Engineering **921** (2020) 012009 doi:10.1088/1757-899X/921/1/012009

Samples 33 and 32 have the smallest initial values of the dynamic Young modulus (20.73 GPa and 23.03 GPa). This may indicate either the initial presence of microcracks or strong change in mechanical properties (decrease in strength, plasticity) under the influence of water – the so-called water-induced compaction [16]. The initial values of porosity (14.11% and 15.60%) and permeability ($6 \cdot 10^{-3}$ μm^2 and $8.5 \cdot 10^{-3}$ μm^2) of these samples are also among the largest. These factors may explain the intense decrease in permeability with a decrease in pore pressure.

In contrast, for samples 26 and 35 having the largest initial values of the dynamic Young modulus (51.63 GPa and 44.14 GPa) and the smallest values of porosity (12.03% and 12.9%) and permeability ($0.3 \cdot 10^{-3}$ μm^2 and $3.8 \cdot 10^{-3}$ μm^2), with different saturating fluids, the permeability changed slightly with a decrease in pore pressure.

Among the kerosene-saturated samples, the largest amplitude of permeability variations was observed for samples 34 and 36 having the initial values of the dynamic Young modulus 25.47 GPa and 32.47 GPa, and the porosities of 13.76% and 14.21%, respectively.

For Tournai samples, permeability graphs vs. pressure are given in Fig. 3.

• Similar to the Bashkir formation, the influence of saturation is noted for the Tournai samples.

Samples 2 and 3 saturated with water demonstrated similar trends. A decrease in permeability was recorded for the primary decrease in pressure. With the following increase in pressure, permeability did not significantly increase but remained at approximately the same level or continued to decrease. A relatively sharp increase in permeability was noted in the pressure range of 12-14 MPa, presumably associated with the initial stage of tensile failure (hydraulic fracturing). With the subsequent decrease in pressure, permeability increased by 20-50%.

For sample 5, also saturated with water, similar trends with the samples 2 and 3 were observed starting from the 2nd stage of the initial decrease in pressure and for the subsequent increase in pressure. However, no evidence of tensile failure was noted, which explained no increase and a slight decrease in permeability with the secondary decrease in pressure. The mechanism of the sharp increase in permeability at stage 2 is not obvious.

However, as a common result for the samples 2, 3, and 5, a lesser effect of water saturation on the change of flow characteristics was noted for the Tournai formation in comparison with the Bashkir formation.

For kerosene-saturated Tournai samples, the situation is as follows.

The results for sample 1 are not fully informative because of its destruction at the 3rd stage of pore pressure reduction. However, the fact of destruction under the action of compressive stresses is worth mentioning, because this is the key mechanism of the cyclic geomechanical treatment.

Sample 6 also failed to fully complete the experiment due to technical issues. However, after the initial decrease in permeability, an increase was recorded at stages 3-4 under the action of compressive stresses which corresponds to the expected geomechanical effect of intrinsic shear failure. Unfortunately, since the change in pressure at the subsequent stages did not correspond to the experimental program, it is impossible to quantify the final effect on this sample.

The most interesting results were obtained for sample 4 saturated with kerosene and studied under the correct axial loading conditions corresponding to the actual in-situ overburden stress of the Tournai formation. For this sample, a pronounced tendency was observed of a substantial and almost uniform increase in permeability starting from the 3rd stage of the experiment. This can be explained by the formation of microcracks due to intrinsic shear (compression) failure and their subsequent opening under the influence of increasing pore pressure. At the pore pressure value of about 14 MPa, a sharper increase in permeability can be noted, which is presumably associated with additional tensile failure (hydraulic fracturing). With a subsequent decrease in pressure to 11 MPa, permeability somewhat decreased, probably due to compression of tensile cracks, but remained at higher values than for the same pressures during the previous stage of pressure increase. With the further decrease in pressure, the permeability only increased, which can be associated with the secondary development of microcracks under compression.

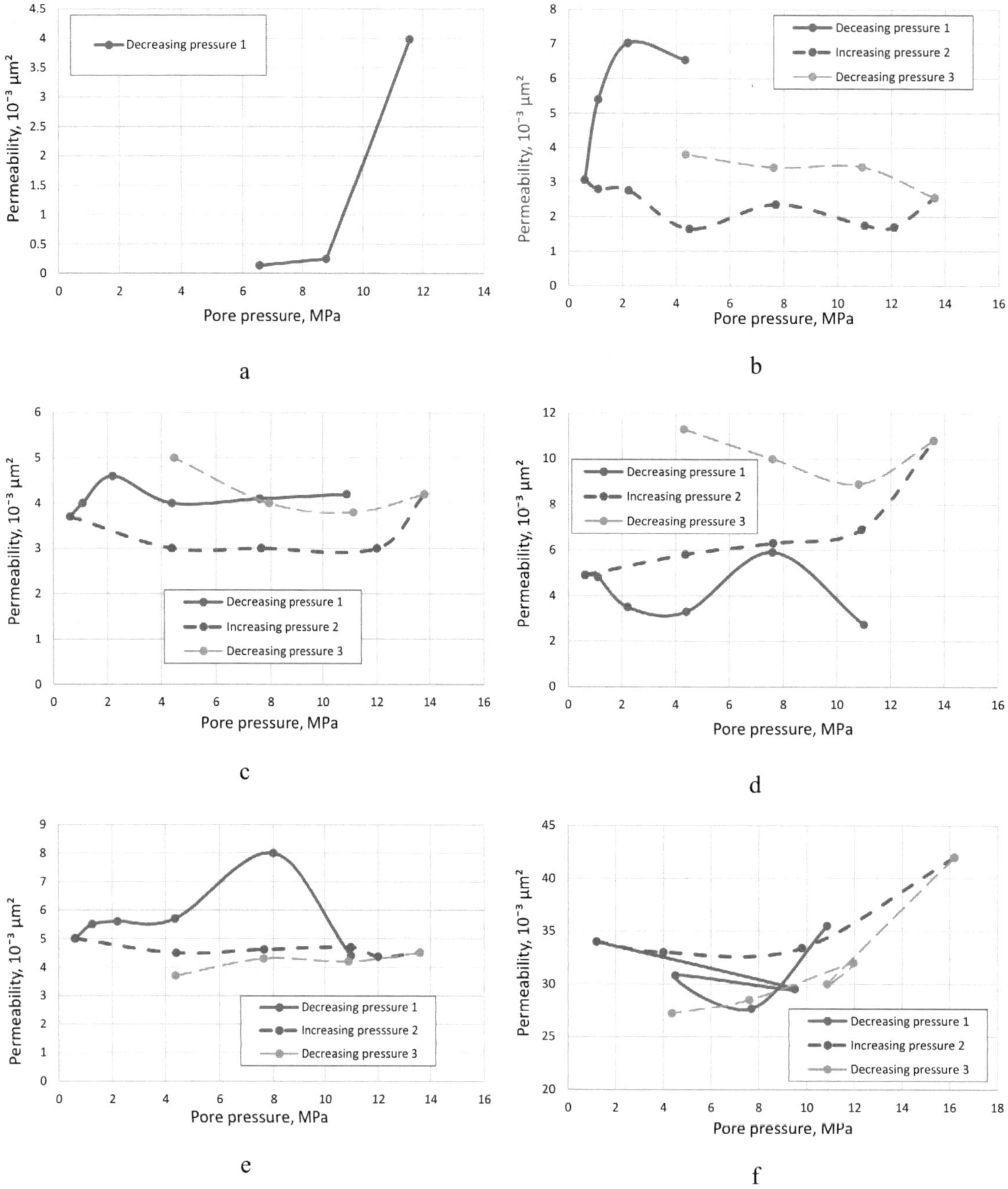

Figure 3. Experimental graphs of permeability vs. pressure for Tournai samples. Samples: a – 1, b – 2, c – 3, d – 4, e – 5, f – 6

Thus, for samples saturated with kerosene, a manifestation of destruction of the internal structure due to compressive stresses was observed, which can lead to the subsequent development of microcracking and an increase in permeability.

An overall increase in permeability for sample 4 compared to the initial value is ~3.5 times, which shows the potential effectiveness of the cyclic geomechanical treatment under the actual stress state of the Tournai formation.

• Different trends of permeability behavior with a secondary decrease in pore pressure were recorded. For samples 2, 3, and 4, with evidences of the onset of tensile failure, stabilization or increase in permeability was observed. For sample 5, with no signs of hydraulic fracturing, a decrease in permeability was noted during the secondary reduction of pore pressure.

• Similar to the Bashkir samples, one can note a relation between the specifics of variation of the flow characteristics with pressure and the initial values of the elastic moduli for the Tournai samples.

The minimum values of the initial dynamic Young modulus are 29.32 MPa for sample 6 and 33.94 GPa for sample 1, and the maximum value is 42.78 GPa for sample 4. Among the samples with a complete experiment, lowest values correspond to samples 5 (34.93 GPa) and 2 (35.88 GPa). It can be assumed that these rocks initially contain microcracks or other defects of the internal structure. This is confirmed by a more intense decrease in permeability, compared to other samples, with a decrease in pressure, especially for sample 2. Also, the initial presence of microcracks with the reduced values of Young modulus can explain the complete destruction of sample 1 and partial destruction of sample 6 at an early stage of the experiment.

The highest values of the initial Young modulus were observed for samples 3 (40.88 GPa) and 4 (42.78 GPa). With a decrease in pore pressure, they demonstrated no significant decrease in permeability. And for sample 4, the mechanism of microcracks creation by the cyclic geomechanical treatment was found to be very efficient.

4. Conclusions

From the presented experimental data and analysis, the following conclusions can be drawn.

- The change in permeability for the carbonate reservoirs of the Bashkir and Tournai formations with a change in pore (fluid) pressure can be significant and depends both on saturating fluid and the initial flow and elastic properties (i.e., on the internal structure) of the rock samples.
- Saturation with water leads to a more intense decrease in permeability than saturation with kerosene. This may be due to the effects known as water induced compaction [16]. However, since fresh water was used in the experiments, non-geomechanical reasons for the permeability reduction due to interaction of water with carbonate rock cannot be excluded. For both the factors, further studies are required. The effect of water is more significant for the Bashkir formation than for the Tournai formation.
- For kerosene-saturated samples, a positive effect of cyclic geomechanical treatment was obtained. For the Bashkir formation, the increase in permeability was about 20% (with axial load overstated). For a Tournai sample studied under the correct loading conditions, 3.5-times increase in permeability was recorded. For several samples, the experimental program was not fully completed due to their destruction, but evidence of microcracking under decrease in pore pressure was also recorded.
- Extra analysis showed that pore pressure required for tensile failure (hydraulic fracturing) is reduced, compared to theoretical estimates, after primary pressure reduction in the experiment. This effect was most pronounced for water saturated Bashkir samples.

The results of the study show that geomechanical effects associated with cyclic changes of pore pressure are significant and can lead to creation of microcracking and increase in permeability in carbonate reservoirs. These effects should be taken into account in petroleum reservoir simulation and can form a basis for treatment methods to improve productivity of oil and gas wells.

Acknowledgements
The results presented in the paper were obtained under the Research Contracts of ASOI and TatNIPIneft funded by PJSC Tatneft, and under the State Research Contract of OGRI RAS (topic AAAA-A19-119022090096-5).

References

[1] Zoback M D 2007 *Reservoir geomechanics* (New York: Cambridge University Press) p 505

[2] Terzaghi K 1943 *Theoretical Soil Mechanics* (New York: John Wiley) p 510
[3] Fatt I 1958 Compressibility of sandstones at low and moderate pressures *AAPG Bulletin* **42** 8 p 1924-57
[4] Zimmerman R, Haraden J and Somerton W 1985 Effects of pore pressure and confining pressure on pore and bulk volume compressibilities of consolidated sandstones *ASTM Special Technical Publication* 24-36
[5] Biot M A 1962 Mechanics of deformation and acoustic propagation in porous media *Journal of Acoustic Society of America* **28** 168–91
[6] Tiab D and Donaldson E C 2003 *Petrophysics: theory and practice of measuring reservoir rock and fluid transport properties: Second edition* (Burlington: Gulf Professional Publishing) p 889
[7] Beattie C I, Boberg T C and Mcnab G S 1991 Reservoir simulation of cyclic steam stimulation in the Cold Lake oil sands *SPE Reservoir Engineering* **6** 2
[8] Zakirov S N 1998 *Development of gas, gas-condensate and oil-gas-condensate deposits* (Moscow: Struna) p 628
[9] Khristianovich S A, Kovalenko Yu F, Kuinich Yu V and Karev V I 2000 Increasing productivity of oil wells by the method of "geo-loosening" *Oil and gas Eurasia* 2 p 90-4
[10] Karev V I 2010 *Influence of stress-strain state of rocks on the flow process and well rates - Doct Tech Sci Thesis* (Moscow: IPMech RAS)
[11] Kovalenko Yu F 2012 *Geomechanics of oil and gas wells – Doct Phys-Math Sci Thesis* (Moscow: IPMech RAS)
[12] Fjær E, Holt R M, Horsrud P, Raaen A M and Risnes R 2008 *Petroleum related rock mechanics. Second edition* (Elsevier) p 492
[13] Zakirov S N, Drozdov A N, Zakirov E S et al 2018 Technical and technological aspects of geomechanical reservoir treatment *Neftegaz.RU Magazine* 6 p 24-9
[14] Zakirov S N, Drozdov A N, Alekseyev B G and Kolobanov A V 2018 Natural evidences of geomechanical processes *Subsurface Management – XXI Century* 3 p 72-7
[15] Eaton B A 1969 Fracture gradient prediction and its application in oilfield operations *Journal of Petroleum Technology* **246** p 1353–60
[16] Sylte J E, Thomas L K, Rhett D W, Bruning D D and Nagel N B 1999 Water induced compaction in the Ekofisk field *SPE Annual Technical Conference and Exhibition (Houston)* SPE-56426-MS

IOP Conf. Series: Materials Science and Engineering **921** (2020) 012010 doi:10.1088/1757-899X/921/1/012010

The formation of nanoparticles in the process of ultrasonication of the pure water

D D Ivanova[1], M V Gorbachevskii[2], A A Novikov[2] and V N Kuryakov[3]
[1] D.I. Mendeleev University of Chemical Technology of Russia, 9, Miusskaya Street, Moscow, 125047, Russia
[2] Gubkin University, Department of Physical and Colloid Chemistry, 65/1 Leninsky prospect, Moscow, 119991, Russian Federation
[3] Oil and Gas Research Institute of the Russian Academy of Sciences, 3, Gubkin Street Moscow, 119991, Russian Federation

E-mail: vladimir.kuryakov@gmail.com

Abstract. Ultrasonication is a very convenient method for the preparation of various dispersions. However, it is often not clear what influence the ultrasonication procedure is on the purity of the prepared dispersion. In this paper, we show that the prolonged ultrasonication of the pure water results in dispersion formation. We characterized the nanoparticles in the resulted dispersion by the TEM, DLS, conductivity, and zeta potential measurements. This work points to the possible source of contamination of the sonicated dispersions.

1. Introduction

Ultrasonic equipment is widely used in various applications, for example, for dispersing particles of various nature in a liquid, homogenizing, and emulsions preparation, for cleaning surfaces (ultrasonic baths for cleaning), for welding and therapeutic applications.

In the ultrasonic treatment technique, sound waves with frequency range equal or more than 20 kHz are used. Rapid changes of pressure in liquid formed by a vibrated activator (horn) lead to the formation of small vapor-filled cavities in places where the pressure is relatively low. When subjected to higher pressure, these cavities, called "bubbles" or "voids," collapse and can generate a shock wave that is strong very close to the bubble, but rapidly weakens as it propagates away from the bubble.

Ultrasonic equipment is used for fine grinding of solids or liquids, for transferring substances into a dispersed state (dispersion) with the formation of a sol under the action of ultrasonic vibrations. Particles, drops, or aggregates are destroyed as a result of repeated exposure to their surface of pulsed loads accompanying the closure of cavitation bubbles [1]. Besides, it is also possible grinding due to the collision of the powder particles resulting from their random movement under the influence of ultrasound.

In [2-4], the authors obtained stable aqueous dispersions of *n*-alkanes without the addition of surfactants using ultrasonication (22 kHz, 75 W). The dispersion particles had an average size (radius) of about 100 nm and a zeta potential of up to -50 mV. In this paper, we study the effect of ultrasonication on pure water when a submersible titanium activator (horn) is used. The observed dispersion formation may explain the high stability of ultrasonicated *n*-alkane dispersions without the addition of surfactants. This kind of information can be interesting in applications related to the production of various types of dispersions.

2. Experimental

Samples for studies were prepared from deionized water produced by the Simplicity UV (Millipore, Bedford, MA, USA). The volume of each sample was 10 mL. For sample preparation, we used an ultrasonicator UZDN-A (22 kHz, 75 W, Russia) with a titanium submersible activator. Three samples were prepared. For two samples (sample G1 and sample G2), the sonication was carried out in glass vials. One sample was prepared in a plastic cuvette (sample P1). Sample P1 in a plastic cuvette and sample G1 in one of the glass vials were prepared by successively sonicating ten times for 30 seconds. Moreover, before each of the ten sonications, the scattered light intensity, particle size utilizing dynamic light scattering (DLS), conductivity, and particle zeta potential were measured.

For the second sample G2 in a glass vial, the preparation was performed by single sonication for 300 seconds, and the same measurements were performed as for the two samples described above. Thus, we could study both the effect of the vessel's material in which the sonication took place and the effect of the sonication procedure. For all three samples, a change in time of such properties as scattered light intensity, particle size (DLS), conductivity, and zeta potential of the particles was studied. Before each measurement, the samples were intensively mixed so that the particles settled between measurements were redispersed. Sample G1 was also examined by transmission electron microscopy 40 days after preparation.

For analysis by transmission electron microscope (TEM), 5 μL of dispersion was dropped on a formvar TEM grid (Ted Pella, Redding, CA, USA) and dried at room temperature. TEM micrographs were obtained using a JEM-2100 electron microscope (JEOL, Japan) at an accelerating voltage of 200 kV and additionally using a LEO912 AB OMEGA (Carl Zeiss, Germany). Energy-dispersive X-ray spectroscopy analysis (EDX) was performed with JED-2300 (JEOL, Japan) analysis station.

Scattered light intensity, conductivity, particle size, and zeta potential were measured using Photocor Compact-Z equipment (Russia). With a scattering angle of 90 degrees for measurements of scattered light intensity and particle size and a scattering angle of 20 degrees for zeta potential measurements. The laser wavelength was 654 nm; the laser power was 30 mW. Each measurement was carried out at least three times and the average value of the obtained results was found.

3. Results and discussion

Figure 1 shows the results of measurements of scattered light intensity, conductivity, particle size, and zeta potential for samples G1 and P1 before each of the ten sonications. Figure 1 shows that with an increase in the sonications of samples G1 and P1, an increase in the scattering intensity in both samples is observed. In the sample G1, the scattering intensity grows linearly with each sonication treatment, and in the sample P1, the dependence has the form of a growing exponential. The intensity of the scattered light in sample G1 after ten sonications is 2.5 times greater than in sample P1. The conductivity of sample G1 increases with each sonication and, after ten treatments, is approximately 2.5 times higher than the initial value. For sample P1, the conductivity remains the same within the accuracy of the measurements. No particles were observed by the DLS method in pure water. After 1-2 sonications in samples G1 and P1, 80-90 nm particles were detected by DLS. Within the limits of measurement accuracy, the particle size does not change after repeated sonications. However, the reproducibility of size measurements is low, which may be due to the low concentration of such particles and the broad size distribution of such particles in the samples. The zeta potential of these particles does not vary with the number of sonications. For particles in sample G1, the zeta potential is about -40 mV, for particles in sample P1 about -20 mV. One can say that the sonicating vessel material affects the change in the physicochemical properties of the sample.

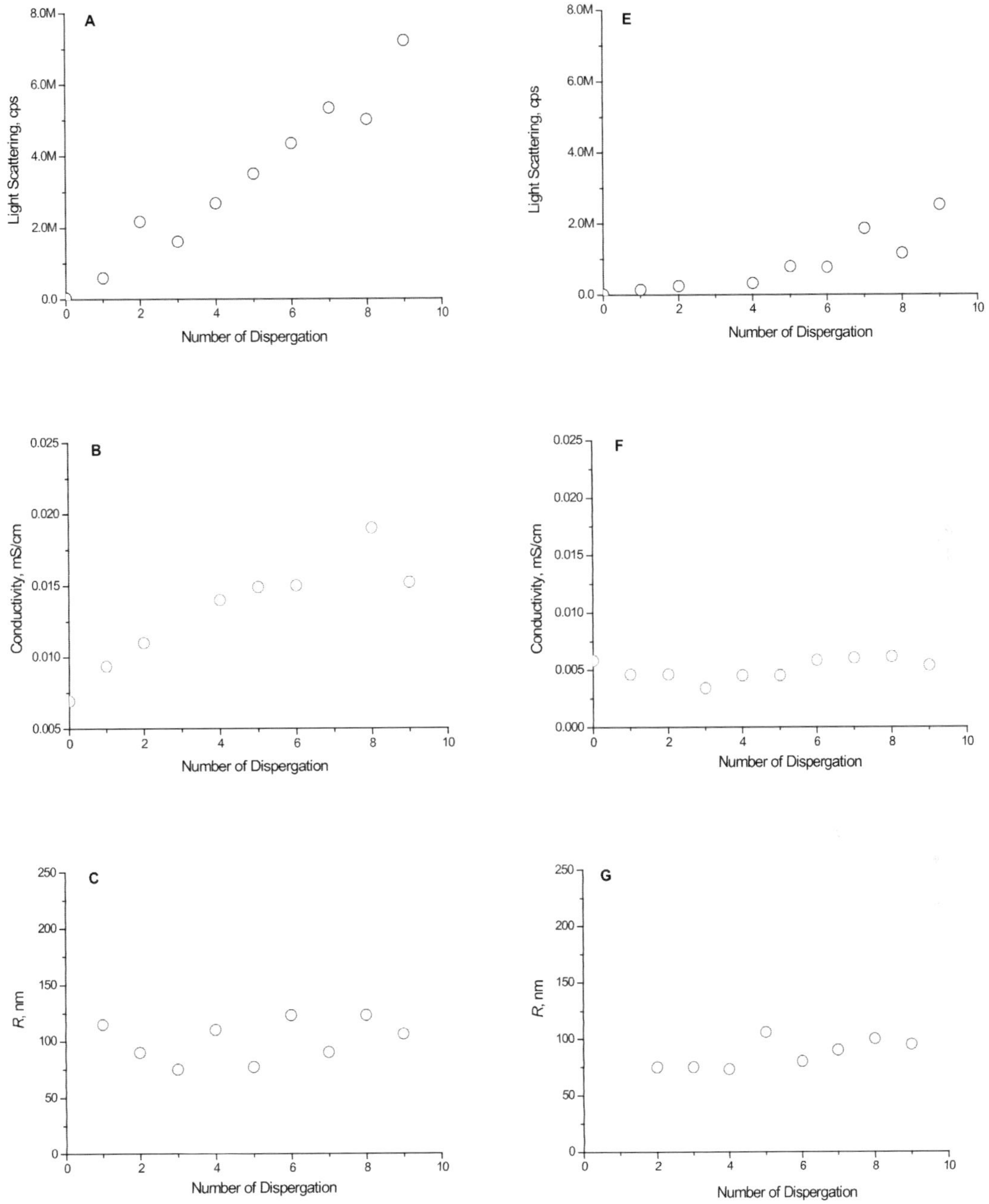
A
Light Scattering, cps
Number of Dispergation
E
Light Scattering, cps
Number of Dispergation
B
Conductivity, mS/cm
Number of Dispergation
F
Conductivity, mS/cm
Number of Dispergation
C
R, nm
Number of Dispergation
G
R, nm
Number of Dispergation

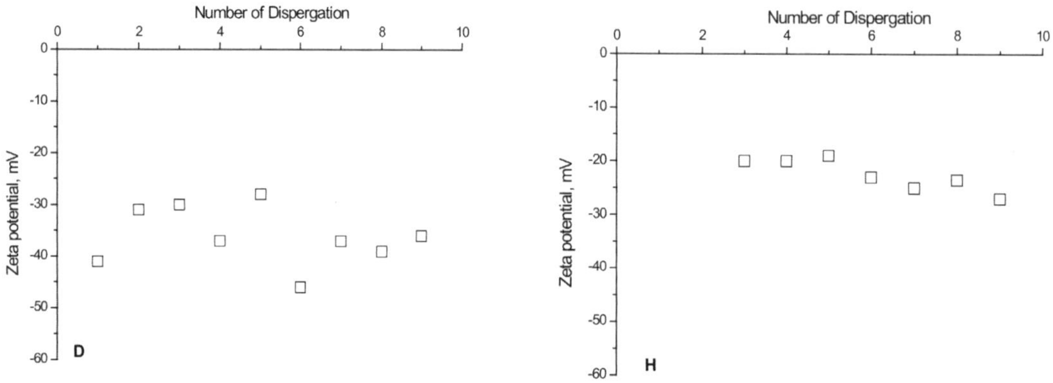

Figure 1. The dependence of the intensity of the scattered light, conductivity, radius and zeta potential of particles on the number of sonications. A, B, C, D and E, F, G, H for samples G1 and P1, respectively.

After ten sonications over one day, samples G1 and P1 were used to measure the scattered light intensity, conductivity, particle radius, and zeta potential over several weeks. The results of these measurements are presented in Figure 2.

It can be seen from Figure 2 that, 2-3 days after dispersion, the scattering intensity in samples G1 and P1 decreases 2–3 times and subsequently does not change significantly. This decrease may be because, after sonication, the samples are not in equilibrium, and in the first 2-3 days, they come to an equilibrium state. Such equilibrium establishment times are close to equilibrium establishment times in a system containing nanobubbles and are associated with the lifetime of such nanobubbles [5-8]. In water, nanobubbles can be visualized using Cryo-EM [9], and their size and concentration can be measured using various experimental techniques (DLS, NTA) [10]. At the same time, the particle radius measured by the DLS method during the observation period increased from 100-130 nm to 250-300 nm in both samples.

Samples G1 and P1 are characterized by an increase in conductivity after several days and stabilization at the level of 0.05 mS/cm and 0.03 mS/cm, respectively. The zeta potential in the studied samples decreases monotonously, and the observation time changes approximately twofold over time. An increase in the average particle size and a decrease in the zeta potential modulus indicates low stability of this dispersion.

Figure 3 shows the results of measurements of scattered light intensity, conductivity, particle radius, and zeta potential for sample G2 within 28 days after sample preparation. Samples G1 and G2 were prepared in the same glass vials. The sonication time is the same for 5 minutes, but sample G1 was sonicated ten times for 30 seconds, and sample G2 was sonicated once for 300 seconds. Comparing the results of studies of samples G1 and G2 (Figs. 1, 2, and 3), it can be seen that the properties of these samples differ slightly in the first days after preparation, but subsequently, after establishing equilibrium values, they become close in their values. The intensity of the scattered light is notably different in these samples immediately after preparation, but after the observation time, it stabilizes at around 1.5×10^6 counts per second (cps). The samples' conductivity increases by about three times; the particle radius increases from 90–100 nm to 200–250 nm, and the zeta potential decreases by about two times in absolute value.

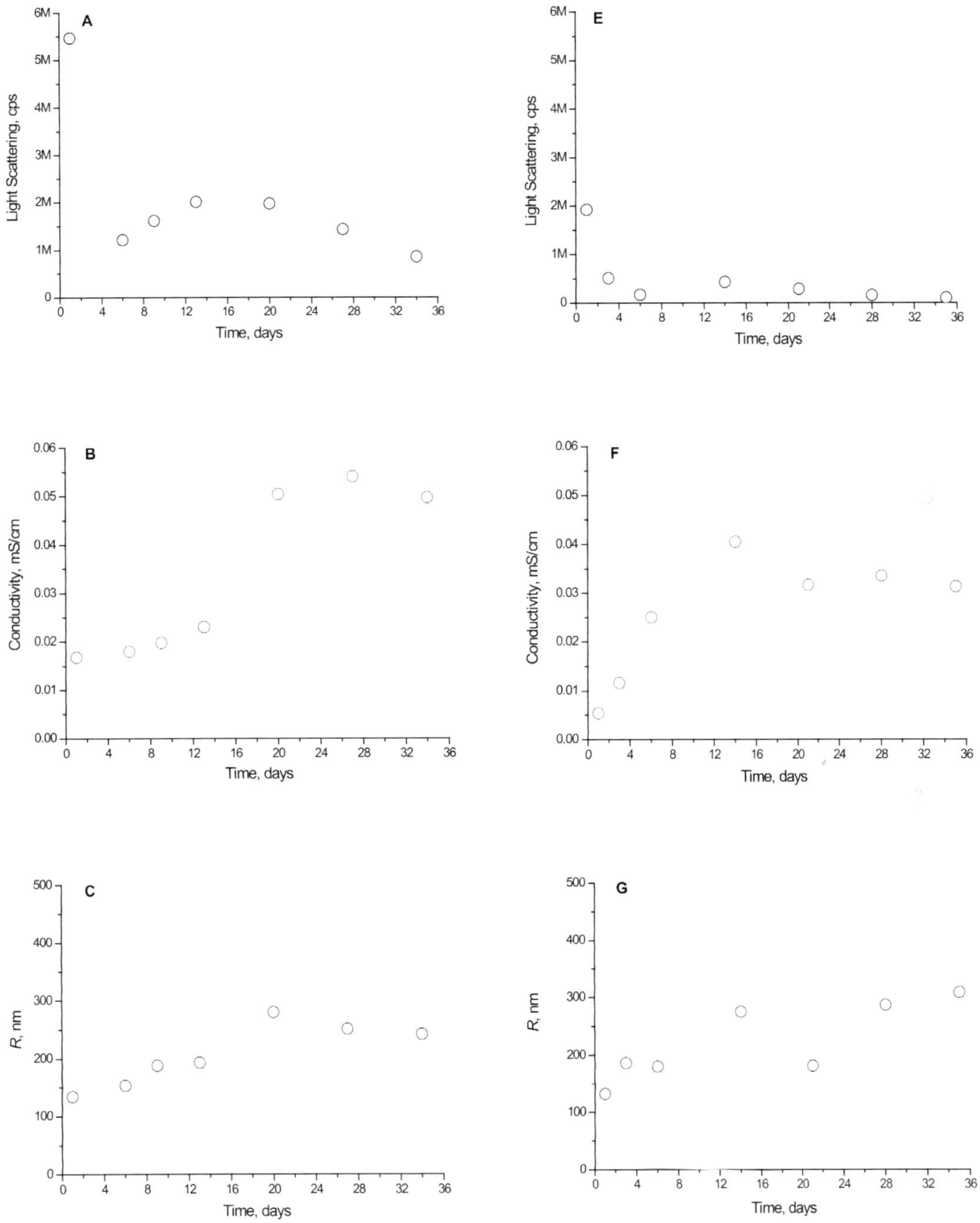
A
Light Scattering, cps
Time, days
E
Light Scattering, cps
Time, days
B
Conductivity, mS/cm
Time, days
F
Conductivity, mS/cm
Time, days
C
R, nm
Time, days
G
R, nm
Time, days

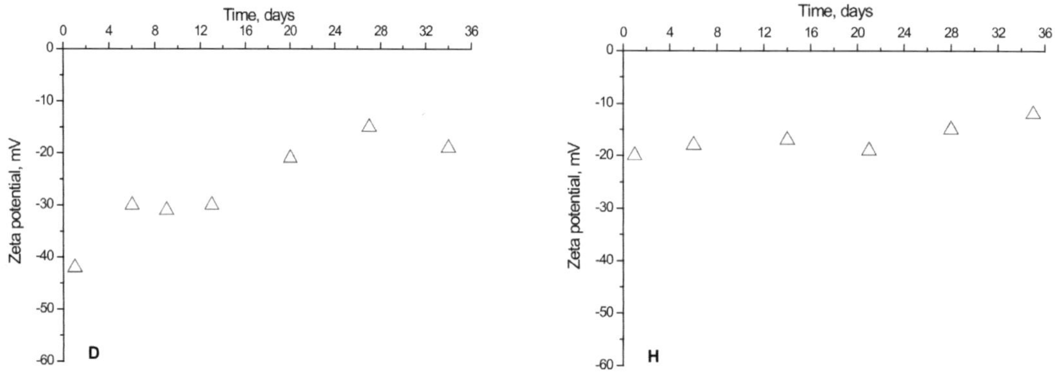

Figure 2. Dependence of scattered light intensity, conductivity, particle radius and zeta potential on the time elapsed after sonication. A, B, C, D and E, F, G, H — for samples G1 and P1, respectively.

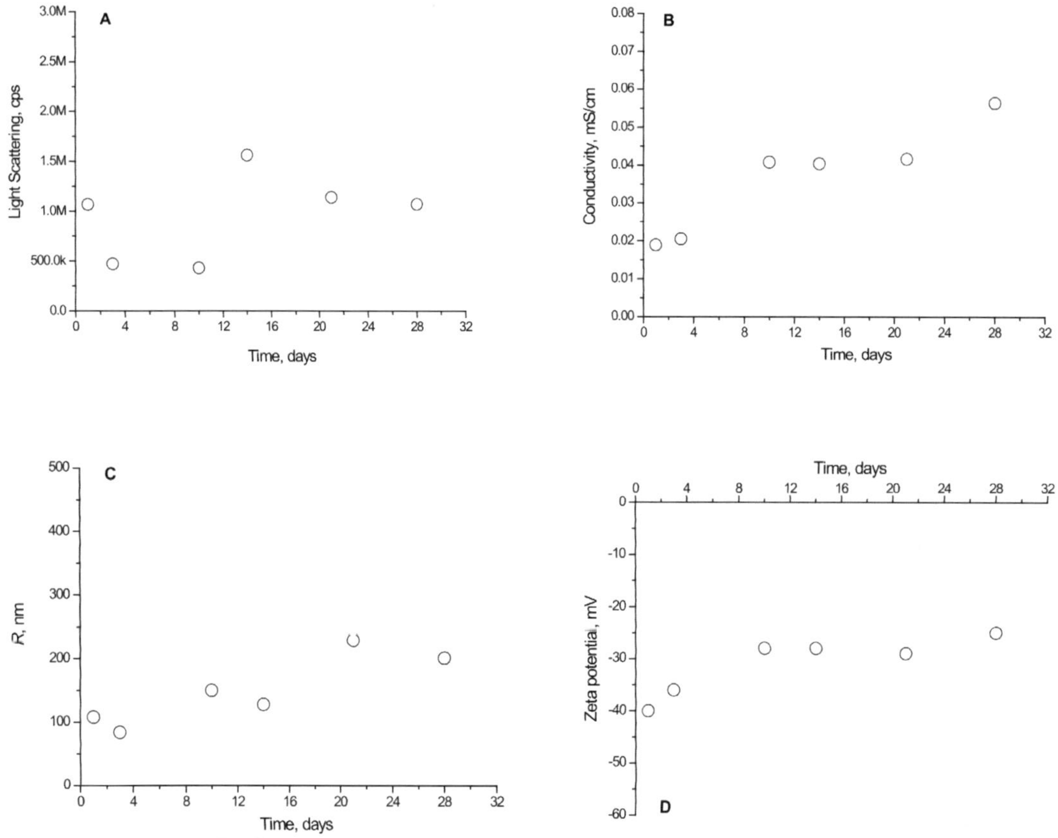

Figure 3. The time dependence of the scattered light intensity, conductivity, particle radius and zeta potential for sample G2.

Figure 4 shows microphotographs obtained using TEM for sample G1 40 days after preparation. Microphotographs clearly show that particles with sharp edges, not spherical, are present in the studied sample. This form is typical for particles obtained by mechanical grinding of a large piece of material. EDX analysis of the detected nanoparticles showed that the nanoparticles have titanium in their composition. The Kα line of titanium (4.508 keV) was present in all the obtained EDX spectra. Also present were Cu lines (TEM grid material). Lines of other metals were not uniquely identified. These titanium-containing particles appeared in the sample, breaking off from the surface of the titanium horn of the sonicator. From the analysis of microphotographs, one can conclude that the sizes (diameter) of these particles lie in the range of 50–300 nm, which is consistent with the results obtained for these particles by the DLS method. With an increase in the sonication time, the formation of these particles can explain the increase in light scattering intensity.

In Figure 4A, in addition to the particles described above with a size of 50–300 nm, one can notice that there are light and dark areas on microphotographs. TEM microphotographs of these regions with maximum magnification did not allow for reliable identification of individual particles. The dark areas in this figure indicate the presence in the sample of some substance forming a film upon drying. In this work, we were not able to determine what kind of substance it is. It likely provides an increase in the conductivity of the samples.

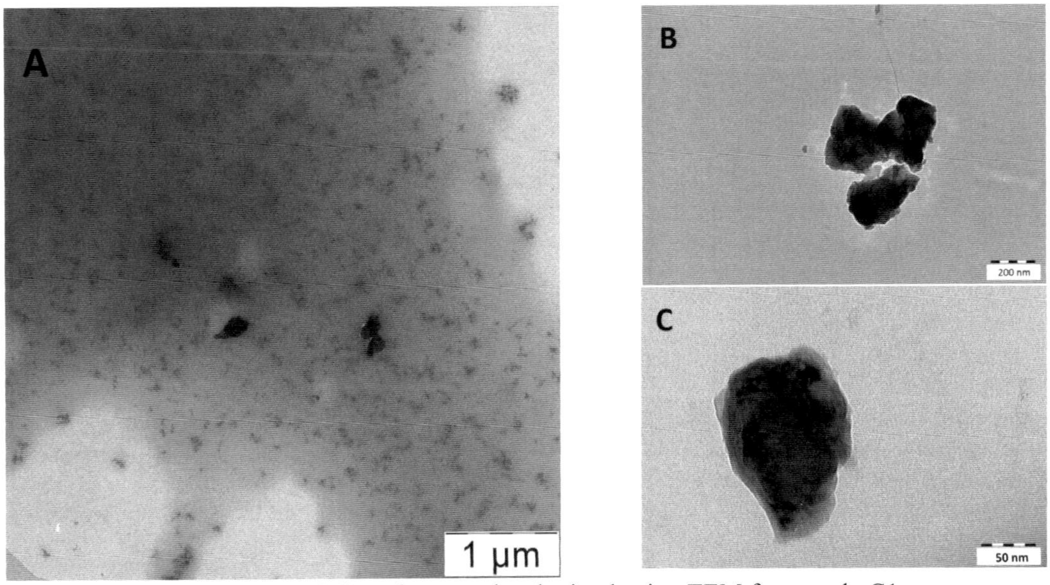

Figure 4. Microphotographs obtained using TEM for sample G1

4. Conclusions

The formation of the nanoparticles in pure water after sonication with a titanium submersible activator (horn) was studied.

It was shown that the treatment of pure water by ultrasonication leads to the formation of titanium-containing nanoparticles in it with a size of 50–300 nm with a zeta potential of around -30 mV and an increase in the conductivity of water. Most likely, this process is due to the erosion of the submersible horn surface. It should be noted that the sonicated dispersions can be contaminated by the eroded horn material, which could affect particle size, conductivity, and zeta potential of the particles in the prepared dispersion.

Acknowledgments

The work of D.D.I. was funded by RFBR, project number 20-35-80007. Electron microscopy and elemental analysis studies were performed by A.A.N. and M.V.G. and were funded by the Ministry of

Science and Higher Education of the Russian Federation within the State Assignment (project 0768-2020-0007).

5. References

[1] Kobus Z 2008 Influence of physical properties of liquid on acoustic power of ultrasonic processor *TEKA Kom. Mot. Energ. Roln.* **8**
[2] Kuryakov V N and Ivanova D D 2019 Crystallization Behavior of Pure n-Alkane (n-Nonadecane) in a form of Nanoemulsion *Int. J. Nanosci.* **18** 1940032
[3] Kuryakov V N and Ivanova D D 2019 Determination of melting point of n-alkanes by means of light scattering technique *J. Phys. Conf. Ser.* **1385** 12045
[4] Kuryakov V N, De Sanctis Lucentini P G and Ivanova D D 2018 Tricosane (C23H48) and Octacosane (C28H58) mixture phase transition insight via Light scattering techniques *IOP Conference Series: Materials Science and Engineering* vol 347
[5] Kikuchi K, Tanaka Y, Saihara Y, Maeda M, Kawamura M and Ogumi Z 2006 Concentration of hydrogen nanobubbles in electrolyzed water *J. Colloid Interface Sci.* **298** 914–9
[6] Kikuchi K, Ioka A, Oku T, Tanaka Y, Saihara Y and Ogumi Z 2009 Concentration determination of oxygen nanobubbles in electrolyzed water *J. Colloid Interface Sci.* **329** 306–9
[7] Bunkin N, Suyazov N, Shkirin A, Ignat'ev P and Indukaev K 2009 Cluster structure of stable dissolved gas nanobubbles in highly purified water *J. Exp. Theor. Phys.* **108** 800–16
[8] Bunkin N, Suyazov N, Shkirin A, Ignatiev P and Indukaev K 2009 Nanoscale structure of dissolved air bubbles in water as studied by measuring the elements of the scattering matrix *J. Chem. Phys.* **130** 134308
[9] Li M, Tonggu L, Zhan X, Mega T and Wang L 2016 Cryo-EM Visualization of Nanobubbles in Aqueous Solutions *Langmuir* **32**
[10] Kobayashi H, Maeda S, Kashiwa M and Fujita T 2014 Measurements of ultrafine bubbles using different types of particle size measuring instruments *Proceedings of SPIE - The International Society for Optical Engineering* vol 9232 p 92320U

IOP Conf. Series: Materials Science and Engineering **921** (2020) 012011 doi:10.1088/1757-899X/921/1/012011

On experimental determination of the elastic and strength properties of anisotropic rocks

V I Karev, Yu F Kovalenko and Yu V Sidorin

Ishlinsky Institute for Problems in Mechanics RAS, Moscow, Russia
wikarev@ipmnet.ru

Abstract. Influence of anisotropy of elastic and of strength properties on the stress-strain state near wells and openings is considered. The choice of parameters characterizing both types of anisotropy is discussed. A method of experimental determination of the parameters characterising anisotropy of both elastic and strength properties from the direct measurements using true triaxial loading systems is described.

1. Introduction

Wide groups of rocks exhibit directional variation of properties that may be described in terms of transversal isotropy, inherent to layered structures. Among these properties are permeability and elasticity, characterized by 2-nd rank and 4-th rank tensors, respectively. In case of transversal isotropy, these tensors have 2 and 5 independent parameters. Such important properties as strength, or transition from elastic to inelastic deformation in general case, do not possess tensor nature. To describe these properties a combined criterion based on existence of two mechanisms of fracture and two alternative extensions of Hill's yield criterion is used.

2. Anisotropy of elasticity

For transverse isotropy, which is the case for the majority of sedimentary rocks, the Hooke's law involves five independent elastic constants [1], however for the 2-D problem in a plane x_1x_3 normal to the plane of symmetry x_1x_2 the number of independent constants β_{ij} reduces to four

$$\begin{aligned} \varepsilon_{11} &= \beta_{11}\sigma_{11} + \beta_{13}\sigma_{33} \\ \varepsilon_{33} &= \beta_{13}\sigma_{11} + \beta_{33}\sigma_{33} \\ 2\varepsilon_{13} &= \beta_{55}\sigma_{13} \end{aligned} \qquad \beta_{ij} = \begin{cases} s_{ij} & \text{for plane stress} \\ s_{ij} - s_{i2}s_{j2}/s_{22} & \text{for plane stain} \end{cases} \tag{1}$$

Here ε_{ij}, σ_{ij} are components of strain and stress tensors; β_{ij}, s_{ij} are compliance coefficients for plane stress and plane strain, respectively.

It was shown [1] that for any problem with boundary conditions given in terms of stresses the stress distribution depends on two dimensionless combinations of elastic constants. However, it is desirable to distinguish the most essential parameter. Usually the ratio of modules (compliances) in x_1 and x_3 directions are chosen [2]. Although this choice appears to be natural it will be shown to be not optimal.

The values of attraction are those characterizing deviation of stress state comparing to that in isotropic body under the same boundary conditions. Since there could be no universal parameter appropriate for arbitrary geometries and boundary conditions, it is reasonable to consider simple typical problems: primarily, stress concentration at a well (cylindrical opening) contour.

2.1. Horizontal well in layered rock under hydrostatic stress

The problem of a cylindrical hole which axis coincides with one of the principle axes of elasticity tensor of orthotropic media was solved by [1]. Using that solution, the stress concentration at a well contour bored along bedding in transverse isotropic massif stressed hydrostatically $\sigma_{xx}^{\infty} = \sigma_{zz}^{\infty}$ may be written as follows

$$f_0(\theta) = \frac{\sigma_{\theta\theta}}{\sigma_{zz}^{\infty}} = 1 + \frac{E_\theta}{E_1}\left\{n - k + n(k-1)\cos^2\theta + \left[(k+1)^2 - n^2\right]\sin^2\theta\cos^2\theta\right\} \tag{2}$$

Here E_θ is the modulus of elasticity in the tangent direction along contour; $E_1 = \beta_{11}^{-1}$ is the modulus of elasticity in x_1 direction.

$$\frac{E_\theta}{E_1} = \left(\sin^4\theta + m\sin^2\theta\cos^2\theta + k^2\cos^4\theta\right)^{-1} \tag{3}$$

Angle θ is measured from x_1 axis to x_3 axis (figure 1). Parameters k,m,n are determined as follows

$$k^2 = \frac{\beta_{33}}{\beta_{11}},\ m = \frac{2\beta_{13} + \beta_{55}}{\beta_{11}} = \frac{2s_{13} + s_{55}}{s_{11}} - \left(\frac{s_{12} - s_{13}}{s_{11}}\right)^2,\ n^2 = 2k + m \tag{4}$$

Contribution of the last term in the second formula of (4) is usually negligible; hence the difference between parameters characterizing plane strain and plane stress is not essential.

According to (2) the extreme concentrations are reached at points corresponding to $\theta = 0, \pi$ and $\theta = \pm\pi/2$

$$f(0) = f(\pi) = 1 + \frac{n-1}{k} = 1 + \frac{\sqrt{2k+m} - 1}{k} \tag{5}$$

$$f(-\pi/2) = f(\pi/2) = 1 + n - k = 1 + \sqrt{2k+m} - k \tag{6}$$

for $E_1 \geq E_3$ $f_B \geq f_A$. Distribution of stress concentrations along well contour are given on figure 2. It follows from equation (2) that for

$$n - k - 1 = 0 \tag{7}$$

The stress concentration at the contour remains constant and equal to the stress concentration of at the contour of a hole in isotropic media. Condition (7) are written in terms of elastic constants as follows

$$a_{55}^0 \equiv \frac{1}{G"} = \beta_{11} + \beta_{33} - 2\beta_{13} = s_{11} + s_{33} - 2s_{13} - \frac{(s_{12} - s_{13})^2}{s_{11}} \tag{8}$$

The last term appeared here for plane strain is usually negligible, similar to (4). It is followed from the above that the deviation of $a_{55} = G_{13}^{-1}$ from the value determined by (8) is an essential parameter characterizing elastic anisotropy. Therefore the ratio of independent shear modulus $G_{13} = a_{55}^{-1}$ to $G"$ determined by (8) characterizes the degree of anisotropy

$$k_A = \frac{a_{55}^0}{a_{55}} = \frac{G_{13}}{G"} \tag{9}$$

The role of the independent shear modulus $G_{13} = a_{55}^{-1}$ on stress distribution was pointed out by Lekhnitsky [1].

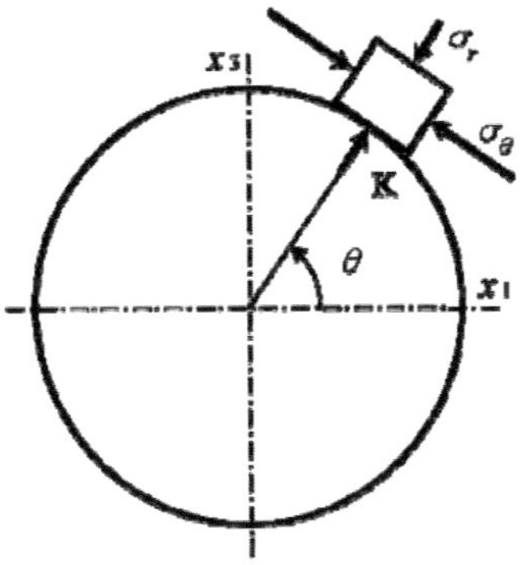

Figure 1. Configuration and stresses.

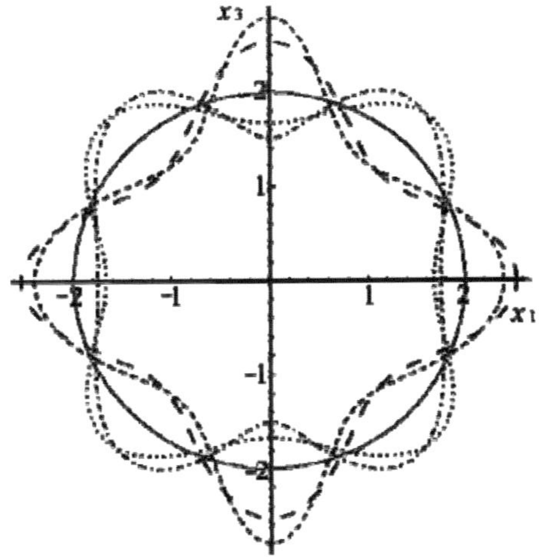

Figure 2. Stress concentrations along the well contour in transverse isotropic rocks for $\beta_{12}/\beta_{11}=-0.2$, $\beta_{13}/\beta_{11}=-0.2$, $\beta_{44}=\beta_{55}=k_A(\beta_{11}+\beta_{33}-2\beta_{13})$: long dashes $k=1$, $k_A=0.5$; short dashes $k=2$, $k_A=0.5$; solid line $k=2$, $k_A=1$; dash-dot line $k=2$, $k_A=2$; dot line $k=1$, $k_A=2$.

For $k_A<1$, which corresponds to $n-k-1>0$, stress concentrations have maxima at points corresponding $\theta=0,\pi$ and $\theta=\pm\pi/2$ and minimum at points K lying in the interval $\pi/4<\theta<\pi/2$. The exact values of the minimum are obtained by equating to zero of the derivative of $f(\theta)$, determined by (2). For $k_A>1$ maximum transits to minimum. Influence of parameter k_A on the stress concentration on the well contour is illustrated by figure 3.

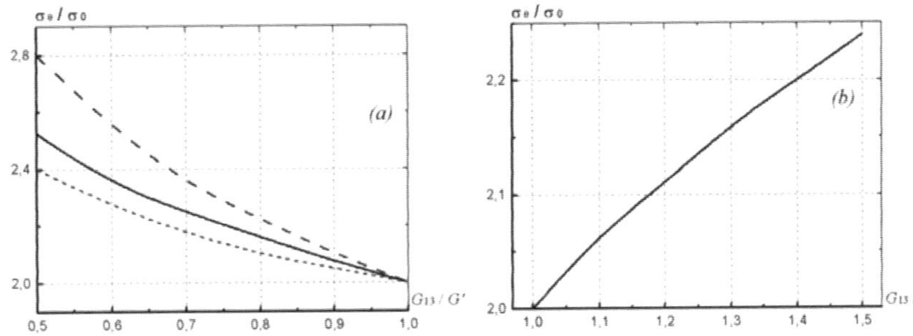

Figure 3. Stress concentrations: (a) f_A и f_B for $k_A\le 1$, $\beta_{12}/\beta_{11}=-0.2$, $\beta_{13}/\beta_{11}=-0.2$, $\beta_{44}=\beta_{55}=k_A(\beta_{11}+\beta_{33}-2\beta_{13})$: solid line $f_A=f_B$ for $k=1$; dashed line f_B for $k=2$; dash-dot line f_A for $k=2$, (b) f_K for $k_A\ge 1$.

Note, that deviation of stress concentration from the isotropic case even for equality of principle values of modules ($k=1$). Note also, that combination of elastic constants (8) corresponds to one of particular cases of elastic symmetry considered by Saint-Venant [3]. Equation (8) was pointed out in [2], and it was stated that it yields the correct value with the accuracy within 10% for 45 out of 47 rocks, that can be treated as transversely isotropic.

2.2. Horizontal well in layered rock under non equi-component stress

IOP Conf. Series: Materials Science and Engineering **921** (2020) 012011 doi:10.1088/1757-899X/921/1/012011

Let one of the principle stresses (σ_{zz}) acts along the vertical axis, two other are supposed to be equal

$$\sigma_{xx} = \sigma_{yy} = \lambda \sigma_{zz} \tag{10}$$

where λ is the coefficient of lateral pressure (the case of inequality of all three principle values and arbitrary inclination of their direction to the hole axis was considered in [4]. The stress concentration at the contour is [1]

$$f(\theta) = \frac{\sigma_{\theta\theta}}{\sigma_{zz}^{\infty}} = f_0(\theta) - (1-\lambda) f_1(\theta), \tag{11}$$

where $f_0(\theta)$ is given by (2), and $f_1(\theta)$ is the stress concentration due to uniaxial stress applied along horizontal axis x_1

$$f_1(\theta) = \frac{E_\theta}{E_1}\left\{-k\cos^2\theta + (1+n)\sin^2\theta\right\} \tag{12}$$

Extreme values of this function corresponds to points at $\theta = 0, \pi$ and $\theta = \pm\pi/2$ for all possible values of elastic parameters

$$f_1(0) = f_1(\pi) = -\frac{1}{k},\ f_1(-\pi/2) = f_1(\pi/2) = 1+n \tag{13}$$

For this case the relative value of the independent shear modulus (9) is no longer the drastic parameter determining the stress concentration.

Stress concentrations at points $\theta = 0, \pi$ and $\theta = \pm\pi/2$ are obtained by substitution (5), (6), (13), into (11)

$$f(0) = 1 + \frac{n-1}{k} + (1-\lambda)\frac{1}{k} = 1 + \frac{n-\lambda}{k} \tag{14}$$

$$f(\pi/2) = 1 + \frac{n-1}{k} - (1-\lambda)(1+n) \tag{15}$$

It is seen from (14), (15) that, similar to isotropic case reducing the stress along horizontal x_1 axis comparing to hydrostatic stress distribution results in increase in stress concentration at points $\theta = 0, \pi$ and decrease of stress concentration at points $\theta = \pm\pi/2$. It is found from analysis of properties of 39 rocks [2], that for not very strong deviation of initial stress state from hydrostatic state the influence of this deviation is not very essential.

2.3. *Inclined well*

To find the role of parameter $k_A = G_{13}/G'$ on distribution of stresses at the contour of a well inclined to the principle axes of elasticity tensor (inclined well in transverse isotropic massif with horizontal plane of isotropy) FEM calculations have been carried out, which revealed that for satisfied condition (8) the stress concentrations differ from those corresponding to isotropic massif $f_0(\theta)$ by less than 1%. Such a discrepancy is most likely due to numerical errors inherent to the method.

2.4. *Summary*

The carried analysis of elastic solution [1], and study of rocks [2] suggest that Lame solution for a hole in isotropic media in most cases is a very good approximation for the problem of a well in anisotropic massif. However, in rare cases of rocks, for which condition (8) is strongly violated for the rock in question, using Lame solution leads to underestimation of the stress concentrations.

The most dangerous is the case of $k_A = G_{13} / G' > 1$, since the maximums of stress concentration correspond to zones, within which the shear stresses act along planes of isotropy, and these planes are usually the weakest [5–8].

Thus, on the one hand, for the majority of rocks, parameter $k_A = G_{13} / G'$ is close to unity [2], which allows using Lame solution, on the other hand, for rare rocks with parameter k_A being far from unity this parameter becomes the most essential for determining stress state.

3. Anisotropy of strength

Contrary to elasticity and filtration, strength, or more generally transition from elastic to inelastic deformation, although being directional sensitive, do not posses tensor properties. Physical nature of dependence of strength on direction lies in existence of planes of weakening associated with layered structure or initial directional crack system. Various criteria of strength accounting for anisotropy were suggested (e.g. [5, 6]).

A variant of anisotropic strength criteria was considered in [8] based on assumption of existence of two mechanisms of fracture: according to Drucker-Prager criteria in massif

$$\frac{1}{\sqrt{6}}\sqrt{(\sigma_1-\sigma_3)^2+(\sigma_1-\sigma_2)^2+(\sigma_2-\sigma_3)^2}-B(\sigma_1+\sigma_2+\sigma_3)-A>0 \qquad (16)$$

and according to Coulomb-type law on planes of weakening. If the planes of weakening have normal coinciding with axis x_3, the law may be written as

$$\sqrt{\sigma_{13}^2+\sigma_{23}^2}=\tau_c\sigma_{33}\,\mathrm{tg}\,\rho_c \qquad (17)$$

Here σ_1, σ_2, σ_3 are the principle stresses; A, B, τ_c, ρ_c are model parameters.

Also two forms of generalization of Hill's yield criterion that account for influence of compressive stresses are considered for description of elastic-plastic transition of rock deformation.

3.1. Generalized Hill's criterion

According to the original Hill criterion [9], the traditional value of the second invariant of stress deviator used as the critical value in the Von Mises yield criterion is replaced with the general quadratic form of components of stress tensor, so that the criterion is written as follows

$$\begin{aligned}F&=G^0_{(23)}(\sigma_{22}-\sigma_{33})^2+G^0_{(13)}(\sigma_{11}-\sigma_{33})^2+G^0_{(12)}(\sigma_{11}-\sigma_{22})^2+\\&+2L^0_{(23)}\sigma_{23}^2+2L^0_{(13)}\sigma_{31}^2+2L^0_{(12)}\sigma_{12}^2-1=0\end{aligned} \qquad (18)$$

Here $G^0_{(ij)}$, $L^0_{(ij)}$ are material constants.

To account for the influence of compressive stresses two generalizations were suggested. Thus, according to [10] the criterion was appended with terms accounting influence of compressive stress

$$\begin{aligned}F&=G^C_{(23)}(\sigma_{22}-\sigma_{33})^2+G^C_{(13)}(\sigma_{11}-\sigma_{33})^2+G^C_{(12)}(\sigma_{11}-\sigma_{22})^2+\\&+2L^C_{(23)}\sigma_{23}^2+2L^C_{(13)}\sigma_{31}^2+2L^C_{(12)}\sigma_{12}^2+\left(B^C_{(1)}\sigma_{11}+B^C_{(2)}\sigma_{22}+B^C_{(3)}\sigma_{33}\right)-1=0\end{aligned} \qquad (19)$$

Here $B^C_{(ij)}$ are additional material constants characterizing influence of compressive stresses. An alternative variant of generalization was suggested in [11] differing from (19) by exponent 0.5

$$\begin{aligned}F&=G^L_{(23)}(\sigma_{22}-\sigma_{33})^2+G^L_{(13)}(\sigma_{11}-\sigma_{33})^2+G^L_{(12)}(\sigma_{11}-\sigma_{22})^2+\\&+2L^L_{(23)}\sigma_{23}^2+2L^L_{(13)}\sigma_{31}^2+2L^L_{(12)}\sigma_{12}^2+\left(B^L_{(1)}\sigma_{11}+B^L_{(2)}\sigma_{22}+B^L_{(3)}\sigma_{33}\right)-1=0\end{aligned} \qquad (20)$$

IOP Conf. Series: Materials Science and Engineering **921** (2020) 012011 doi:10.1088/1757-899X/921/1/012011

Constants $G^L_{(ij)}$, $L^L_{(ij)}$ $B^L_{(ij)}$ have similar meaning as in (19). In case of transversal isotropy with the isotropy plane having normal n_3 the number of constants in (18) reduces to five for both models

$$G_{(13)} = G_{(23)}\,,\ L_{(13)} = L_{(23)}\,,\ L_{(12)} = G_{(13)} + 2G_{(12)}\,,\ B_{(1)} = B_{(2)} \tag{21}$$

For isotropy

$$G_{(13)} = G_{(23)} = G_{(12)} = G\,,\ L_{(13)} = L_{(23)} = L_{(12)} = 3G\,,\ B_{(1)} = B_{(2)} = B_{(3)} = B/3\,, \tag{22}$$

criterion (18) reduces to Drucker-Prager criterion. Questions related to inelastic deformation of anisotropic rocks are considered in [12, 13]. For permeable rocks in all the above expressions the stresses should be understand as the effective stresses.

3.2. Experimental determination of the model parameters

The parameters involved have been experimentally obtained for various types of rocks for 8 oil and gas fields using the true triaxial loading facility – Triaxial Independent Loading Test System (TILTS), developed in IPMech RAS [8, 12, 13]. Experiments were carried out on cubic specimens ($40 \times 40 \times 40$ mm). Two types of the loading programs were used: uniaxial loading with lateral pressure, and the generalized shear [8, 12, 13].

According to the loading program of the first type the specimens were subjected to comprehensive pressure up to a certain value (2, 10 or 20 MPa), and then loaded along one of the axes up to the level of elastic/plastic transition in the displacement controlled mode.

During the second loading program the specimens subjected to hydrostatic compression up to the level, corresponding to in situ stress state, then loading was being continued along one of the axes, accompanied with unloading along another axis, the stress along the third axis was being preserved constant; the average stress was being conserved during this stage. The stress state constructed in this program correspond to the state near the well. The tests were carried for two orientations of the minimal stresses: along the layering and normal to the layering, which corresponds to two different points at the contour of horizontal well.

3.3. Influence of strength anisotropy on the stability of wells

Boundaries separating regions of elastic and inelastic regions in the near-well zone for condition of Feodorovskoe field [14] are presented in figure 4.

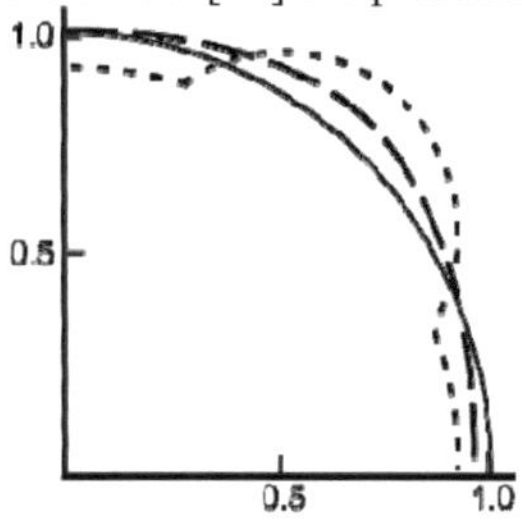

Figure 4. Boundaries separating elastic and inelastic regions in the near-well zone for condition of Feodorovskoe field. Dashed line – modified Hill criterion (20); dotted line – combined criterion (16), (17); solid line – the well contour.

4. Conclusions

The influence of elastic anisotropy on stress distribution on the contour of wells and openings has been investigated. By analyzing analytical and numerical elastic solutions, it is shown that for a cylindrical opening (well) oriented at an arbitrary angle to the axis of isotropy in transversely isotropic rocks loaded by hydrostatic stresses, the deviation of the stress distribution along its contour from the similar distribution in isotropic media is governed by a single parameter – the ratio of the independent shear modulus to a particular combination of elastic constants.

The analysis of existing experimental data has demonstrated that for great majority of rocks (about 95%) the value of this parameter ensures deviation of stress concentration from the isotropic media for the case in question to be within 10%; therefore for this group the parameter may be set equal to unity at first approximation, thus reducing the number of parameters involved. For the relatively small remaining group of rocks this particular parameter become drastic for determining stress concentration, while widespread parameter – ratio of the maximal and minimal compressive modules – has almost negligible influence on the stress concentration for that case. In case of non-hydrostatic initial stresses the deviation of stress concentration is governed by two dimensionless combinations of elastic constants.

Two forms of generalization of Hill's yield criterion accounted for influence of compressive stresses are considered for description of elastic-plastic transition at rock deformation. In case of transversal isotropy both models are characterized by 5 parameters. The values of these parameters have been experimentally obtained using the true triaxial loading facility TILTS for rocks from various oil and gas fields.

The methodic of experimental determining the parameters from the direct measurements using a true triaxial loading system is described.

Acknowledgment
The work has been financially supported by state assignment (No. AAAA-A17-117021310371-9).

References

[1] Lekhnitsky S G 1977 *Theory of elasticity of anisotropic body* (Moscow: Nauka) p 416
[2] Batugin S A and Nirenburg R K 1972 Approximate dependence between elastic constants of rocks and anisotropy parameters *J. Mining Sci.* **7(1)** pp 7–11
[3] Saint-Venant A J-C B 1863 Memoire sur la distribution des elasticites autour de chaque point d'un solide ou d'un milieu de contexture quelconque, particulierement lorsqu'il est amorphe sans etre isotrope (Deuxieme article) *J. Mathematiques Pures Et Appliques* **2(8)** pp 257–430
[4] Mimouna A and Prioul R 2017 Closed-form approximations to borehole stresses for weak transverse isotropic elastic media *Int. J. Rock Mech. and Mining Sci.* **98** pp 203–216 doi 10.1016/j.ijrmms.2017.06.002
[5] Goodman R E 1980 *Introduction to rocks mechanics* (New York: John Wiley and Sons) p 561
[6] Jaeger J C, Cook N G W and Zimmerman R W 2007 *Fundamentals of rock mechanics* (Malden, MA: Blackwell Publishing) p 489
[7] Zobak M D 2007 *Reservoir Geomechanics* (Cambridge: Cambridge University Press) p 464 doi 10.1017/CBO9780511586477
[8] Karev V I, Klimov D M, Kovalenko Y F and Ustinov K B 2016 Fracture of sedimentary rocks under a complex triaxial stress state *Mechanics of Solids* **51(5)** pp 522–526
[9] Hill R 1948 A theory of the yielding and plastic flow of anisotropic metals *Proc. Roy. Soc. London A* **193** pp 281–297
[10] Caddel R M, Raghava E S and Atkins A G 1973 A yield criterion for anisotropic and pressure dependent solids such as oriented polymers *J. Materials Sci.* **8** pp 1641–1646
[11] Lui C, Huang Y and Stout M G 1997 On the asymetric yield surface of plastically orthotropic materials: a phenomenological study *Acta Mater.* **45(6)** pp 2397–2406
[12] Karev V I, Kovalenko Y F and Ustinov K B 2017 Modeling Deformation and Failure of Anisotropic Rocks nearby a Horizontal Well *J. Mining Sci.* **53(3)** pp 425–433
[13] Karev V I, Klimov D M, Kovalenko Y F and Ustinov K B 2018 Modelling of mechanical and filtration processes near the well with regard to anisotropy *J. Phys.: Conf. Series* **5** 012039
[14] Karev V I, Kovalenko Y F and Ustinov K B 2018 *Modeling of geomechanical processes in the vicinity of oil and gas wells* (Moscow: IPMech RAS) p 529

IOP Conf. Series: Materials Science and Engineering **921** (2020) 012012 doi:10.1088/1757-899X/921/1/012012

Monitoring of mechanisms of formation of corrosive aggressiveness of soils and grounds of underground gas pipeline routes and on its basis assessment of service life of gas pipes in various soil-climatic zones

O I Khudyakov[1*], O V Reshotkin[1], D V Demin[2] and M A Velikotskiy[3]

[1] Institute of Physicochemical and Biological Problems in Soil Science, Russian Academy of Sciences, Pushchino 142290, Russian Federation
[2] Institute of Basic Biological Problems, Russian Academy of Sciences, Pushchino 142290, Russian Federation
[3] Faculty of Geography, Lomonosov Moscow State University, GSP-1, Leninskie Gory, Moscow 119991, Russian Federation

*E-mail: oix@rambler.ru

Abstract. The experience of studying the problem of corrosive aggressiveness of soils and grounds in terms of improving the reliability of Gazprom's pipeline systems operation is summarized and presented. Soil factors causing surface corrosion of gas pipelines in soil conditions are studied. Based on the obtained results, it was suggested to evaluate not only the corrosive aggressiveness of soils and the gas pipe lifetime, but also the corrosive resistance of pipe metal in corrosive environments. Special attention is paid to the stress corrosion cracking formation mechanism.

1. Introduction

Corrosion aggressiveness of soils and grounds is currently a topical problem in modern gas pipeline construction and operation. Metal corrosion in underground conditions is the main reason for wear and tear of gas transportation systems. The actual service life of pipelines is determined primarily by the rate of metal corrosion. For steel pipelines built in 60-80s of the XX century, the standard service life was determined by 33 years [1]. By now, this design term has come to an end, but the pipelines continue to operate. Therefore, monitoring of corrosion aggressiveness of soils and grounds, and on this basis working out of the technology reducing and preventing development of corrosion processes, is a reliable basis for prolongation of service life of the iron pipelines laid underground.

The problem of determining the corrosive aggressiveness of different soils and grounds in relation to gas pipe steel is an urgent scientific and technical task. The rate of metal corrosion in various soils and grounds differs by its magnitude in dozens and even hundreds of times. The idea of corrosion aggressiveness of soils and grounds and its determinants is necessary when solving issues related to the development of technology to extend the life of gas pipelines, forecasting the sections of the highest corrosion rate, identifying the causes of stress corrosion cracking (SCC), justification of the method of pipe protection, etc. At the same time, it is important to have a quantitative characteristic of corrosive aggressiveness of soils and grounds in order to develop and apply technologies for the most reliable

IOP Conf. Series: Materials Science and Engineering **921** (2020) 012012 doi:10.1088/1757-899X/921/1/012012

protection of pipeline routes from corrosive destructive action. However, many criteria of an estimation of the corrosion aggressiveness connected with physical and chemical characteristics of soils and grounds, are reliable only within the limits of certain physical-geographical regions and their application in other regions is ineffective.

Relevance of this work is determined by the increased scale of economic and environmental disasters caused by the long-term operation of transcontinental gas pipelines, which is ultimately due to the lack of reliable methods to assess the interaction in the system "aggressive external environment of the containing thickness – pipeline". From the point of view of practice, it seems relevant to study the corrosive aggressiveness of each zonal soil and to develop on this basis the technology to reduce and prevent the corrosive aggressiveness of soils and grounds is not only an economic but also an environmental necessity.

2. Objects and methods

The objects of the study were the soils of the zonal series, including tundra, northern, middle and southern taiga, forest-steppe, steppe, dry steppe and semi-desert. Corrosion aggressiveness of zonal soils in relation to pipe metal was determined in the zonal series. In addition, the corrosion resistance of gas pipe metal in soils of the zonal series was determined. In laboratory conditions in soil samples determined the ionic composition of water extraction, particle size distribution, electrical conductivity, electrical resistance, hydrogen Clark Index, pH water in the soil sample with different degrees of moisture, magnetic susceptibility. Assessment of corrosive aggressiveness of the soil was determined by the most aggressive indicator.

3. Results and discussion

Types of corrosion aggressiveness of zonal soils

The introduction of a gas pipe into any zonal soil creates favorable conditions for the development of various types of corrosive aggressiveness. Hydrological, chemical, physicochemical, biological and other types of corrosive aggressiveness stand out among them.

3.1. Hydrological corrosive aggressiveness of soils

The initial mechanism of all types of corrosive activity in soils is the organization of a trench and laying of a gas pipe. Violation of soil integrity as a bio-inert matter of nature during trenching and gas pipe laying irreversibly changes soil properties and regimes. A special role in this is played by the natural process of constant saturation of the soil with the moisture of the near-pipe space. The pushing mechanism of soil saturation with water around the gas pipe is the gradient between the gas pipe temperature and the ground temperature. Condensation saturation of the water goes up to the full moisture content of the near-pipe soil. The moisture of the near-pipe soil is retained by capillary forces, forming anaerobic conditions for the development of anaerobic sulfate-reducing microorganisms. The presence of moisture in the soil has a significant impact on corrosion of pipe metal.

$Fe + H_2O = Fe\,(OH)_2$

Water is a component of soil electrolytes which interact with metal and create damage to gas pipes. We judge the qualitative composition of cations and anions of electrolytes by the water extract of each soil. The qualitative composition of the water extract determines the chemical aggressiveness of soils to the metal of the gas pipe.

3.2. Chemical corrosive aggressiveness of soils

The first data on the assessment of the effect of water extract ions on metal corrosion were noted in the works [2-4]. It was found that the most aggressive are Cl^- and SO^{2-}_4 ions. Their presence above 0.1% is an essential criterion of high corrosive aggressiveness of soil. First of all, chlorine and hydrogen ions destroy the protective film and thus accelerate the corrosion of metals. According to [5-8] the mechanism destroying the protective film is not clear. According to [1], the content of chlorine ions exceeds 0.1%, which causes greater corrosion of soils and natural waters in relation to steel. According to [9], the

IOP Conf. Series: Materials Science and Engineering **921** (2020) 012012 doi:10.1088/1757-899X/921/1/012012

chlorine ion content in river water is more than 5 mg/l, which increases its corrosion activity in relation to metals, destroys protective films and accelerates corrosion of low-alloy and medium-alloy steels. Not only anions but also cations play an active corrosive role in water extraction of soils and grounds. Interaction of water extract ions with metal of a gas pipe has stage character. The first stage is characterized by the interaction of ions with water. In soils, each anion interacting with water cation forms acid, forming around the gas pipe corrosive acid conditions with respect to the metal of the gas pipe.

$H^+ + Cl^- = HCl$
$2H^+ + SO^{2-}_4 = H_2SO_4$
$2H^+ + CO^{2-}_3 = H_2CO_3$
$H^+ + HCO^-_3 = H_2CO_3$

Interacting in soil with water ion each cation forms in soils around the gas pipe corrosive alkaline conditions in relation to the gas pipe metal.

$2OH^- + Ca^{2+} = Ca(OH)_2$
$2OH^- + Mg^{2+} = Mg(OH)_2$
$OH^- + Na^+ = NaOH$
$OH^- + K^+ = KOH$

The second stage is characterized by interaction of acids and alkalis with the pipe metal, forming chemical damages of the gas pipe metal.

$2HCI + Fe = FeCI_2 + 2H^+$
$H_2SO_4 + Fe = FeSO_4 + 2H^+$
$H_2CO_3 + Fe = FeCO_3 + 2H^+$
$Ca(OH)_2 + Fe = Fe(OH)_2 + Ca^{2+}$
$Mg(OH)_2 + Fe = Fe(OH)_2 + Mg^{2+}$
$2Na(OH) + Fe = Fe(OH)_2 + 2Na^+$
$2K(OH) + Fe = Fe(OH)_2 + 2K^+$

The percentage of each anion from their sum characterizes its contribution to the overall acidic and alkaline corrosive aggressiveness of soils and grounds (table 1). By the sum of alkaline ions and each of them the coefficient of corrosive aggressiveness in grey forest soil reaches 0.6 and is characterized as increased. Corrosion aggressiveness as high on acid potential is noted in three samples.

Table 1. Anionic composition of water extraction of grey forest soils (Albic Luvisols).

№	Mg-eq/100 g soil					Coefficient				Corrosion aggressiveness by anion coefficient
						Acid		Alkaline		
	Cl	SO_4	CO_3	HCO_3	Sum	Cl	SO_4	CO_3	HCO_3	
1	0.125	0.057	0	0.10	0.282	0.44	0.20	0	0.36	Low
2	0.120	0.343	0	0.68	1.143	0.11	0.30	0	0.59	High
3	0.125	0.042	0	0.05	0.217	0.58	0.19	0	0.23	High
4	0.125	0.393	0	0.75	1.268	0.09	0.31	0	0.59	High
5	0.125	0.111	0	0.35	0.586	0.21	0.19	0	0.60	High

Analyzing the ion composition of the water extract, it can be unequivocally said that each ion of the water hood in the interaction with water forms in soils and grounds corrosive conditions with respect to the metal gas pipe.

Soil saturation of near-pipe space with water to the category close to the total porosity is the initial mechanism for the development of corrosive physical and chemical processes (table 2).

Table 2. Corrosion aggressiveness of soils.

Type of water regime	Soils	Ground temperature, °C			Duration vegetative period with $t > 0°C$, month			Moisture, %	Ions SO_4	pH water (80-120 cm)	Electrical resistivity, Ohm-m	rH2	Pipe stability coefficient	Gas pipe lifetime, years
		0-5	5-10	> 10	< 3	3-6	> 6							
Soils of the tundra and arctic areas of the polar belt														
Automorphic	Arctic tundra	*			*			32-44	*	5.8	<20	>28	1.6-1.8	64-72
	Tundra soddy	*			*			32-44	*	5.6	<20	>28	1.7-1.9	68-76
Semi-hydromorphic	Tundra soddy gley	*			*			42	*	6.4	<20	>28	1.6-1.7	64-68
	Tundra peaty and peat-gley	*			*			42	*	6.2	<20	>28	1.7-1.9	68-76
Hydromorphic	Boggy tundra	*			*			42	*	6.4	<20	>28	1.9-2.2	76-88
Alluvial	Alluvial soddy tundra	*			*			22-32	*	6.2	<20	>28	1.4-1.6	56-64
	Alluvial boggy tundra	*			*			20-26	*	6.0	<20	>28	1.6-1.8	64-72
Soils of the steppe areas of the subboreal belt														
Automorphic	Chernozems			*			*	<24	*	6.5-7.0	>20	>28	1.0	40
Soils of semi-desert and desert areas														
Automorphic	Brown semi-desert			*			*	<32	*	7.6-8.1	>20	>28	0.7-0.8	28-32
	Grey-brown semi-desert			*			*	<32	*	7.7-7.8	>20	>28	0.7-0.8	28-32
	Solonetz semi-desert			*			*	<32	*	9.6-10.6	>20	>28	0.6-0.7	24-28
	Solonchaks semi-desert			*			*	<32	*	9.8-10.7	>20	>28	0.5-0.6	20-24

3.3. Physicochemical corrosive aggressiveness of soils

Physicochemical corrosiveness in soils and grounds is determined by redox processes called oxidation-reduction potential (ORP). In the zonal series, each soil type is characterized by a certain ORP value. In sod-podzolic soils, the ORP value varies from 550 to 750 mv. In chernozems ORP varies from 400 to 600 mv, and in serozems it varies from 350 to 450 mv [10]. Introduction of a gas pipe into the soil changes the character of water and temperature regimes, density, aeration, water permeability and other properties that lead to changes in ORP, pH, Clark Index and other properties. By the example of grey forest soil, it is shown that measurement of ORP in 5 samples in 10 replicates of grey forest soil indicates

IOP Conf. Series: Materials Science and Engineering **921** (2020) 012012 doi:10.1088/1757-899X/921/1/012012

the development of gley process near-tube space in soils, at which ORP value was below 200 mv (table 3).

Table 3. Oxidation-reduction potential (ORP) values (mv).

Sample number	Replicates									
1	180	140	165	170	230	160	110	115	90	120
2	80	90	100	110	130	95	90	75	110	130
3	120	130	80	90	115	110	110	130	100	110
4	130	140	90	110	120	100	115	120	90	80
5	100	100	120	130	110	140	150	90	90	100

According to the assessment scale, the corrosive aggressiveness of the near-pipe space soil is characterized as strong and in some places average.

Assessment scale of aggressiveness of ORP (mv) in soils:

Negative values	Very high
Under 100	Strong
From 100 to 200	Mean
From 200 to 400	Weak
Above 400	Non-corrosive

The concentration of hydrogen potential in the near-tube grey forest soil was calculated based on ORP and pH (acidic or alkaline values). At rather low ORP values rH2 did not exceed 23–25 units, which characterizes grey forest soil near-tube conditions as corrosive. It was found out that each zonal soil is characterized with different combination of corrosive-aggressive processes. It is very characteristic that combination of processes in each zonal soil changes depending on change of temperature or hydrological factor in seasonal cycle.

For example, let us give an example of complex development of corrosive-aggressive influence of grey forest soil on metal of gas pipe. Analysis of factors determining corrosive aggressiveness of soils shows that the most indicative integral factor characterizing corrosive aggressiveness of soils and grounds is the redox potential of soils and grounds. The main mechanism of ORP dynamics is change of trench ground watering in seasonal cycle.

On the one hand, anaerobic conditions contribute to the development of ORP, which itself as a process, dramatically increases the corrosive aggressiveness of soils and grounds, involving in the physical and chemical processes iron pipe itself. On the other hand, ORP characterizes aerobic and anaerobic conditions in soils. In this regard, the recovery regime can be judged on the development and vital activity of anaerobic sulfate-reducing bacteria. Sulfate-reducing bacteria under anaerobic conditions restore sulfates to hydrogen sulfate and sulphuric acid, which interact with iron, while emitting atomic hydrogen, which is stored in the structure of the metal pipe. The development of sulfate-reducing bacteria in soils requires strictly defined environmental conditions. The most important physiological feature of sulfate-reducing bacteria is their ability to develop under strictly anaerobic (complete absence of oxygen, oxides and oxygen-containing ions) conditions.

The analysis of the soil on which the pipe lies is very indicative in this respect. It is a dove-colored gleyed loam, whose ORP varies in the range from -80 to -96 mv at five measuring points. Low ORP values of the loam mean a high degree of its corrosive aggressiveness, which manifests itself in the pipe section with broken insulation tape.

The chemical composition of the protective crust above the corrosion cavity in the metal of the pipe taken under the broken insulation film included biological corrosion products, including oxides of divalent iron (FeO), hydroxides of iron $Fe(OH)_2$, iron sulphide (FeS), mineral siderite ($FeCO_3$). At

IOP Conf. Series: Materials Science and Engineering **921** (2020) 012012 doi:10.1088/1757-899X/921/1/012012

selection of samples hydrogen sulfide (H_2S) has been defined. Under the protective film the depression was located along the pipe. The width of the cavity was from 1.6 to 2.3 mm. In the head part of the cavity, in terms of gas movement in the pipe, the cavity width increased to 2.9-3.4 mm. The length of the cavity was 2.4 cm. The interaction of hydrogen sulfide with iron in the pipe forms iron sulphide and two hydrogen atoms. This process is called hydrogenation (saturation of pipe metal with atomic hydrogen), which dramatically reduces the corrosion resistance of pipes and their service life, leads to the development of SCC. Determination of gas pipelines trench corrosion aggressiveness by various methods allows to monitor corrosion activity of soils and grounds and on this basis to estimate service life of gas pipelines in different zonal climatic conditions and in different geochemical crust of weathering due to modern warming tendency.

The stated above shows that the problem of assessment of corrosion aggressiveness to metals of natural factors participating in underground corrosion of pipelines in Russia is a multifactorial system. In such a system, the initial mechanism in the development of metal corrosion is the condensation process of moisture accumulation in the near-pipe space. Saturation of near-pipe space with water leads to the development of anaerobic conditions as a medium for sulfate-reducing bacteria.

First of all, this complexity is connected with the fact that the rate of deep corrosion of pipes is predetermined not only by natural factors, but also by the state of insulation coating of pipes and operating modes of electrochemical protection stations.

According to numerous physicochemical data of soils and grounds very often in soils with the same physicochemical indicators, different rates of deep corrosion are revealed. Such facts can be explained only by different efficiency of electrochemical protection. Only a few pits can be distinguished from the total number of samples, where high rates of deep corrosion of metal and low values of protective potential are observed. These are the most representative sections of pipeline natural-technical systems for determining corrosion aggressiveness to metals of soils and grounds. The aforesaid allows to assume, that estimation of corrosion aggressiveness to metals of soils and grounds on speed of deep corrosion of metal of pipes in the conditions of work of stations of electrochemical protection is not quite correct. The complexity of corrosion processes lies in the fact that a single corrosion process runs simultaneously on anode and cathode. At the anode section, metal atoms are oxidized, i.e. electrons are emitted and transformed into ionic state, while at the cathode section ions (hydrogen, oxygen, iron) involved in the corrosion process are restored.

Secondly, the difficulty of determining the corrosion aggressiveness to metals of soils and grounds by the rate of deep corrosion is due to the inaccuracy of determining the rate of deep corrosion itself. For exact determination of the deep corrosion rate it is necessary to have 15–20-year monitoring (3–4 scientific field stations in each natural observation zone with the interval of 5 years).

Thirdly, the task of determining corrosion activity of soils is complicated by insufficient development of electrochemical theory of corrosion as applied to underground conditions. Corrosion products, formed in underground conditions and in conditions of cathode protection stations in operation, are poorly investigated. Many issues related to depolarization on cathode and anode pipeline sections are under discussion. The electrochemical theory of corrosion aggressiveness to metals of soils and grounds in the permafrost zone is especially poorly grounded. It is necessary to notice, that around pipes with "warm gas" in frozen soils a halo of thawing is formed and, in these conditions, (in the near-pipe space) electrochemical corrosion can occur on the usual (more southern) scheme. It is more difficult to estimate corrosion aggressiveness of frozen ground for pipes with "cold gas" (gas condensate or any other product).

In the literature there is an opinion that chemical reactions in the Far North are extremely slowed down due to weak water dissociation at low temperatures. The work of recent years has convincingly proved that perennially frozen soils of the permafrost zone are not a chemical rest zone and there are active processes of chemical and biological weathering, neoplasm and migration of substances. The main factors of the redox reaction in the permafrost zone are: unfrozen water, hydrogen ions and atomic oxygen. Dissolved gases, ions of salts and acids are concentrated in the film water. Chemical reactions in film water do not fade even at low temperatures. According to [11], even at low temperatures of soils from 0 to -50°C the energy of redox processes increases. Therefore, it is possible to assume that potential corrosion activity of soils and grounds of the permafrost zone, despite high specific electrical resistance, is much higher than that of soils and grounds of the chernozem zone. In this connection it is possible to consider as rather debatable record in GOST (Russian National Standard) No. R 51164-98, paragraph 3.6: "Main gas pipelines, which wall temperature is lower than 268 K" (-5°C), are not subject to electrochemical protection...". These questions are not theoretically grounded and require further research, theoretical and experimental substantiation. According to our observations, when the soil freezes, water in the soil solution passes from liquid to solid, to the form of ice. Decrease of dissolving volume of water in soil solution at ice formation is accompanied by increase of concentration of soil solution, thus chemical reactions of interaction of ions of soil solution with pipe metal in such conditions develop according to laws of reactions in saturated solutions.

4. Conclusions

The obtained materials allow making a number of general conclusions about the mechanism of corrosive aggressiveness of soils in relation to gas pipe metal in different soil-climatic zones. On their basis in the future it is possible to carry out an in-depth analysis of corrosive aggressiveness to metals of soils and grounds by geographical zones in the rank of groups of natural boundaries. On the other hand, this aggressiveness is realized in soils and grounds in relation to metal in conditions of broken pipe insulation. In this case, we must pay attention to the corrosion resistance of the gas pipe metal. In the zonal series of soils, we take chernozem soil as the soil standard, where the corrosion resistance of the gas pipe metal is taken as a unit, and the accident-free service life of the metal gas pipe is taken for 40 years under the condition of undisturbed insulation, the absence of factory cracks on the gas pipe and the quality of the weld.

Table 2 shows the data on corrosive aggressiveness of zonal soils by physical and chemical parameters. To the south from the chernozem zone soils of alkaline crust of weathering are formed. In alkaline soils the pH value of water extraction is more than 7.0. Such soils have high corrosive aggressiveness, therefore in them low corrosive resistance of pipe metal in relation to chernozem does not exceed 0.5–0.8 (table 2).

In alkaline conditions, the organ-iron film is soluble. Besides in soils with pH of water extraction above 8.2 there is carbonate ion CO_3^{2-} which at interaction with sodium ion forms sodium carbonate, very strong corrosion destroyer of metal. This dramatically reduces the accident-free service life of the gas pipe, which is 24–32 years at soil temperatures near the pipe above 10°C and undisturbed insulation.

$2Na^+ + CO^{2-}_3 = Na_2CO_3$

To the north of the chernozem zone, acidic crust soils of weathering are formed. Laboratory studies showed that organic matters play important role in soil aggressiveness. When fragment of gas pipe was placed in suspension of tundra zone soils organo-iron film was formed on the surface of sample. This film was the area where corrosive tundra soil and pipe metal surface were divided. In acidic environment organo-iron film is stable. Besides, in soils of circum-pipe space of Arctic and Antarctic regions the temperature does not exceed the first biological minimum (5.0°C), in conditions of which corrosive aggressiveness of chemical, physicochemical and biological processes sharply decreases, that promotes increase of accident-free service life of gas pipe in 1.4–2.2 times in comparison with chernozem (table 2). Based on the results obtained, it can be concluded that the service life of gas pipes in corrosive soils of alkaline and acidic weathering crusts can be extended, relative to the service life of gas pipes in chernozems, 2–4 times, if the gas temperature does not exceed the temperature of the first biological minimum – 5°C. At such temperature corrosion aggressiveness of biological, chemical and physical-chemical processes is sharply suppressed. Therefore, we recommend pumping gas through the pipe at a temperature no higher than 5°C, which will lead to the extension of accident-free operation of gas pipes.

Acknowledgements
The work was carried out within the framework of the State Assignment No. 0191-2019-0046.

IOP Conf. Series: Materials Science and Engineering **921** (2020) 012012 doi:10.1088/1757-899X/921/1/012012

References

[1] Krylov G V, Stepanov D A and Ugryumov R A 2001 *Anticorrosive protection of the trunk pipelines Reference guide* (St. Petersburg: Nedra) p 190 (in Russian)
[2] Pritula V A 1934 *Determination of soils corrosion aggressiveness* (Moscow–Leningrad) p 135 (in Russian)
[3] Pritula V A 1945 *Cathodic protection of pipelines against soil corrosion* (Leningrad–Moscow) p 85 (in Russian)
[4] Pritula V V 1975 *Corrosion protection for Canada and Alaska underground pipelines* (Moscow: VNIIOENG) p 64 (in Russian)
[5] Evans Yu R, Akimov G V, Timashov N D, Krasnoyarskiy V A 1925 *Soda formation at contact of iron, solutions of glauber salt and air in connection with iron rusting process* 97–107 (in Russian)
[6] Evans Yu R 1962 *Corrosion and oxidation of metals* (Moscow: Mashgiz) p 856 (in Russian)
[7] Akimov G V 1945 *Theory and methods of metal corrosion research* (Moscow–Leningrad) p 396 (in Russian)
[8] Timashov N D 1959 *Theory of corrosion and metal protection* (Moscow: USSR Academy of Sciences Publishing House) p 592 (in Russian)
[9] Krasnoyarskiy V V, Nosov R P, Frankel G Ya 1963 *Corrosion and protection of metal structures of hydraulic structures* (Moscow–Leningrad) p 200 (in Russian)
[10] 1969 *Soil Science* ed I S Kaurichev (Moscow: Kolos) p 544 (in Russian)
[11] Tyutyunov I A 1961 Physical and chemical change of rocks in the Far North *Physical and chemical processes in freezing and frozen rocks* (Moscow: USSR Academy of Sciences Publishing House) 7–27 (in Russian)

IOP Conf. Series: Materials Science and Engineering **921** (2020) 012013 doi:10.1088/1757-899X/921/1/012013

Study on rheological properties of clayless drilling fluids influenced by fractional composition of carbonate weighting agents

E L Leusheva and V A Morenov

Petroleum Department, Saint Petersburg Mining University, Vasilevsky Island, 21th Line, 2, Saint Petersburg, 199106, Russia

E-mail: Leusheva_el@pers.spmi.ru

Abstract. Study of the influence of weighting-bridging agents on technological parameters of drilling fluids is an important aspect in the development of washing fluid compositions, especially when drilling-in productive formations. At selection of weighting agents, it is often assumed that these substances are inert and do not affect the structural and rheological parameters of the fluid, but only increase the density (due to higher specific gravity) and lower the filtration parameters (due to an increase in the solid phase content). Paper presents experimental study to assess the effect of the fractional composition of carbonate weighting agents on structural and rheological parameters of a biopolymer drilling fluid used to drill-in a reservoir. Carbonate substances with an average particle size of 5 to 150 microns were used as weighting-bridging agents. Based on the data obtained, dependences of the main structural and rheological parameters of the drilling fluid on the size of the weighting-bridging agent's fraction are revealed.

1. Introduction

In the process of drilling oil and gas wells in conditions of high reservoir pressures, occurrence probability of various complications increases. To prevent the occurrence of complications during drilling, such as oil, gas and water backflows, cavings and collapses of the wellbore walls and etc., it is necessary to create pressure in the well that exceeds the hydrostatic pressure. The primary method of creating the necessary pressure is to increase the density of the drilling mud. For this purpose, special weighting materials are used.

Currently more than 20 weighting materials are used in world market. Barite and carbonate weighting agents are most often used. However, more and more materials are being created for weighting. Sometimes it is required to increase the density of the drilling mud to 2500 kg/m^3 and higher, at the same time maintaining structural-rheological and filtration parameters. For this purpose, more thorough and comprehensive analysis of weighting agents' composition and properties is required, as well as their effect on the properties of washing fluids.

Several scientists, such as Abrams, Kaeuffer, Vickers and Ishbaev, studied the properties of bridging agents and their influence on the process of formation drilling-in [1, 2, 4, 5]. In their research, they considered various theories on the fractional composition of the weighting agent particles and their interaction with the pore volume of the formation. Ishbaev investigated ways to reduce the contamination of the productive reservoir with a filtrate and a solid phase of drilling solution. Dick and Heinz determined the optimum content of the bridging agents in the drilling mud to seal the permeable zones [8].

Aim of this article is to assess the effect of the carbonate weighting agent dispersion on the rheological parameters of the drilling fluid.

2. Theories of the bridging agent's fractional composition selection

Drilling-in with a water-based solution results in contamination of productive reservoir and reduction in oil recovery rates. Basically, the pollution is caused by penetration of the solid particles and mud filtrate into the pore volume. Quick and efficient formation of low-permeable filter cake leads to reduced contamination of the oil reservoir (Figure 1).

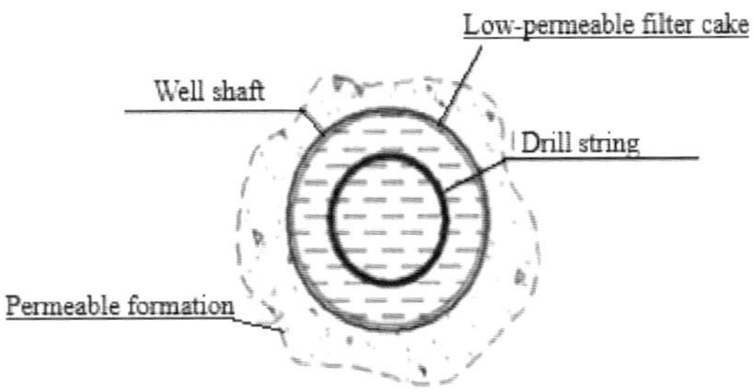

Figure 1. Formation of filter cake on the wellbore walls

The permeability of the filter cake, which is formed on the wall of the wellbore, is significantly influenced by the grain composition and the amount of the bridging agent contained in the solution. Fractional calcium carbonate is most often used as a bridging agent since it is easily soluble in acid with subsequent hydrochloric acid treatment.

If the colmatage mixture consists of particles that are significantly larger in size than the pore openings, they will not be able to form an effective filter cake. Some of them will be carried away by the flow of drilling mud, and channels will form between the particles, through which the drilling mud is easily filtered (Figure 2, a). Particles that are much smaller than the pore openings will penetrate the pore volume of the formation unhindered and contaminate it (Figure 2, b). Optimal selection of particles by their size in the bridging agent leads to the formation of a filter cake with minimal penetration of the filtrate and solid particles into the formation (Figure 2c).

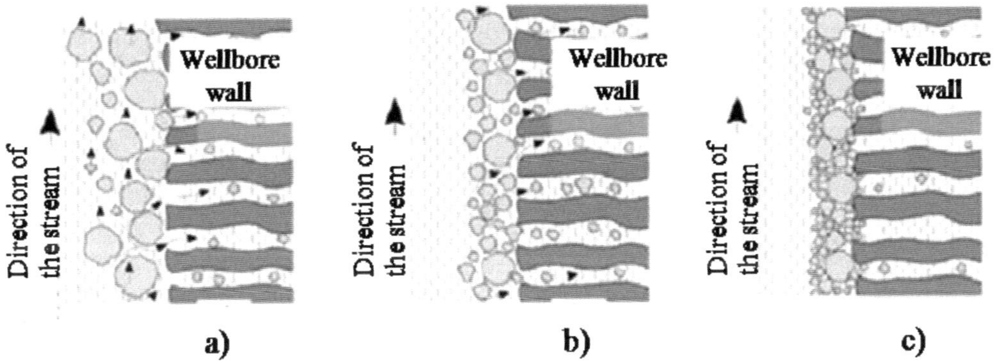

Figure 2. Influence of the bridging agent's fractional composition on the filtration properties of the drilling mud

Studies of Dick and Heinz revealed that most effective sealing of permeable zones is done when the content of the bridging agent in the drilling mud is about 60-90 kg/m^3 [1, 8].

Over the years, Abrams' criteria were used to calculate the fractional composition of the bridging agent. According to these criteria, the size of particles capable of forming arched bridges in the pore volume of the permeable formation should be equal to or greater than 1/3 of the average pore size of the formation, and their content should be at least 5% of the solid particles volume in the drilling mud. These criteria allow to determine only the particle size, which is necessary to start colmatation, but do not determine the requirements for particle size distribution in the mixture to ensure their effective packaging [2].

Along with the selection of the fractional composition of the bridging agent according to Abrams' criteria, a method aimed at the coverage of the largest range of particles ("Shotgun") is widely used. The essence of this method consists in mixing bridging agents of different fractional composition in proportions providing the widest range of particle sizes. This method is usually applied when the specific characteristics of the formation are unknown, and is not always effective [3].

Another method for determining optimal fractional composition of the bridging agent is based on the Ideal Packing Theory (IPT) developed by Kaeuffer [4]. Initially, it was used in the paint industry for the ideal pigment-dimensional distribution of particles in paints, and later found application in the drilling industry. According to this method, the bridging agent forms an ideal package, if the particle size distribution ensures effective clogging of all pores, including the pores formed between the solid particles. Ideal packaging is achieved if the total distribution of particles in the mixture is directly proportional to the square root of the particle size, i.e. graphically represented as a straight line (ideal mixture) (Figure 3) [1].

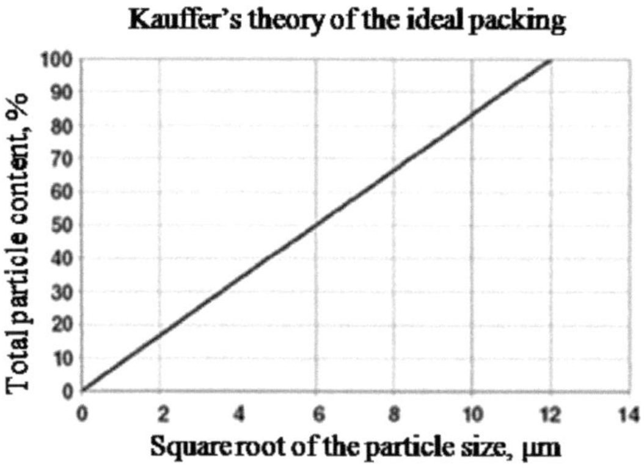

Figure 3. Graphical representation of the Kaeuffer's theory of the ideal packing

Drilling mud with a bridging agent, which has fractional particle distribution in accordance with the theory of ideal packing, is capable of forming the filter cake with a minimum permeability.

One of the most recent methods for selecting fractional composition of the bridging agent is the "Vickers Method" (Figure 4). The authors of this method established that in order to increase the effectiveness of colmatation, it is necessary that the fractional composition of the drilling fluid colmatage mixture should correspond to the size of the reservoir pore throats [5, 7]. This correspondence is represented by five target criteria, which carry basic information on the porosity of the particular reservoir and are defined by the following:

D50 – the size of the main pores of the formation, ± 1\3 of their size, μm;

D10 – is the size of the smallest pores of the formation, μm;

D25 – 1/7 of the size of the main pores, μm;

D75 – < 2\3 of the total porosity, μm;

D90 is the size of the largest pores of the formation, μm.

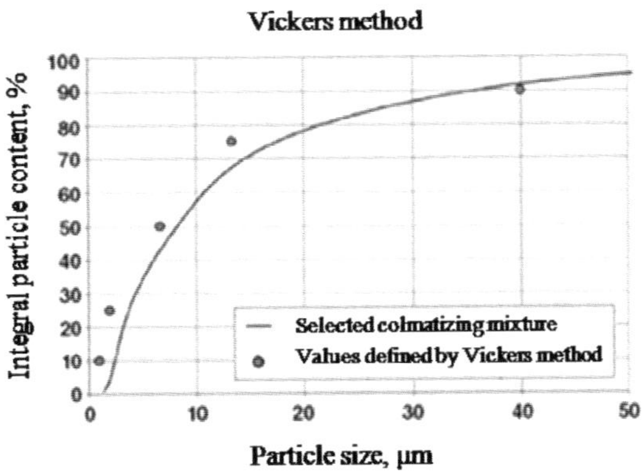

Figure 4. Graphical representation of the Vickers method

3. Selection of drilling mud for research

Current requirements to drilling muds for drilling-in productive reservoirs exclude the possibility of deep penetration of drilling mud filtrate, particles of the solid phase and other elements into the formation, that can reduce its permeability [6, 8]. Therefore, the drilling mud which is used for drilling-in productive reservoirs must comply with the following points:

• The filtrate of the washing fluid should not cause the swelling of clay particles, increase in hydrophilicity of the rocks and the amount of physically bound water in the pores of the formation;

• The composition of the filtrate should be such that, when it penetrates into the formation, no physical or chemical interactions occur, which can lead to creation of insoluble precipitation;

• The grain composition of the washing fluid solid phase must correspond to the structure of the productive formation pore volume. Surface tension at the interface of the filtrate and hydrocarbon content in the formation should be minimal;

• Fluid loss at the borehole bottom thermobaric conditions should be minimal, density and rheological properties should be so that the differential pressure during drilling of the productive reservoir would be close to zero;

• The degree of mineralization and salt composition of the filtrate should be close to degree as at reservoir, and the osmotic pressure is minimal.

At present, several types of solutions are used at drilling-in productive formations: thin clay solutions, clayless (biopolymer) solutions, petroleum based solutions [10]. The latter type of solution is used more often at present when drilling reservoirs, since the degree of mineralization and the composition of such fluid filtrate comply with formation liquids. However, such solutions have several drawbacks:

1. Expensive preparation;
2. Difficulties with preparation and regulation of parameters;
3. Solutions are environmentally unsafe systems.

According to mentioned above, two types of water-based solutions were chosen for the experiments: a biopolymer solution, the thin clay of which was BARAZAN (xanthan gum), and a thin clay solution, thin clay - bentonite powder, Duovis.

The composition of the solutions and the required amount of reagents for their preparation are presented in Tables 1 and 2.

Reagents, presented in the tables, have following functions: BARAZAN - biopolymer, viscosifier; KCl - salt, heaver; PAC HV and PAC LV - polyanionic cellulose of high and low viscosity, used for filtration and rheology control; NaOH – caustic soda, pH control; Na_2CO_3 - calcined soda, hardness control; Bentonite - clay, viscosifier, base of clay drilling mud; Modified starche for filtration control; Duovis - polymer, viscosifier.

Table 1. Component composition of biopolymer solution

Drilling mud	
Component	Yield, kg/m^3
BARAZAN	4
KCl	50
PAC HV	3
PAC LV	3
NaOH	1,1
Na_2CO_3	0,5

Table 2. Component composition of thin clay solution

Drilling mud	
Component	Yield, kg/m^3
Bentonite	30
NaOH	1,1
Na_2CO_3	0,5
PAC HV	1,7
PAC LV	1,7
Modified starche	10,8
Duovis	2

Density (ρ), as well as the main rheological parameters, such as static shear stress (SSS) calculated after 10 seconds (SSS_{10s}) and 10 minutes (SSS_{10min}) of rest, dynamic shear stress (DSS), plastic viscosity (PV) and filtration rate in 30 minutes (F_{30}) of the prepared solutions are presented in Table 3.

As can be seen in Table 3, the biopolymer solution has a higher initial slurry density than the thin clay solution. This means that in order to achieve the required density, lesser amount of weighting agent is needed.

Table 3. Main parameters of clayless and thin clay solutions

Solution parameter	Clayless (biopolymer) solution	Thin clay solution
ρ, kg/m^3	1050	1020
F_{30}, ml	10-11	10-11
PV, mPa·s	16-18	16-18
DSS, Pa	17-18	16-17
SSS_{10s}, Pa	4-5	5-6
SSS_{10min}, Pa	5-6	7-8

The filtration indices in both solutions were approximately the same: 11 ml and 10.6 ml - in the biopolymer and thin clay solution, respectively.

Rheological parameters of solutions (PV, DSS) also have approximately the same values.

The SSS of the biopolymer solution has lower values both for 10 seconds and for 10 minutes, rather than for thin clay solution.

The SSS values of the biopolymer solution correspond to the parameters of the "ideal" drilling mud to a greater degree according to [11].

The advantage of biopolymer solutions is that they increase the effective viscosity at low shear rates with an insignificant increase in plastic viscosity, which contributes to the enhancement of the carrying and suspending properties of the drilling mud and the reduction of the equivalent density at circulation. Strengthening the viscoelastic properties of drilling agents significantly improves cleaning

IOP Conf. Series: Materials Science and Engineering **921** (2020) 012013 doi:10.1088/1757-899X/921/1/012013

of the wellbore, and also reduces the rate of filtration of the liquid phase into the formation [12, 14, 19, 22].

In addition, the biopolymer solution contains a lesser amount of an insoluble solid phase in the composition than the thin clay solution [13, 15, 17, 23]. Therefore, there is a lower chance of reservoir contamination during drilling.

After the comparison, clayless (biopolymer) drilling mud was chosen for further studies, since it better satisfies current requirements to drilling muds for productive reservoir drilling-in [16].

4. Evaluation of weighting agents' fractional composition influence on drilling mud main parameters

For each experiment, a sample of a solution with a volume of 400 ml was prepared. The density of the base suspension was 1050 kg/m^3 [18, 20]. The required amount of reagents for preparing a solution sample is shown in Table 4. The drilling mud parameters are shown in Table 3.

Each sample of the solution was weighted to a density of 1160 kg/m^3. The required amount of weighting agent was:

$$C_w = \frac{\rho_w(\rho_{ws} - \rho_{cs})}{(\rho_w - \rho_{ws})} \quad (1)$$

$$C_w = \frac{2{,}7 \cdot (1{,}16 - 1{,}05)}{(2{,}7 - 1{,}16)} = 0{,}193 \ \text{g/cm}^3 \rightarrow m_к = 0{,}193 \cdot 400 = 77{,}2 \ \text{g},$$

where C_w - concentration of calcium carbonate, g/cm^3; $m_к$ - mass of calcium carbonate, g.

Table 4. Needed amount of reagents to prepare a solution sample of 400 ml

Drilling mud	
Component	Needed amount, g/400 ml
BARAZAN	1,6
KCl	20
PAC HV	1,2
PAC LV	1,2
NaOH	0,44
Na_2CO_3	0,2

The mass content of the heaver in the solution was 19% when weighted with calcium carbonate.

To draw the rheological curves of the solutions under consideration, the readings of the shear stresses were taken at different shear rates. The data obtained is presented in Table 5.

Table 5. Data for constructing the rheological curves of the prepared solutions.

Solution / Shear stress at different shear rates	3	6	$1\cdot10^2$	$2\cdot10^2$	$3\cdot10^2$	$6\cdot10^2$	n	K
Base	9	11	33	44	53	70	0,471	2,990
(D50) 5 μm	11	14	42	55	68	90	0,470	3,885
(D50) 10 μm	12	15	45	60	72	96	0,485	3,764
(D50) 50 μm	13	15	46	61	73	97	0,485	3,764
(D50) 60 μm	11	14	44	59	70	93	0,474	3,930
(D50) 100 μm	11	14	44	58	67	87	0,440	4,537
(D50) 150 μm	13	15	43	56	64	81	0,415	4,780

IOP Conf. Series: Materials Science and Engineering **921** (2020) 012013 doi:10.1088/1757-899X/921/1/012013

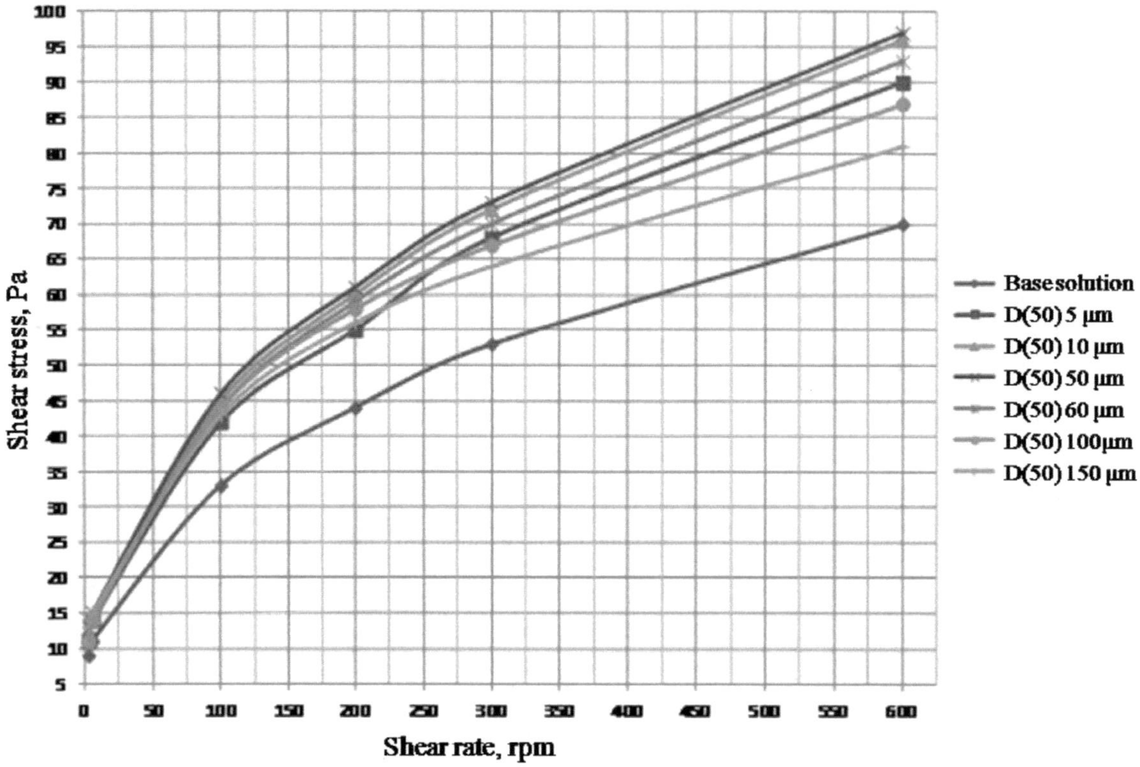

Figure 5. Rheological curves of solutions, constructed according to the Herschel-Balkley model

In accordance with the Herschel-Bulkley model, the rheological curves of the prepared solutions were constructed (Figure 5) [24].

Also in the course of the study following parameters of solutions were investigated: density, plastic viscosity, dynamic shear stress, static shear stress for 10 seconds and for 10 minutes, static filtration, dynamic filtration, data obtained for these experiments will be used for future research.

Proposed rheological model is described by the coefficients of nonlinearity and consistency. Figures 6 and 7 show the graphs of the change in the indices of nonlinearity (n) and consistency (K) as a function of the change in the average particle size of the weighting agent in the solution [25].

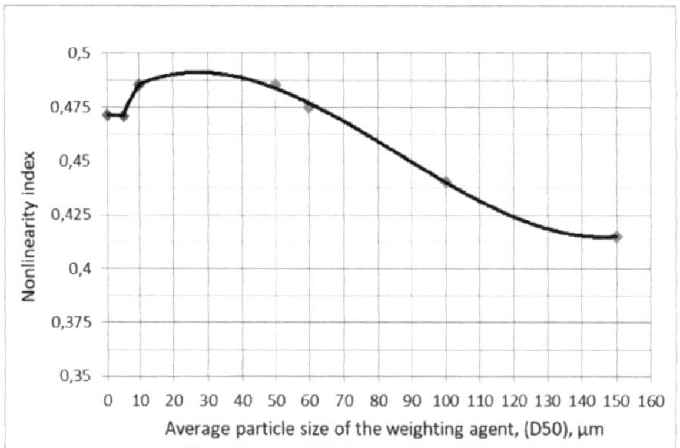

Figure 6. Graph of change in the nonlinearity index

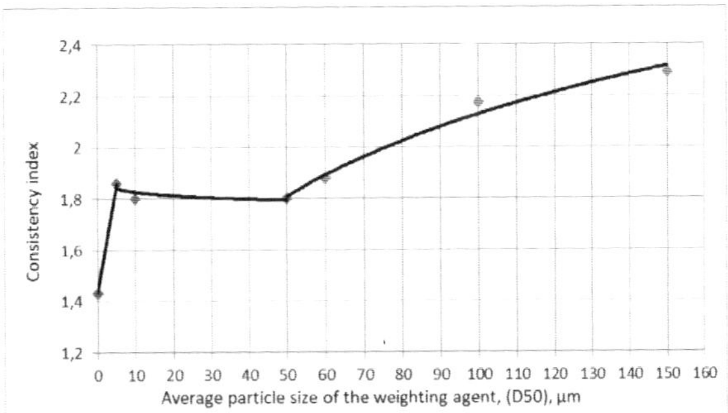

Figure 7. Graph of the change in the consistency index

Based on the experimental data obtained, the following statements can be made:

- introduction of a calcium carbonate into a biopolymer solution with an average particle size of up to 50 µm leads to a general increase in the resistance of the solution to the flow;
- figure 5 shows that the rheological curve corresponding to a solution weighted by calcium carbonate with a fractional composition (D50) of 50 µm has the highest shear stresses;
- introduction of fractions with a size (D50) of 10 µm also leads to an increase in the total rheology of the solution;
- drilling mud weighted with calcium carbonate with an average particle size of 150 µm has the lowest index of flow resistance;
- with the rise in the dispersity degree of the weighting agent, the amount of fine fractions increases, the total surface area of the material increases, which contributes to the structure-forming and increase of drilling mud rheology.

5. Conclusion

At present, a wide variety of weighting materials are used to weight the drilling muds at well drilling. Most common are carbonate and barite weighting agents.

The degree of dispersion and the fractional composition of the weighting agents affect their weighting ability, process of structure-formation in weighted drilling muds, and also the rheological properties of the solutions. For example, with an increase in weighting agent's degree of dispersion, its weighting ability decreases, and the growth of structural-rheological parameters is observed.

The correct choice of the colmatizing composition at drilling productive reservoir ensures creation of a thin low permeable filter cake on the wellbore wall and minimizes the penetration of the filtrate and the solid phase of the solution into the formation.

The introduction of calcium carbonate into the drilling mud with an average particle size of 10 to 50 µm leads to a general increase in the rheology of the solution and, as a consequence, to an increase in resistance to the flow of the solution. With the increase in the degree of dispersion of the weighting agent the amount of fine fractions increases, as well as the total surface area of the material and adsorption of the weighting agent particles by water and polymers that facilitates the process of structure-formation.

Conducted research showed that the rheological parameters of the drilling biopolymer solution are mostly influenced by the particles of the carbonate bridging agents with an average particle size (D50) of 50 µm, and less influenced by the largest particles of the weighting agents under consideration with the size (D50) of 150 µm.

In order to improve the quality of the reservoir drilling-in, in addition to the filtration characteristics of the drilling mud, it is also necessary to take into account its rheological parameters. Increased shear stress leads to an increase in pressure losses, which in turn causes a decrease in an overall drilling efficiency. Optimum selection of bridging agents in the composition of drilling muds will allow quick and efficient drilling-in of the formation.

Acknowledgments

Project is done within the frame of State scholarship for support of young scientists - SP-4606.2018.1 and SP-3147.2018.1.

References

[1] G. Ishbaev, M. Dilmiev, A. Khristenko, A. Mileyko Bridging theories of particle size distribution // Drilling and Oil. – 2011. - №6. – 16-18 p.;
[2] A. Abrams. The Influence of Fluid Viscosity, Interfacial Tension, and Flow Velocity on Residual Oil Saturation Left by Waterflood. Society of Petroleum Engineers Journal. Vol. 15, Issue 05, 1975;
[3] A. Abrams. Mud design to minimize rock impairment due to particle invasion. / Journal of Petroleum Technology. Vol. 29, Issue 05,1977, Pp. 586 – 592;
[4] Kaeuffer M. Determination de L'Optimum de Remplissage Granulometrique et Quelques properties S'y Rattachant, presented at Congres de I'A.F.T.P.V., Rouen, Oct 1973;
[5] Stephen Vickers, Martin Cowie, and Tom Jones, Baker Hughes Drilling Fluids, and Allan Twynam. A New Methodology that Surpasses Current Bridging Theories to Efficiently Seal a Varied Pore Throat Distribution as Found in Natural Reservoir Formations, BP [№16]];
[6] Don Whitfill Lost Circulation Material Selection, Particle Size Distribution and Fracture Modeling with Fracture Simulation Software. / Society of Petroleum Engineers. 2008;
[7] Valerie Belton, Stephen Vickers. Use of a Simple Multi-Attribute Value Function Incorporating Visual Interactive Sensitivity Analysis for Multiple Criteria Decision Making. / Readings in Multiple Criteria Decision Aid. pp 319-334;
[8] Keith Fraser Lamb, Brian Frank Harper, Syed Afaq Ali, Michael T. Darring, Hope C. Marine, William Edward Foxenberg Fluid-Loss Pills Control Losses through Sand-Control Screens Without Damage or Cleanup Treatment. / Society of Petroleum Engineers. 2007;
[9] M.A. Dick, T.J. Heinz, C.F. Svoboda, M. Aston Optimizing the Selection of Bridging Particles for Reservoir Drilling Fluids. / Society of Petroleum Engineers. 2000;
[10] Hubbatov A.A., Sharafutdinov Z. Z., Gaidarov M. M-R. Formation of properties of oil fluids / Construction of oil and gas wells on land and sea. 2010, №8, 28-31 p.;
[11] Myasnikov Ya., Gadzhiev S., Ionenko A., Evdokimov I., Losev A., Kronin A. Marcs engineer software predicting tight packing of lost circulation materials for drilling fluids. Drilling and Oil, 2014, №5, 35-37 p;
[12] Leusheva E.L., Morenov V.A. Combined oilfield power supplying system with petroleum gas utilization as an energy carrier. Neftyanoe hozyajstvo - Oil industry, №5, (2015), p. 96-100;
[13] S. S. Tabatabaee Moradi, N. Ivanovich Nikolaev, Considerations of Well Cementing Materials in High-pressure, High-temperature Conditions, International Journal of Engineering (IJE), TRANSACTIONS C: Aspects Vol. 29, No. 9, (September 2016) 1214-1218 ;
[14] V. Morenov, E. Leusheva, Energy Delivery at Oil and Gas Wells Construction in Regions With Harsh Climate, International Journal of Engineering (IJE), TRANSACTIONS B: Applications Vol. 29, No. 2, (February 2016) 274-279;
[15] S. Tabatabaee Moradi, N.I. Nikolaev, Optimization of Cement Spacer Rheology Model Using Genetic Algorithm, International Journal of Engineering (IJE), TRANSACTIONS A: Basics, Vol. 29, No. 1, (January 2016) 127-131;
[16] N.I. Nikolaev, E.L. Leusheva. Increasing of hard rocks drilling efficiency (Russian) - Oil Industry Journal, March 2016, 4p.
[17] S. S. Tabatabaee Moradi, N. Ivanovich Nikolaev, Sedimentation Stability of Oil Well Cements in Directional Wells, TRANSACTIONS A: Basics Vol. 30, No. 7, (July 2017) 1105-1109;
[18] V. Morenov, E. Leusheva, Development of Drilling Mud Solution for Drilling in Hard Rocks, International Journal of Engineering (IJE), TRANSACTIONS A: Basics Vol. 30, No. 4, (April 2017) 620-626;

[19] Morenov V.A Utilization of associated petroleum gas as an energy carrier (Primenenie popytnogo neftyanogo gaza v kachestve energonositelya). Naychno-tehnicheskie vedomosti SPBGPU - Scientific Technic News of SPBSPU, №3-2, (2012), 61-65;
[20] Nikolaev N.I., Leusheva E.L. Development of drilling fluids composition for efficiency increase of hard rocks drilling. Zapiski Gornogo instituta. 2016. Vol.219, p. 412-420;
[21] S. Tabatabaee Moradi, N.I. Nikolaev, E. Leusheva, Improvement of cement properties using a single multi-functional polymer, International Journal of Engineering (IJE), TRANSACTIONS A: Basics, Vol. 31, No. 1, (January 2018) 174-180;
[22] Belsky A.A., Dobush V.S., Malarev V.I. Electro Steam Thermal Complex Powered by Wind-Driven Generator for the Treatment of the Oil Formation's Bottomhole Area. Journal of Physics: Conference Series. 2020, vol. 1441(1), 012020. DOI: 10.1088/1742-6596/1441/1/012020
[23] Strizhenok A.V., Korelskiy D.S. Improvement of the system of industrial environmental monitoring of atmospheric air in the area of anthropogenic arrays impact. Journal of Physics: Conference Series. 2019, vol. 1384(1),012052. DOI: 10.1088/1742-6596/1384/1/012052;
[24] Huang, X., & Garcia, M. (1998). A Herschel–Bulkley model for mud flow down a slope. Journal of Fluid Mechanics, 374, 305-333. doi:10.1017/S0022112098002845;
[25] Nikolaev N.I., Haoya L. Results of Cement-to-Rock Contact Study. Zapiski Gornogo instituta. 2017. Vol. 226. P. 428-434. DOI: 10.25515/PMI.2017.4.428.

IOP Conf. Series: Materials Science and Engineering **921** (2020) 012014 doi:10.1088/1757-899X/921/1/012014

Method of signal identification based on wavelet analysis.

V A Loginov and A V Elnikov

Budgetary institution of higher education of the Khanty-Mansiysk Autonomous Okrug - Ugra "Surgut State University"

loginovva93@gmail.com; andreyeln@mail.ru

Abstract. The article presents the results of the development of a method for identifying the source of vibroacoustic impact on the sensitive element of the fiber-optic perimeter protection system of industrial facilities. The methodology was tested on the identification of signals from the following sources of vibroacoustic impact: impact over a certain period of time, or shaking the fence (t-impact); short-term impact type impact (δ-exposure). In the course of the work, polygon tests of the perimeter security system were conducted, a library of signals was compiled, and characteristic features of signals of various types of impact on the sensitive element of the system were identified. The main signal processing tools are the Fourier and wavelet transforms.

1. Introduction

The identification method was developed during the surveys of signals of the fiber-optic perimeter security system. This method allows you to distinguish the signals of various vibroacoustic sources. The following types of effects on the sensitive element of the system acted as vibroacoustic sources:

- Exposure over a period of time, or shaking the fence (τ-exposure);
- Short-term exposure type impact (δ-exposure).
- At the initial stage of development the technique was focused on identifying signals using the Fourier transform.

2. The Fourier transform

The Fourier transform allows you to transfer the signal from the time domain to the frequency domain, that is, to obtain the spectrum of this signal. On account the analysis of the various the spectra of signals types, it is possible to identify the frequency components characteristic of the specific vibroacoustic sources. For a discrete signal, which is a lattice function and, as a rule, defined on a finite period of time (measurement time), the Fourier transform takes the form of the so-called discrete Fourier transform:

$$X_k = \sum_{k=0}^{N-1} x_n e^{\frac{-2\pi i}{N}kn} \quad (1)$$

Figure 1 shows the τ-exposure signal and figure 2 shows its Fourier spectrum. It is possible to distinguish harmonics inherent to τ-exposure signals in the range of 5-40 Hz. However, the most characteristic harmonics are 6 Hz and 25 Hz, which are present in all the signals presented above.

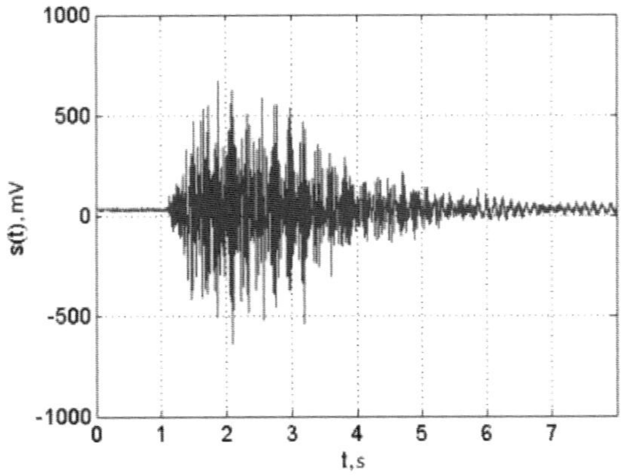

Figure 1. Signal of τ-action

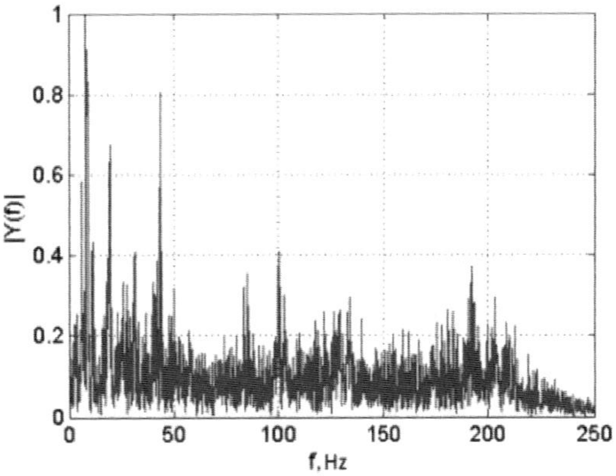

Figure 2. Fourier spectrum of τ-action signal

Figure 3 shows us the δ-exposure signal and figure 4 shows its Fourier spectrum. The harmonics lying in the range of 160-180 Hz are typical for δ-exposure signals. In the low-frequency range, harmonics with an amplitude above the noise components are absent, unlike t-signal signals.

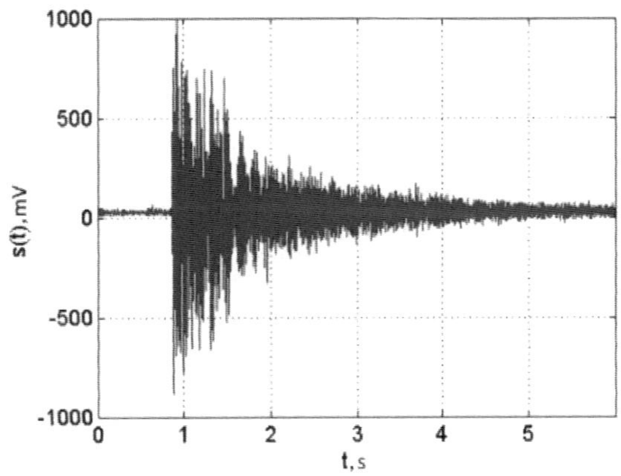

Figure 3. Signal of δ-action

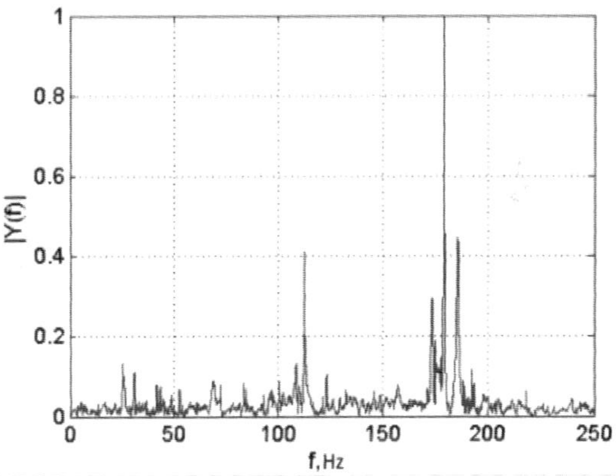

Figure 4. Fourier spectrum of δ-action signal

However, the Fourier transform provides global information about the investigated signals and does not allow considering local features of the signals (distribution of harmonics in time).

3. The wavelet transform

At the end of the last century, a new and important direction appeared in the theory and technique of processing signals, images, and time series, called the wavelet transform (WP), which is well suited for studying the structure of non-stationary (time-varying) processes. Wavelets are special functions in the form of short waves (bursts) with a zero integral value and with localization along the axis of an independent variable (t or x), capable of shifting along this axis and scaling (stretching / compression).

The wavelet transform (WT) of a one-dimensional signal is its representation in the form of a generalized series or Fourier integral over a system of basis functions:

$$\psi_{ab}(t) = \frac{1}{\sqrt{a}}\psi(\frac{t-b}{a}) \tag{2}$$

constructed from the parent (source) wavelet Ψ (t), which has certain properties due to time-shifting operations (b) and changing the time scale (a). The multiplier 1/a ensures that the norm of these functions is independent of the scaling number a.

Let's consider the result of the wavelet transformation of the previously presented signals.

For the wavelet transform of the system signals, the CWT (...) function was used in the Matlab environment. Wavelet transform implies the choice of the mother wavelet used. In preparation for the wavelet analysis, the following wavelets of various orders were considered: the Haar wavelet, the Mexican hat wavelet, the Meyer wavelet, the Morlet wavelet, and the Gaussian wavelet.

The most detailed spectrum was obtained using 8th order Gauss as the mother wavelet (Figure 5).

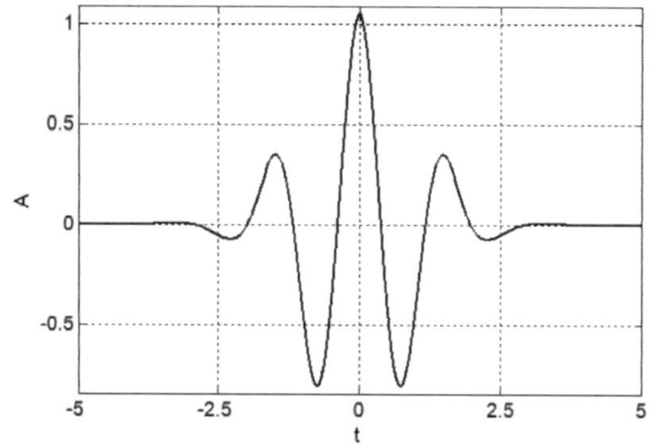

Figure 5. 8th-order Gaussian wavelet (ordinate axis - amplitude; abscissa axis - time in s)

The wavelet spectrum is a three-dimensional graph of the dependence of the harmonic amplitude (displayed in color) on the frequency scaling factor and time in s. It is also worth noting that you must first multiply the initial signal by the window function so that "false" harmonics do not occur at the beginning and end of the signal presence interval. As a window (weighted), the Hann function was selected with a length of 5000 samples in accordance with the length of the signals (Figure 6).

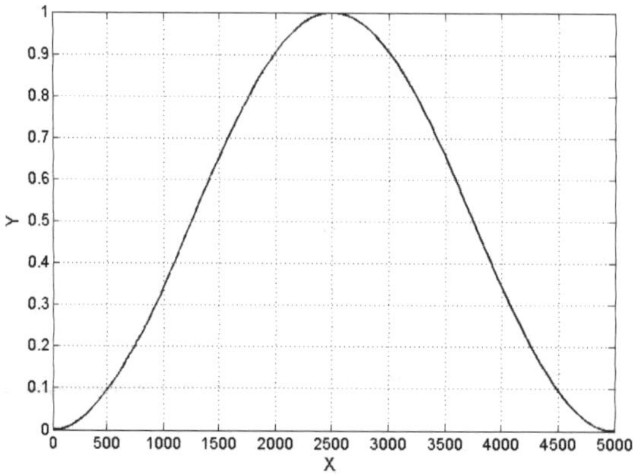

Figure 6. Hann's window function (ordinate axis - amplitude; abscissa axis - reference number).

Figure 7 shows us the wavelet spectrum of the τ-exposure .. In this case, in the period from 2.8 s to 7 s, harmonics lying in the frequency range from 10 to 80 Hz make the greatest contribution.

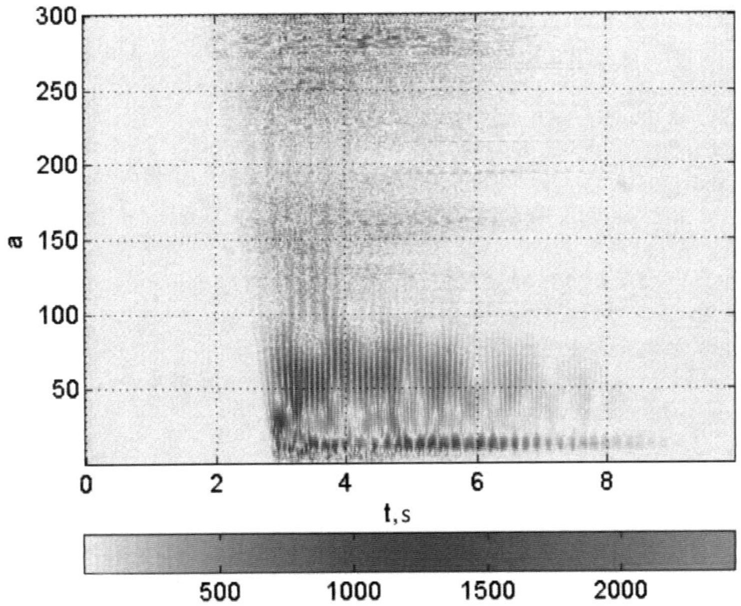

Figure 7. Wavelet spectrum of the τ-exposure signal.

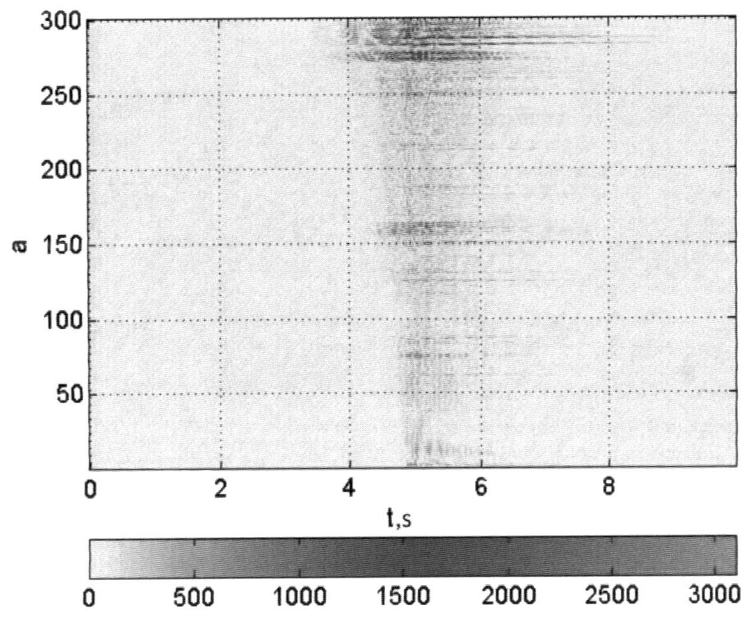

Figure 8. Wavelet spectrum of the δ-exposure signal.

Figure 8 the wavelet spectrum of δ-exposure is presented. In this case, the largest contribution is made by harmonics lying in the range of 250 to 300 Hz.

Based on the obtained data, it is possible to form an algorithm for the method of identification of the signal source of the vibro-acoustic monitoring system.

τ-exposure signals are characterized by low frequencies up to 80 Hz, while δ- exposure signals are characterized by higher frequencies, ranging from 250 Hz to 300 Hz. Based on this information, an algorithm for identifying the vibroacoustic source was formed.

The algorithm consists of the following steps:

- Wavelet transformation;
- Separation of the wavelet spectrum in the region of a fixed length in time and width in frequency;
- Calculation of the average value of the harmonic amplitude in each region;
- Selecting areas with the maximum value and converting them to a two-dimensional spectrum;
- Analysis of the obtained spectrum.

The regions formed by dividing the spectrum have a time length of 0.2 s and a frequency width of 10 Hz. The grid of division of regions is presented in figure 9.

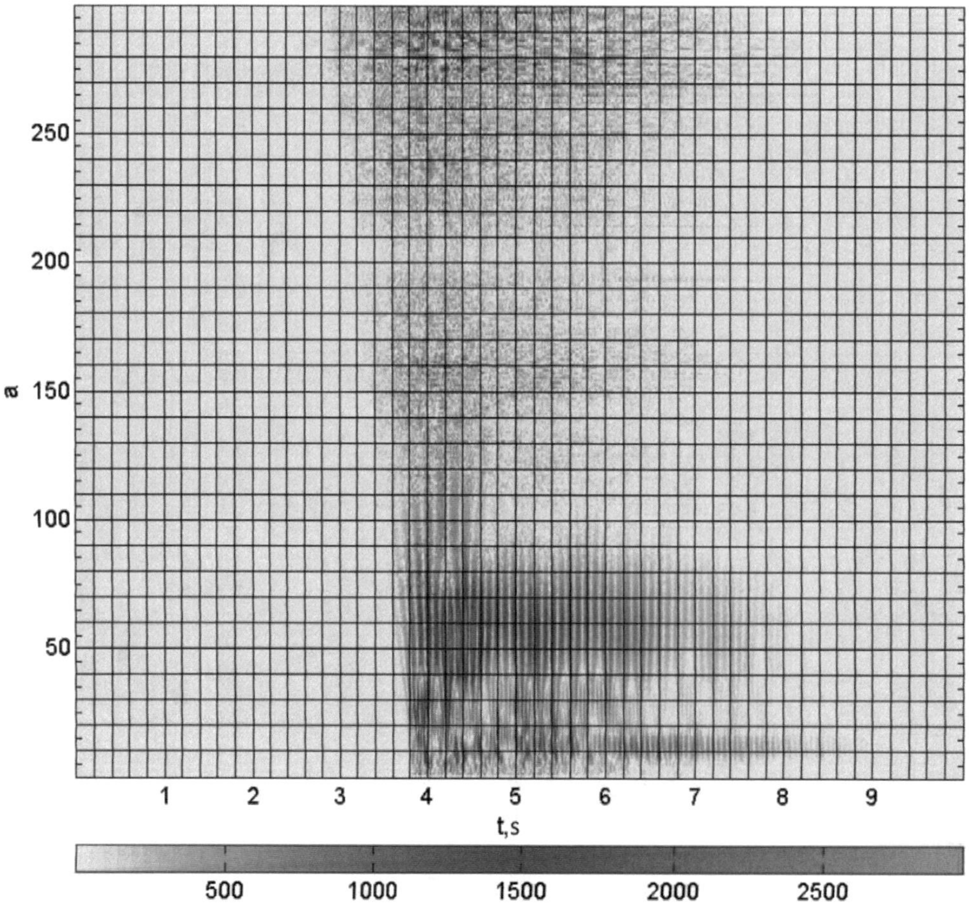

Figure 9. Grid for dividing the wavelet spectrum into regions.

IOP Conf. Series: Materials Science and Engineering **921** (2020) 012014 doi:10.1088/1757-899X/921/1/012014

After calculating the average value of harmonics amplitudes in each region 4 regions stand out with the maximum value are allocated and the frequency coordinates of the Central points of these regions are saved. The resulting frequency values are converted to a spectrum. The next step is to analyze the spectrum. The type of vibroacoustic signal source is determined by the harmonics belonging to one of the ranges: from 0 Hz to 80 Hz (τ-impact); from 250 Hz to 300 Hz (δ-impact).

In figure 10 shows the spectra of τ-exposure and δ-exposure. Some harmonics of the spectrum of the δ-exposure signal are not included in the previously determined frequency range from 250 Hz to 300 Hz, however, Figure 10 shows us that the main contribution is made by harmonics belonging to this range.

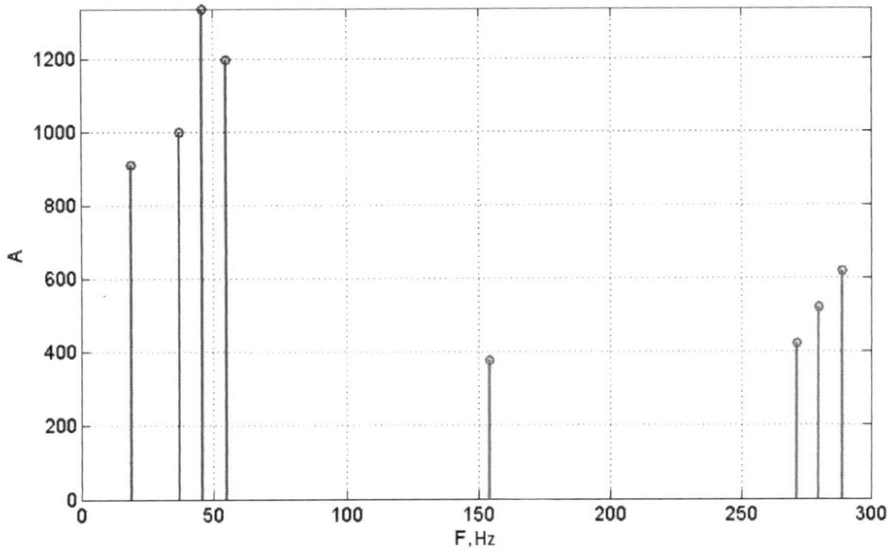

Figure 9. The spectrum of τ-exposure signals (blue) and δ-exposure (red) obtained as a result of the processing algorithm.

As a result of all the work carried out, a methodology was identified for identifying a vibro-acoustic signal source, which makes it possible to distinguish between t-effect and d-effect signals. This methodology can exist independently, and can be applied in conjunction with other techniques to increase reliability.

This work was supported by the Russian Foundation for Basic Research (Grant No. 18-47-860014).

References

[1] Loginov V A, Elnikov A V and Sysoev S M 2017 *Fundamentals of the method for identifying the type of impact to the perimetral fiber-optic system* (Surgut: Proceedings in Cybernetics)

[2] Kirillov S N 2007 *Digital signal processing algorithms based on wavelet decompositions in radio engineering devices*

[3] Lyons R 2001 *Understanding digital signal processing*

[4] Solonina A I 2013 *Digital signal processing and MATLAB: tutorial* (St. Petersburg: BVH-Petersburg)

IOP Conf. Series: Materials Science and Engineering **921** (2020) 012015 doi:10.1088/1757-899X/921/1/012015

Polymeric stabilization of drilling drills with low content of solid phase

Yu K Lukanina[1], E V Belenko[2], P A Sakharov[1] and A V Khvatov[1]

[1] Emanuel Institute of Biochemical Physics, Russian Academy of Sciences, Kosygin str., 4, Moscow, 119334 Russia
[2] MIREA - Russian Technological University, 78 Vernadsky Avenue, Moscow, 119454 Russia

E-mail: hvatovanatoliy@gmail.com

Abstract. This article is devoted to identifying the relationship between the structure-forming properties of polymer stabilizers of clay suspensions (drilling fluids) and the structure of their macromolecules based on thermodynamic representations of the stability of polymer-dispersed systems. New types of polymer-micellar gels combining high structural and pseudo-plastic properties and optimal surface-active characteristics characteristic of concentrated micellar systems of the "surfactant-solubilizate" type are presented.

The use of polymeric thickeners and stabilizers for washing fluids pursues a variety of purposes, the most important of which are: reducing the filtration rate, ensuring sedimentation stability of drilling dispersion, controlling the rheological and structural-mechanical characteristics of the system, as well as implementing special technological requirements, for example, thermal stability, resistance to biodegradation, salt resistance, etc. Polymer reagents differ in the chemical nature of the monomers forming the macromolecular chain, the size of the molecular mass and the degree of ionization of the functional groups. The most common synthetic polymers are polysaccharide and polyacrylic series. In addition, starch derivatives, lignosulfonates, humates, as well as microbial polysaccharides (XC biopolymers) are often used as stabilizing agents. The esterified derivatives of cellulose (carboxymethyl cellulose (CMC), polyanionic cellulose (PAC), hydroxyethyl cellulose (HEC), carboxymethyloxyethyl cellulose (CMOEC), etc.) are polyelectrolytes containing hydroxyl, macromolecular, other these polymers are divided into anionic, cationic, amphoteric and nonionic. The solubility of nonionic compounds in water is due to the presence in their composition of hydroxyl and ether polar functional groups resistant to hardness salts.

An important non-ionic stabilizer is the starch reagent, which is the main component of cereal seeds (corn, wheat, rice) and tuber crops (potatoes and tapioca). Hydrolysis of starch produces glucose, as well as a small amount of fatty and phosphoric acids. The carbohydrate part of starch contains two polysaccharides - amylose and amylopectin. Amylose consists of long chains of α-glucose, the molecular weight of which varies from 10 thousand to 100 thousand (Figure1).

IOP Conf. Series: Materials Science and Engineering **921** (2020) 012015 doi:10.1088/1757-899X/921/1/012015

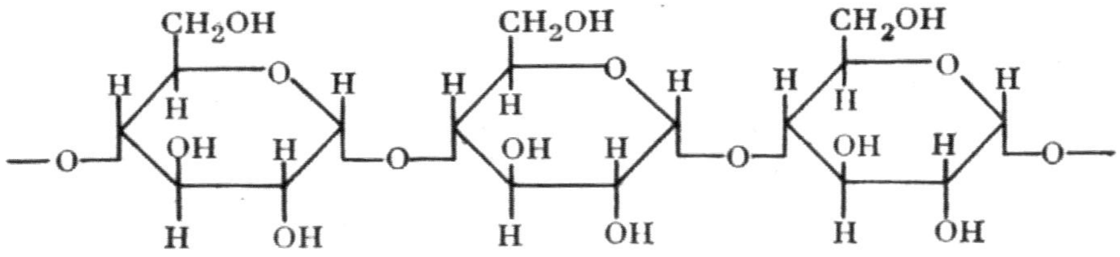

Figure 1. Amylose.

Amylopectin contains a mixture of branched chain molecules (Figure 2).

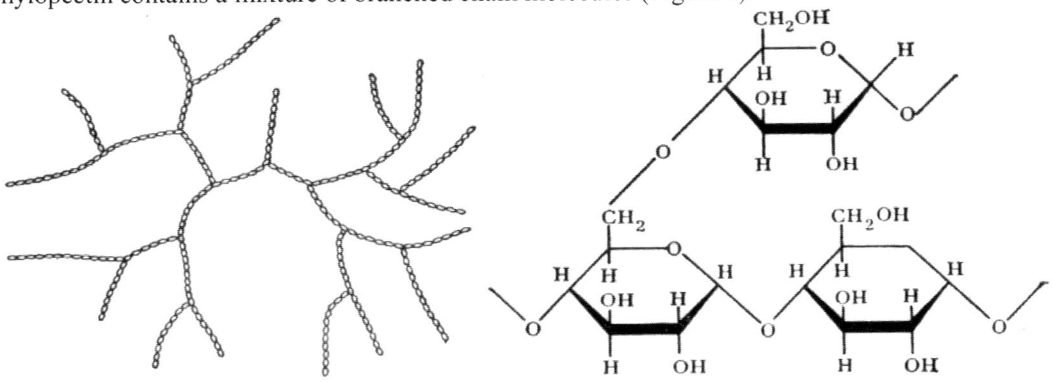

Figure 2. Amylopectin.

The molecular weight of amylopectin varies from 40,000 to 100,000. Starch is used in drilling fluids only to reduce the filtration rate and is susceptible to enzymatic decomposition by many microorganisms (yeast, mold, bacteria).

In the chemical structure of biopolymers, a β-linked backbone containing D-glucose, D-mannose and D-glucuronic acid with one side chain for every 8 sugar residues and one 4,6D-glucose side chain for every 16 sugar residues takes part [1] (Figure 3).

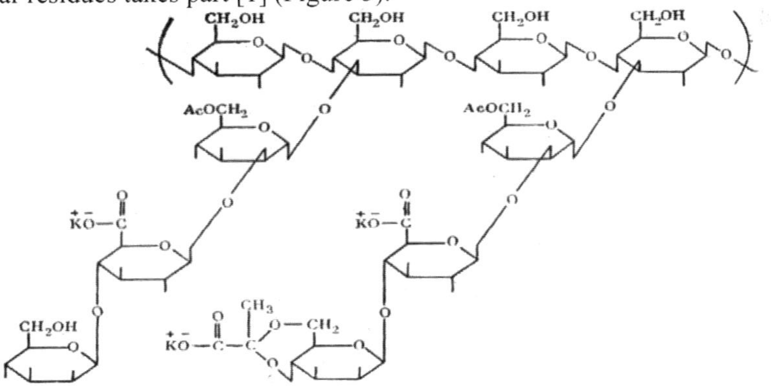

Figure 3. Chemical structure of biopolymers.

The biopolymer macromolecule contains carboxyl, carbonyl and hydroxyl groups, as a result of which it is characterized by a high thickening ability, giving the drilling fluid highly pronounced pseudoplastic properties and resistance to polymineral aggression.

The success of using low-charged and non-ionic polymers (starch, biopolymer, HEC) in low-clay and non-clay drilling fluids is associated with ensuring the aggregate stability of the system, as well as with the formation of a three-dimensional structural grid. The stabilization of the disperse system and structure formation are interrelated processes in which the polymer provides the implementation of

IOP Conf. Series: Materials Science and Engineering **921** (2020) 012015 doi:10.1088/1757-899X/921/1/012015

structural-mechanical, sorption-solvate and electrostatic factors of system stability. There are several mechanisms of structure formation with the participation of a nonionic polysaccharide:
- in aqueous solutions of biopolymers, the tendency toward macrostructuring is very pronounced as a result of the formation of supramolecular spatial networks consisting of spiral structural units that are contacted by means of hydrogen and Van der Waals bonds. The biopolymer macromolecules, in addition to polar functional groups, also carry anionic carboxylate groups, the source of which is the structural units of pyruvic acid and glucuronic acid, located on the side branches inside the main (d) glucose helix. It is this shielding of the charged sections of the biopolymer macromolecule that apparently explains the unique reversibility and shear strength of hydrated macromolecular structures, which provide the solid, pseudoplastic nature of the rheological behavior of biopolymer solutions. This, in particular, is confirmed by the fact that the maximum solubility of biopolymers is observed at pH = 4–6, while the solubility region of conventional polyanionic stabilizers lies at pH> 8 [2];
- non-ionic polymers (XC-polymer, starch) form spatially structured systems when interacting with dispersed particles of high-grade bentonite in polymer-bentonite drilling fluids. An important property of such solutions is their pseudoplasticity, characterized by a pronounced dependence of the effective viscosity on shear rate;
- polymer structured solutions are obtained by complexation using polyvalent metal ions, characterized by a large charge and a small ionic radius. Metals whose ions have a large hydration capacity include Cr, Al, Ca, Zr, Ti. To enhance the thixotropic properties of biopolymer solutions, trivalent chromium salts are introduced into them, the effect of which is due to the ability to form multinuclear complex compounds, for example, aquacomplexes, in the inner sphere of which water molecules and hydroxyl groups are contained. The coordinating ability of cations depends on their ionic potential (ratio of ion charge to its radius) [3], in accordance with the value of which complexing agents are located in the following order: Al> Fe> Cu> Ni> Mn. The thixotropic properties and viscosity of polymer solutions containing one or another complexing agent are directly proportional to its ionic potential. Thus, Al3 + cations are capable of very tightly coordinating the hydroxyl and carboxyl groups of polysaccharide derivatives in the pH range of <6, while a three-dimensional gel structure of the Ecorish solution is formed on the basis of the complex of iron (III) salts and HEC [4].

The requirement to control the structural and mechanical characteristics of drilling fluids is also presented to polyanions, which are less prone to complexation, however, form adsorption layers on the surface of clay particles, characterized by a long-range order.

The purpose of this article is to identify the relationship between the structure-forming properties of polymer stabilizers and the structural features of their macromolecules based on thermodynamic ideas about the stability of polymer-dispersed systems that can perform useful work. If, as a result of reagent treatment, the mud filtration rate increases, then this means that negative work has been performed on the system (ΔAf <0). Moreover, regardless of the change in the rheological properties of the system, the structural work performed by the system is also negative (ΔAs <0). If the filtration rate decreases, then the system itself does a positive job: ΔAf> 0, ΔAs> 0.

The gradient of the resulting operation of the mud system can be found as follows:

$$\mathbf{grad\ A} = \mathbf{T} \cdot \mathbf{l}$$

This is the scalar product of the unit vector **l** and the tensor **T**, which is the result of the product of the gradients of the work produced by the system:

$$\mathbf{T} = \mathbf{grad\ A_s} \bullet \mathbf{grad\ A_f} = \begin{bmatrix} a_{11} & a_{12} & a_{13} \\ a_{21} & a_{22} & a_{23} \\ a_{31} & a_{32} & a_{33} \end{bmatrix} \mathbf{i_m j_n} = \mathbf{i_m j_n} T_{mn}$$

Having calculated the value of **grad A**, the differential value of the resulting system operation dA is determined. Drilling fluid, being an open system, can perform positive work only due to loss of free energy:

$$dG < -dA,$$

где: $dG < 0$.

Therefore, in order for the system to perform positive work, it is necessary to have a supply of free energy, the change of which is associated with the positive work of the system:

$$dA = -k \cdot dG$$

The coefficient k characterizes the completeness of the conversion of free energy into useful work dA. In an isothermal irreversible process, when a certain amount of energy δQ is transferred to the system, entropy production is observed in accordance with the ideas of R. Clausius (for an isolated system), which is expressed by "uncompensated" heat (δQ`):

$$\delta Q = TdS - \delta Q`$$

Chemical treatment of the drilling fluid, which is an isobaric-isothermal process, should be accompanied by a decrease in the free energy level of the system ($\Delta GS < 0$). Suppose that as a result of introducing a polymer component, the system does a positive job ($\Delta A > 0$), accompanied by a decrease in the filtration index. The system before polymer processing ($\Delta_1 A$) and after it ($\Delta_2 A$) is expressed by the equations:

$$\Delta_1 A = -k \cdot \Delta_1 G = -k \cdot (\Delta_1 H - T \cdot \Delta_1 S)$$
$$\Delta_2 A = -k \cdot \Delta_2 G = -k \cdot (\Delta_2 H - T \cdot \Delta_2 S)$$

During the production process, the enthalpy of the system does not change ($\Delta 1H = 0$, $\Delta 2H = 0$), therefore, the operation of the drilling system before and after chemical treatment has an entropy character:

$$\Delta_1 A = -k \cdot T \cdot \Delta_1 S$$
$$\Delta_2 A = -k \cdot T \cdot \Delta_2 S$$

Define the work performed by the system as a result of chemical treatment:

$$\Delta A = \Delta_2 A - \Delta_1 A = k \cdot T \cdot (S_1^{1} - S_1^{2}),$$

where: $S1^1$ - level of system entropy before chemical treatment;

$S1^2$ is the entropy level of the system after chemical treatment.

The magnitude of the positive operation of the drilling fluid system is greater, the lower the entropy ($S1^2$) of the system after chemical treatment. Thus, chemical treatment of the drilling fluid with a general decrease in the free energy level of the system should lower its entropy. This conclusion allows us to formulate the basic requirements for polymer reagents as structural agents and reducing agents for filtering washing fluids.

The introduction of the polymer into the drilling fluid must be accompanied by an exothermic thermal effect ($\Delta H < 0$), which ensures the solubility of the reagent in the dispersion medium, while the entropy of the system should decrease ($\Delta S < 0$). Therefore, polymer macromolecules must contain a large number of hydrophilic functional groups, both non-ionic type (polyoxyethylene chains) and those capable of dissociation. The optimal polymer should be a polyanion or polycation, in the macromolecules of which the charged fragments alternate with nonionic (hydroxyl, hydroxyethyl, carbonyl) groups.

The requirement of a negative enthalpy of dissolution of the polymer reagent determines the choice of polymers capable of chemisorption on the surface of the particles of the dispersed phase. Moreover, the higher the molecular weight, the higher the affinity of the polymer to the surface of the adsorbent. In drilling fluids, the dispersed phase is mainly represented by bentonite particles, the crystal lattice of which is formed by two structural elements consisting of 3 layers. One structural element consists of two outer silicon-oxygen tetrahedral networks with silicon atoms in the centers. Between the two silicon-oxygen tetrahedrons is an aluminum-oxygen octahedral network. Adjacent packets face each other with the same layers. The weak bond between the layered packets and the possibility of water and some cations penetrating between them explains the high ion-exchange ability of bentonite clays and their strong swelling in water. Particles of a clay dispersed phase have a lamellar structure: the faces of three-layer packets are negatively charged (as a result of dissociation of surface silanol groups), and the edges are positive. As a result, polyanions are prone to chemisorption on the edges of aluminosilicate packets, and polycations are adsorbed on their faces and enter the interlayer space. Polyanions increase the negative charge of dispersed particles, helping to stabilize the system, while polycations neutralize the clay surface, causing the development of coagulation. With an excess of polycation concentration, recharging of the surface of a dispersed particle can take place, acquiring a positive potential, which will ensure the implementation of the electrostatic stability factor of the system.

Effective polymer treatment of the drilling fluid is accompanied by a decrease in the entropy of the system. This means that macromolecules must be integrated into the hydrogen bonds of the existing coagulation structure, reducing the concentration of "free" water. The state of the aqueous phase in the drilling fluid system can be judged by the change in its relative hydrophilicity, which is directly dependent on the time of spin-spin relaxation of water protons, determined by the method of nuclear magnetic relaxation (NM relaxation) [5, 6]. The introduction of the polymer should be accompanied by an increase in the fraction of water protons that are tightly coordinated in the adsorption layers on the surface of the dispersed phase (spin-spin relaxation time T_2 of the order of 10^{-2} s), as well as an increase in the fraction of "exchange" water in ion hydration shells (T_2 of the order of 10^{-1} s). The requirement of low entropy also determines the choice of polymers with a maximum molecular weight and a limited number of macromolecular conformations. Therefore, polymers of regular structure, the macromolecules of which form stable spiral structures, are most in demand.

Let us dwell in more detail on the micellar structures of nonionic surfactants. It is known that gels based on concentrated solutions of nonionic surfactants are of a viscoplastic nature. In such solutions, macromicellar aggregates interact at close distances due to the formation of hydrogen, van der Waals, or ionic bonds, forming a cellular structure. The pseudoplastic nature of concentrated emulsions is due to an elongated elongated form of supramolecular aggregates, which are easily oriented in the direction of shear rate. Micellar solutions used in drilling and oil production practice have a low concentration of micelle-forming surfactants, not more than 5-10 times higher than the critical micelle formation concentration (CMFC). Typically, such solutions include oil-soluble solubilizate (hydrocarbons, higher fatty acids), which leads to the formation of axisymmetric micellar aggregates and, accordingly, the development of structure-forming and pseudoplastic properties [7]. Unfortunately, compared to polymer (crosslinked ionic bonds or biopolymer) structures, micellar gels are characterized by irreproducible rheology under shear or thermal stresses. Therefore, the development of new types of hydrogels — polymer-micellar, combining high structural and pseudoplastic properties characteristic of crosslinked polymer systems, and optimal technological characteristics (surface-active, lubricating, antifiltration, inhibitory) characteristic of concentrated micellar systems of the "nonionic surfactant-solubilizate" type is highly relevant. ".

As the polymer base of the polymer-micellar system (hydrogel), we will choose derivatives of polysaccharides bearing hydroxyethyl and carboxymethyl groups — CMOEC, and as nonionic surfactants — polyalkylene glycol (PAG) —a random copolymer of ethylene and propylene oxides characterized by a hydrophilic-lipophilic balance value of 3. Tall oil (TM), in which the content of fatty acids is 60%, acts as a hydrophobic solubilizate. We studied the structure formation in the ternary system "polymer (CMOEC) - PAG - TM", in which PAG micellar complexes acquire a surface charge due to the introduction of a neutralizing alkaline agent (HA) into the system. The main structure-forming element in such a system (dispersed phase) is, apparently, the hydrophobic surface of the polysaccharide skeleton, the hydrophilic cavities inside which are occupied by mixed PAG micelles and TM fatty acids, which form hydrogen and ionic bonds with functional groups (hydroxyethyl and carboxylate anions) of CMOEC .

The compositions of the obtained polymer-micellar hydrogels with the same content of TM acids are given in Table 1. As can be seen, an increase in the mass fraction of the alkaline agent allows one to achieve a decrease in the content of the polymer stabilizer (CMOEC) and micelle former (PAG) at a constant concentration of hydrophobic solubilizate (TM).

Table 1. Compositions and technological properties of polymer-micellar hydrogels.

№ experience	CMOEC, %	PAG, %	TM, %	HA, %	Filtration rate, cm^3 / 30 min	Clay minerals initial wetting rate	Coefficient of friction

1	0,29	0,64	17,5	0,63	4	1,7	0,026
2	0,2	0,52	17,5	0,88	4	2,3	0,026
3	0,145	0,32	17,5	1,35	6	2,3	0,020

The results of the study of the technological characteristics of the obtained hydrogels are presented in Table 1. As can be seen, an increase in the mass fraction of HA in the polymer-micellar hydrogel leads to a slight increase in the filtration rate and a decrease in the initial rate of swelling of clay minerals in the hydrogel medium. On the other hand, by increasing the degree of neutralization of TM fatty acids, system No. 3 has the best lubricating properties.

The rheological properties of the gels were investigated on a FANN viscometer company Baroid (USA). Experiments show that as a result of an increase in the concentration of an alkaline agent, there is a decrease in the viscosity characteristics, as well as the yield strength of the gel system. However, with an increase in the mass fraction of HA, the static shear stress of the gels increases, which is apparently due to the structural effect of neutralized fatty acids. The effect of temperature on the rheological characteristics of polymer-micellar gels was also studied. Upon heating, there is a gradual decrease in the viscosity and structural characteristics, as well as a slight decrease in the pseudoplastic properties of the gel, which, however, remain at a high level even at 80 ° C.

It should be noted that the mass fraction of fatty acids in the composition of HMs significantly affects the properties of polymer-micellar hydrogels. The effect of the mass fraction of fatty acids in the composition of HM on the technological properties of hydrogel solution No. 3 was studied (Table 1). Table 2 presents the technological parameters of such hydrogels, differing only in the content of alkaline agent.

Table 2. Investigation of the influence of the mass fraction of fatty acids in HM on the technological properties of polymer-micellar gel based on CMOEC.

The content of HA,%	Mass fraction of fatty acids in TM, %	Effective viscosity, MPa·s	Plastic viscosity, MPa·s	Static shear stress, dPa	Consistency Index (Ostwald-Weil Model)	Degree indicator (Ostwald -Weil model)	Coefficient of friction	Filtration parameter, cm / 30 min
1,35	60	119	43	85/90	28	0,29	0,02	6
0,89	79	140	60	30/35	12,7	0,35	0,007	6
0,22	97	150	65	16/18	7,3	0,45	0,007	5,5

As can be seen, an increase in the concentration of fatty acids in the composition of tall oil can reduce the concentration of alkaline agent necessary to ensure complete colloidal solubility of the hydrophobic solubilizate. It is easy to notice that an increase in the content of fatty acids in TMs (Table 2) affects the properties of hydrogels in a similar way to a decrease in the concentration of an alkaline agent at a constant concentration of fatty acids (Table 1): there is an increase in the viscosity characteristics of the system with a simultaneous decrease in structural and pseudoplastic properties, as well as lubricating and anti-filtration indicators are improved.

In the course of the studies, it was established that it is possible in principle to obtain structured hydrogel systems by stabilizing solubilization microemulsions with polysaccharide water loss reducers. Thus obtained polymer-micellar hydrogels combine high structural and pseudoplastic properties characteristic of biopolymer solutions, as well as significant inhibitory and lubricating characteristics inherent in saturated emulsion systems. However, the preparation of such emulsions in a drilling environment is rather laborious due to the multi-stage emulsification process and the low efficiency and productivity of the most common drilling dispersants. A modern engineering solution designed to facilitate the management of the properties of polymer-micellar hydrogels and transfer them to the category of real industrial systems is the concept of flushing liquids with a reversible

IOP Conf. Series: Materials Science and Engineering **921** (2020) 012015 doi:10.1088/1757-899X/921/1/012015

dispersed phase. Polymer-micellar hydrogels are a completely new type of dispersed systems, the main objects of reversion in which can be both oligomeric polyalkylene glycols and polymer stabilizers that are subject to conformationally induced fluctuations in the hydrophilic-lipophilic balance. In most reversible emulsion systems, a reversible emulsifier plays an important role, which, depending on the acidity of the medium, acts alternately as a stabilizer of direct and reverse emulsions. In polymer-micellar systems, the role of a reversible emulsifier can be played by nonionic surfactants that provide phase reversal without changing the ionic strength, as well as the acidity of the medium. For example, the management of the properties of a hydrogel emulsion, including polysaccharide derivatives (CMC, CMOEC, PAC, etc.) and hydrophobic solubilizate (unsaturated higher fatty acids, polypropylene glycols, triglycerides, etc.), is provided by the introduction of an invert emulsifier with HLB = 3-8, which leads to to the formation of a reverse emulsion. Translation of the reverse emulsion into a direct one can be easily carried out by adding an ethoxylated surfactant with HLB = 16-18. Consider the effect of the concentration of the polysaccharide reagent on the technological parameters of a reversible hydrogel emulsion. In this case, we select CMC as the polysaccharide base of the polyemulsion, and oleic acid as the hydrophobic solubilizate. The roles of the invert emulsifier and emulsifier of direct emulsions in such a system are performed, respectively, by polyalkylene glycol with HLB = 3 and neonol. At the first stage of the introduction of PAG into the mixture of CMC and oleic acid, we obtain a highly viscous invert emulsion with a practically zero filtration index and one hundred percent inhibitory activity with respect to the swelling of clay minerals. At the next stage, the addition of neonol to an invert hydrogel emulsion leads to the formation of a less viscous direct emulsion with a simultaneous deterioration in filtration and inhibitory properties. If we now reduce the concentration of CMC in the initial mixture with oleic acid, then the viscosity and structure of the newly formed invert hydrogel emulsion are not as high as in the previous case with a higher polymer concentration. In addition, there is a slight decrease in filtration and inhibitory properties. The direct emulsion obtained after the addition of neonol is also characterized by a decrease in the main technological indicators, in comparison with a similar hydrogel emulsion with a high content of polysaccharide reagent. That is, the technological properties of the hydrogel emulsion at different stages of the reversal cycle with a decrease in the concentration of the polysaccharide reagent approach some average values.

Unlike biopolymer washing systems, a hydrogel emulsion is characterized by increased water repellent and surface-active characteristics. In addition, the additional enhancement of hydrophobic interactions in polysaccharide macromolecules due to the use of PAG in the composition of the polyemulsion helps to increase the degree of stabilization of filtration properties in harsh operating conditions. Thus, at the present stage, emulsions based on polymer-micellar hydrogels are the most advanced systems for the construction of horizontal shafts in productive horizons.

Based on the considered example of creating polymer-micellar hydrogels, it can be concluded that the rule of lowering entropy requires polymers to form a unified spatial structural network in the volume of the dispersion medium of the drilling fluid, approaching the liquid-crystalline state. The system can perform the maximum work, the chemical treatment of which contributes to the transition from a coagulation to a condensation structure with minimal entropy. An example is the preparation of a cement mortar, which during the transition to a condensation structure (at the beginning of setting) generates a large amount of heat ($\Delta H < 0$). Structures close to condensation also have various hydrogels, including the above-described biopolymer solutions "crosslinked" with salts of polyvalent metals. To form a spatial network, the polymer macromolecules should not have too high a charge - the molar fraction of ionic groups is preferred from 50 to 70 mol%.

Thus, in the production of effective chemical treatments of the drilling fluid, including polymer reagents, one should be guided by the requirement of providing a significant exothermic effect while reducing the entropy of the system. Optimal polymer stabilizers should have a large molecular weight, an average charge level of macromolecules, a regular macromolecular structure with a periodic alternation of charged and non-ionic functional groups. The advantage is the ability of the polymer to

IOP Conf. Series: Materials Science and Engineering **921** (2020) 012015 doi:10.1088/1757-899X/921/1/012015

chemisorb on the surface of the particles of the dispersed phase, as well as the tendency to form coherent dispersed systems (gels) with a spatial structure.

References

[1] Brouning W C and Chesser B G 1972 Polymerpolyelectrolyte drilling fluid system Petrol Technology pp 1255-1263
[2] Penkov A I, Vakhrushev L P and Fedosov R I 1981 Requirements for solutions without clay and with a low content of solid phase Oil industry №5 pp 21
[3] Sharipov A W, Anderson B A and Mikhaylov K L 1981 Application of silicate-polymer solution for drilling wells Oil industry №4 pp 19-23
[4] Fedosov R I and Koshelev V N 2004 Physico-chemical and structural-mechanical characteristics of polymer hydrogel drilling fluids Regupol and ECORYS NTJ Construction of oil and gas wells on land and at sea №4 pp 24-28
[5] Penkov A I, Vakhrushev L P, Belenko E V, Ostryagin A I and Dzhioev T E 2000 Influence of polyanion electrolytes on the state of adsorption layers of polyalkylene glycols in bentonite suspensions News Of Universities North Caucasus region Natural science № 1 pp 90-94
[6] Penkov A I, Vakhrushev L P, Belenko E V, Ostryagin A I and Koshelev V N 2001 Interpolymer complexes based on copolymers of ethylene and propylene oxides Natural science №2 pp 45-49
[7] Vakhrushev L P, Koshelev V N, Penkov A I, Belenko E V and Lushpeeva O A 2001 Spatially structured water-based clay-free drilling fluids Oil industry №5 pp 46-51

Biodegradation of polymer compositions with pro-oxidants

Yu K Lukanina[1,3], **A A Popov**[1,2] **and A V Khvatov**[1]

[1] Emanuel Institute of Biochemical Physics, Russian Academy of Sciences, Moscow, Russia
[2] Plekhanov Russian University of Economics, Moscow, Russia
[3] julialkn@gmail.com

Abstract. Oxo-biodegradable plastics are widely used as an alternative to well-known polyethylene and polypropylene. thanks to the polyolefin base, they have the necessary set of operating characteristics and are able to be processed with the same parameters as for example polyethylene. The presence of prooxidant in their composition ensures its oxo-destruction after the end of the product's service life with the subsequent biodegradation of the obtained material fragments. The combination of a natural filler and a pro-oxidant additive in polymer compositions makes it possible to obtain a material with a programmed life and destruction

Increasing population growth and improved quality of life lead to increased consumption and large amounts of waste.

Waste generation leads to loss of materials and energy, as well as additional economic and environmental costs for society in connection with the collection, recycling and disposal of waste. The impact of waste on the environment, resources and human health depends on their quantity and nature. Environmental pressures from waste generation and handling include emissions to air, water and soil, which have potential effects on human health. Most of the municipal waste is landfilled. It is necessary to take into account the fact that the volume of household waste increases annually, and territorial opportunities decrease, which leads to a negative impact on the environment. However, only a small part of the waste is recycled.

Over the past several decades, the formation of urban waste has increased dramatically. On average, the per capita amount of waste is 150–300 kg / year, and the annual increase in the amount of garbage is at least 3%, and in some countries, about 10% [1].

Waste can be domestic, industrial and mining. According to the phase state, they are divided into liquid, solid, or a mixture of solid, liquid, and gas phases (the emission of gases into the atmosphere through pipes is considered to be production emission, not waste).

Municipal solid waste (MSW) - waste generated in residential and non-residential premises in the process of consumption by individuals and legal entities, as well as goods that have lost their consumer properties in the process of consumption by individuals or legal entities in order to meet personal and domestic needs [2].

Solid municipal waste is very diverse in composition: food residues, paper, scrap metal, rubber, glass, wood, cloth, synthetic substances. The morphological and granulometric composition of MSW depends on the source of education and economic conditions, the population of cities, as well as the climatic conditions of populated areas.

According to numerous sources [3,4], 90% of MSW can be used to obtain secondary raw materials or additives to fuel resources. According to data published by Rosprirodnadzor, 9 million tons of waste paper, 1.5 million tons of ferrous and non-ferrous metals, 2 million tons of polymeric materials, 20 million tons of food waste and more than 0.5 million tons of glass are lost each year when disposing of unsorted MSW [5].

In Russia over the past 25 years, household waste has become 4 times more. This happened also because of the improvement of packaging culture (paper and cardboard, containers made of glass, as well as of complex composite materials) and the appearance of a large amount of disposable packaging (polyethylene film, polyethylene terephthalate, polyurethane foam, and other polymers).

Comparing the results of the morphological composition determining of municipal waste in Moscow in 2015 with the results of a similar study in 2008, showed an increase in the proportion of polymers and a decrease in the amount of glass waste is observed [6]. This is due to the decision of some food manufacturers to replace the glass packaging on the polymer, as cheaper. It is predicted that by 2025 the volume of polymer waste in Russia will amount to 7 million tons [7].

At the turn of the XX - XXI centuries for Europe, the problem of recycling polymer packaging is in fourth place, second only to outdated cars, worn car tires and demolished building waste [8].

Degradation of plastic waste by various methods (such as biodegradation, compost, oxo-biodegradation and photodegradation) becomes an alternative to solving such problems [9]. As a result of extensive research in this area, several basic strategies have been developed to solve the problem of pollution caused by polyethylene film [10]. First, the introduction of natural biodegradable fillers into the polymer matrix, secondly, the introduction of carbonyl groups directly into the polyethylene chain or into the α-position of short branches by copolymerization, and third, the introduction of a small amount of prooxidants [11]. Undoubtedly, the latter is more economically feasible, since in this case the existing production and processing technologies remain. It should be noted that in most cases a controlled service life is required from polymeric materials before physical degradation occurs; which implies the preservation of operational properties (physico-chemical and mechanical characteristics) of the product during the entire period of storage and operation [12].

Traditionally, natural fillers such as cellulose, starch, chitin, wood flour, etc. are used as additives to accelerate the biodegradation of polymer composite materials. Such natural fillers act as a nutrient medium for microorganisms, thereby increasing the contact surface of bacteria / micromycetes with polyolefins (not only from the surface, but from the volume) [13].

In the last decade, pro-oxidant additives have proven themselves well as additives to polyolefins [14]. Their composition usually includes stearates or carboxylates of metals with transition valency. They act as catalysts for the oxidation processes of (thermo-, photo-) polymers, which leads to preliminary breaking of the polymer chain, the accumulation of oxygen-containing groups and fragmentation of the material with subsequent biodegradation. According to ASTM D-5488-94d, biodegradation is defined as "the process by which materials are decomposed into carbon dioxide, methane, water, inorganic compounds or biomass, and the main mechanism is the enzymatic action of microorganisms."

It should be noted that at the moment there are a large number of works aimed at:

1. study of the influence of various types of natural fillers and their content on the structure and properties of composite materials;
2. selection of pro-oxidants for polyolefins. However, due attention has not been paid to the creation of new composite materials polyolefin / natural filler / pro-oxidant additive and the study of the process of their biodegradation.

Oxo-biodegradable materials imply a two-step process of material degradation. At the first stage, oxo-decomposition of the material occurs, and at the second stage, biodegradation of fragments obtained as a result of oxo-decomposition. The aim of this work is to study the effect on the biodegradation of polyethylene with a natural filler of the introduction of the third component - a pro-oxidant additive and after preoxidized in air .

As the object of study, films of composite materials (table 1) based on low density polyethylene (LDPE) grade 15803-020 were used. The natural filler is cellulose of the Polycel PC brand (average fiber length 100 nm). A pro-oxidant additive that includes LDPE and metal salts of variable valency (cobalt, zinc and zirconium).

IOP Conf. Series: Materials Science and Engineering **921** (2020) 012016 doi:10.1088/1757-899X/921/1/012016

Table 1. Composite materials.

№	The content of components in the mixture, wt.%		
	LDPE	Cellulose	pro-oxidant additive
1	70	30	0
2	97	0	3
3	67,9	29,1	3

The stability of the studied films was determined in the soil, which was made in accordance with GOST 9.060-75. The essence of the method is to determine the change in mass of the samples after exposure for a certain time in a model environment. Soil parameters: humidity 30%, pH 7-8.

All samples were preoxidized in air at 90 ° C for two days. For comparison, the results of the original samples.

Spectrometric studies were carried out using the equipment of the CCU "New Materials and Technologies" of the Institute of Bioorganic Chemistry, Russian Academy of Sciences, Perkin Elmer Spectrum 100, at T = 23 ± 2 ° C in transmitted light in the wavelength range of 4600 – 450 cm-1.

In the course of determining the stability of film samples in soil, the kinetics of mass loss was measured; the data are presented in Figure 1 for the starting materials (a) and after preliminary oxidation (b).

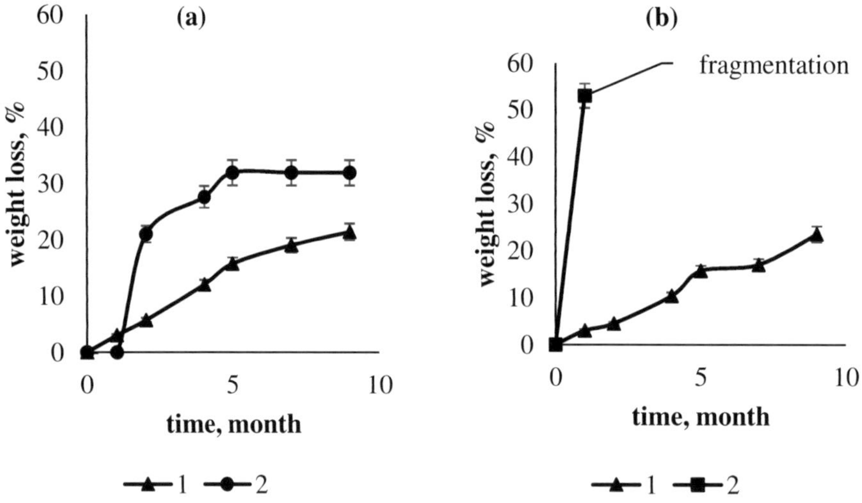

Figure 1. Kinetic curves of mass loss of samples 1 - PE / cellulose, 2 - PE / Cellulose / pro-oxidant additive (a) starting materials, (b) pre-oxidized.

It was previously shown that the introduction of a natural cellulose additive into the polymer matrix accelerates the biodegradation process [15]. But it should be noted that, preliminary acidification of the PE / cellulose composite material does not significantly accelerate the rate of weight loss. The introduction of a pro-oxidant additive in a potentially biodegradable material does not significantly increase weight loss. A preliminary oxidation of the composite material polyethylene / cellulose / prooxidant additive before entering the soil leads to rapid fragmentation of the material (Fig. 1b, curve 2) already in the second month of exposure.

Transition metals act as good prooxidants in polyolefins, so it makes the polymer susceptible to oxidative degradation. Transition metals induce hydroperoxide decomposition and produce free radicals on the polymer chain, which can attack chains and produce low molecular weight products with oxygen-containing groups, such as -COOH, -OH and >C=O groups[16,17]. While antioxidants and UV stabilizers make polyolefins resistant to degradation, prooxidant additives can make them oxo-degradable by catalytic chain breakage and make the polymer more hydrophilic. That is, oxo-biodegradation means a two-step process, including, in order, oxidative destruction, which is usually abiotic, and then the biodegradation of oxidation products [18].

The literature reports that it is possible to achieve various rates of degradation by selecting a balanced combination of antioxidant and pro-oxidant additives [19, 20]. And by selecting a complex of salts of various metals in a pro-oxidant additive, you can control which environmental factor "triggers" the oxidation process to accelerate the subsequent biodegradation of the entire composite material. For example, Fe^{3+} plays an important role in photooxidation, initiating a radical reaction [9]. Mn^{2+} or Co^{2+} are important oxidizing agents that initiate the decomposition of a polymer buried in soil in the absence of light.

In our case, the pro-oxidant additive contains: cobalt, zinc and zirconium, which have a good effect on the biodegradation of the material after preliminary oxidation with atmospheric oxygen, the material is phagmenated after a month of exposure.

It is known that during thermal oxidative degradation, the formation of free radicals O · and · OH occurs first, which turn into thermal decomposition products - ketones, esters, lactones and carboxylic acids [21].

For all samples, after exposure to the soil, changes in the region of 1800-1500 cm^{-1} are recorded (Figure 2), which characterizes the formation of oxygen-containing products of thermal oxidation of polymers.

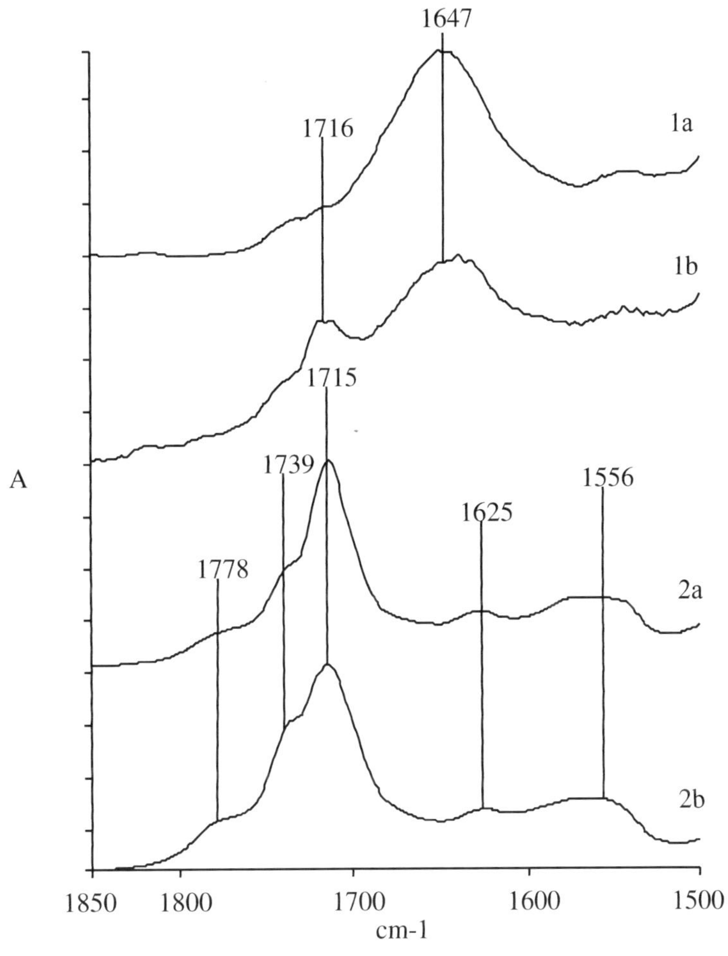

Figure 2. IR absorption spectra of samples as a result of aging in soil, where 1 is PE / cellulose, 2 is PE / Cellulose / pro-oxidant additive (a) starting materials, (b) pre-oxidized.

The absorption spectra for PE / cellulose samples, both for non-oxidized (Figure 2, 1a) and oxidized samples (Figure 2, 2a), have bands of 1715 cm $^{-1}$ — carboxyl groups and 1647 cm $^{-1}$ — terminal vinyl groups [22].

In the absorption spectra of PE / cellulose / pro-oxidant additive samples, intense absorption is recorded in the band region 1715 cm^{-1}, 1780 cm^{-1} - aldehyde groups, 1740 cm^{-1} - ketone groups, and absorption in the region 1650 cm^{-1} - 1530 cm^{-1} - amide I and amide II [22].

In addition to the main difference in the increase in the number of unsaturated oxygen-containing groups in the images containing the pro-oxidant additive, it is necessary to note the presence of absorption characteristic of amide I and amide II. This may indicate the growth and development of micromycetes and soil bacteria in the bulk of the material, i.e. about their penetration into the material and the course of biodegradation processes in the polymer matrix of composites.

The process of biodegradation of polymeric materials or their compositions in nature is not possible to consider in terms of only one factor causing degradation. Traditionally, the biodegradability of prooxidant activated polyolefins has been evaluated in complex media such as soil, sewage sludge or compost. This approach brings several advantages, such as the availability of a very diverse microbial inoculum or a close relationship to actual conditions in nature and / or in waste treatment processes. Understanding the whole complex of the reactions is of great importance for researchers and technologists when creating biodegradable polymeric materials. Thus, the combination of a natural filler and a pro-oxidant additive in polymer compositions makes it possible to obtain a material with a programmed life and destruction. The introduction of a pro-oxidant can significantly accelerate the biodegradation processes (up to several months in the soil) with final fragmentation of the sample under the condition of preliminary acidification in order to form unsaturated oxygen-containing groups.

The work was carried out within the framework of budget financing on the theme of state assignment No. of state registration 01201253305.

References

[1] http://www.nacc.spb.ru/index.php?option=com_content&view=article&id=319:2012-10-11-14-41-20&catid=47:2010-03-24-15-06-27&Itemid=69
[2] http://gkhgroup.ru/musornyy-vopros-iz-tbo-v-tko/
[3] http://www.greenpeace.org/russia/Global/russia/report/toxics/obsor_othodi_msk.pdf
[4] Sister V G 2003 Modern technologies for the disposal and disposal of MSW Moscow: Academy of Public Utilities. K.D. Pamfilova 303
[5] https://www.rpn.gov.ru/sites/all/files/documents/.../analiticheskaya_zapiska2.doc
[6] Zaitsev V and Sotnezov A 2015 Experimentally determining the composition and properties of solid municipal waste is a key step in the development of program activities in the field of municipal waste management Advances in chemistry and chemical technology XXIX 9 21-23
[7] http://www.abercade.ru/research/industrynews/8499.html
[8] www.reda.spb.ru/rep.htm
[9] Singh B and Sharma N 2008 Mechanistic implications of plastic degradation Polym Degrad Stab 93(3) 561-584
[10] Arnaud R, Dabin P, Lemaire J, Al-Malaika S, Chohan S, Coker M and et al. 1994 Photooxidation and biodegradation of commercial photodegradable polyethylenes Polym Degrad Stab 46 211–224
[11] Rabek J F 1995 Polymer photodegradation: mechanisms and experimental methods London: Chapman & Hall 374
[12] Jakubowicz I 2003 Evaluation of degradability of biodegradable polyethylene (PE) Polym Degrad Stab 80(1) 39-43
[13] Koroleva A V, Lukanina Y K., Khvatov A V, Popov AA, Monakhova T V and Huebner M 2013 Oxo-biodegradability of polyethylene blends with starch, cellulose and synthetic

additives New Steps in Physical Chemistry, Chemical Physics and Biochemical Physics 43-57

[14] Lukanina YK, Khvatov A V, Koroleva A V, Popov A A and Kolesnikova N N 2015 Oxo-degrading polyolefin additive Patent RF № 2540273

[15] Lukanina Yu K, Kolesnikova N N, Popov A A, and Khvatov A V 2019 Oxo-degradation of LDPE with pro-oxidant additive IOP Conf. Series: Materials Science and Engineering 525 012104

[16] Singh B and Sharma N 2008 Mechanistic implications of plastic degradation. Polym Degrad Stab 93(3) 561-584

[17] Jakubowicz I 2003 Evaluation of degradability of biodegradable polyethylene (PE) Polym Degrad Stab 80(1) 39-43

[18] Scott G 1999Biodegradable polymers Polymers and the environment Cambridge: Royal Society of Chemistry Paperbacks

[19] Erlandsson B, Karlsson S and Albertsson A C 1997 The mode of action of corn starch and a prooxidant system in LDPE: influence of thermooxidation and UV-irradiation on the molecular weight changes. Polym Degrad Stab 55 237-245

[20] Karlsson S, Hakkarainen M and Albertsson A C 1997 Dicarboxylic acids and ketoacids formed in degradable polyethylenes by zip depolymerization through a cyclic transition state. Macromolecules 30 7721-7728

[21] Ammalaa A, Batemana S, Deana K, Petinakisa E, Sangwana P, Wonga S, Qiang Yuana, Long Yua, Patrickb C and Leong K H 2011 An over view of degradable and biodegradable polyolefins: Biodegradation of polyethylene films with prooxidant additives Progress in Polymer Science 36 1015–1049

[22] Maquelin K, Kirschner C, Choo-Smith L P, van den Braak N, Endtz H Ph and Puppelsa G J 2002 Identification of medically relevant microorganisms by vibrational spectroscopy Journal of Microbiological Methods 51(3) 255-271

IOP Conf. Series: Materials Science and Engineering **921** (2020) 012017 doi:10.1088/1757-899X/921/1/012017

INNOVATIVE APPROACH TO ASSESSING THE CONDITION AND CERTIFICATION OF OIL - CONTAMINATED SITES

E R Mukhametshina and I S Aitov

628616, Russia, Tyumen Oblast, Khanty-Mansiysk Autonomous Okrug – Yugra, Nizhnevartovsk, ul. Lenina 2P, str. 9

mukhametshina.elvina18@gmail.ru

Abstract. The article describes an innovative method of volumetric assessment and certification of oil-contaminated land plots, which allows us to give an accurate assessment of the volume content of certain limiting factors. This technique involves building 3D models, tabular systematization of concentrations and volumes of oil pollution, improving the certification of the above-mentioned areas using special QR codes and parcel codes created by us. The method described by us makes it easier and at the same time increases the objectivity of certification of oil-contaminated land plots, reflects the qualitative and quantitative volume and time parameters of plots, increases the convenience, efficiency and reliability of accounting, data transmission and storage.

1. Introduction

Today, the problem of low efficiency of reclamation works on oil-contaminated sites remains an urgent problem at oil-producing enterprises of our country. The difficulties are caused by the lack of spatially ordered ideas about where and how the pollution is distributed, what their concentration and volume parameters are in the soil. This problem entails an irrational waste of finances. As a rule, the return on the spent funds does not exceed 30%.

To solve these problems, we offer a method for representing pollutants in the form of 3D models, tabular systematization of their concentrations and volumes, creation of QR codes and parcel codes.

Scaling table information determines the color content of three-dimensional models. The described method allows differentiation of plots and their parts into reclamation groups. They focus on the targeted use of allocated funds. Tabular systematization and spatially ordered representations of research results produce a synergistic effect. It manifests itself in obtaining a more accurate assessment of the state of oil-contaminated sites, improving the efficiency of reclamation, and making productive use of finances. And the creation of QR codes and parcel codes allows you to simplify the process of accounting and certification of oil-contaminated land plots.

It is known that there are three main degrees of oil pollution of soils: weak – up to 10%; average – from 10 to 40%; strong – more than 40%.

Low pollution dramatically reduces productivity and leads to the death of about 1/2 of vegetation. With average pollution, stands and shrubs die completely within 10 years. Heavy oil pollution completely destroys the phytocenosis in the period up to 1 year.

IOP Conf. Series: Materials Science and Engineering **921** (2020) 012017 doi:10.1088/1757-899X/921/1/012017

2. Adjusting the location, borders, and description of oil- contaminated sites.
Their topographic survey and photographing were carried out. The division of the parcel into parts: areas and parcel. Forming a hierarchical system of 3 levels for each section. 1 the smallest level – a parcel, as one of the four layers within the boundaries of the allotment; 2 level- an allotment with a set of four parcels; 3 level – the entire site in the aggregate of allotments and parcels.

3. Sampling of soil at the sites. Conducting quantitative and chemical analysis.
Formatting the analysis results in tabular form.

Oil is a liquid natural solution consisting of a large number of hydrocarbons of various structures and high-molecular resinous-asphaltene substances. It contains a certain amount of water, salts, and trace elements. On the one hand, the world's oil fields are distinguished by a huge variety of types (there are no two completely identical oils from different reservoir deposits), on the other - the unity of its composition and structure, similarity in some parameters. The elemental composition of tens of thousands of different individual representatives of oil around the world varies between 3 and 4 % for each element. The main oil-forming elements: carbon (83-87 %), hydrogen (12 - 14%), nitrogen, sulfur, oxygen (1 - 2%, less often 3 - 6% due to sulfur). Tenths and hundredths of a percent of oil are made up of numerous trace elements, the set of which in any oil is approximately the same.

Components of the light fraction, being in soils, water or air environments, have narcotic and toxic effects on living organisms. Normal alkanes with a short carbon chain, which are mainly contained in light fractions of oil, act especially quickly. These hydrocarbons are better soluble in water, easily penetrate the cells of organisms through the membranes, and disorganize the cytoplasmic membranes of the body. Most microorganisms do not assimilate normal alkanes containing less than 9 carbon atoms in the chain, although they can be oxidized. The toxicity of normal alkanes is reduced in the presence of a non-toxic hydrocarbon, which reduces the overall solubility of the alkanes. Due to the volatility and higher solubility of low-molecular-weight normal alkanes, their action is usually not long-term. If their concentration was not lethal to the body, then over time the normal functioning of the body is restored (in the absence of other toxins).

4. The scaling and color values of pH and the content of oil pollution and chlorides.
Calculation of the volume of allotments, deep layers and contamination (table 1).

Table 1. Scale of oil pollution

Content of petroleum products, %	Color
0 - 1,0	green
1,1 - 3,0	yellow
3,1 - 6,0	orange
6,1 - 10,0	light brown
10,1 - 15,0	dark brown
15,1 - 25,0	pink
25,1 - 40,0	light red
> 40,0	dark red

Identical tables were created for scaling and coloring the values and volume reflection of pollutants (table 2) for the concentration of chlorides and acidity in the soils of the oil -contaminated site.

IOP Conf. Series: Materials Science and Engineering **921** (2020) 012017 doi:10.1088/1757-899X/921/1/012017

Table 2. Volume of oil product residues, m3

Location	Volume of oil product residues, m^3				Total, m^3
	parcel «a» (0,1m)	parcel «b» (0,2m)	parcel «c» (0,3m)	parcel «d» (0,5m)	
83	-	3,17 (0,012)	-	-	3,17
84	13,64 (0,078)	4,06 (0,012)	5,72 (0,011)	-	23,42
85	10,79 (0,072)	10,7 (0,04)	-	-	21,49
86	10,96 (0,078)	5,25 (0,019)	-	-	16,21
Total	**35,39**	**23,18**	**5,72**	-	**64,29**

5. Create 3D-models of plots with each parcel displayed in volume and color.
As a rule, the number of parcels within the boundaries of a single plot does not exceed 4, the lower parcel is located no deeper than 2.0 m (in our case, 1.1 m). For reclamation, the top three soil parcels are available: a) - 0-10 cm, b) - 10-30 cm, c) - 30-60 cm; d) - 60-110 cm (hardly accessible) (Fig.1).

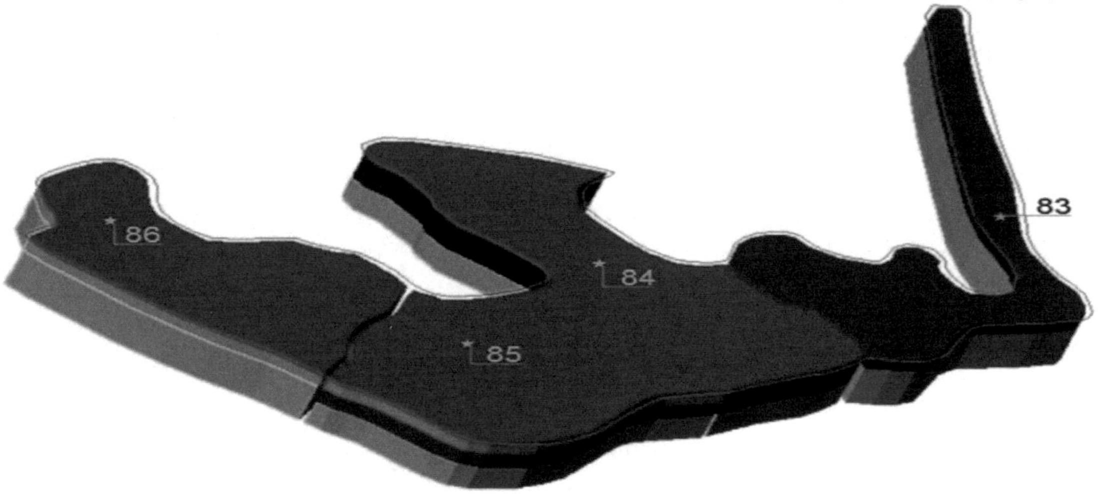

Figure 1. 3D-model of oil contamination of a land plot X-Y-ZZZZ

Almost all chlorides found in soils-NaCl, $CaCl_2$, KCl, $MgCl_2$ - are easily soluble. The chlorides of most trace elements are also soluble, with the exception of silver chloride and mercury chloride. Migration and accumulation of chlorine compounds depend mainly on the temperature and water conditions of the territory. Maximum accumulation is typical for the drainless depressions of arid regions. Negative environmental consequences of excessive chloride content in the soil are possible in the form of deterioration of physical and chemical properties of soils, as well as contamination of ground and surface water.

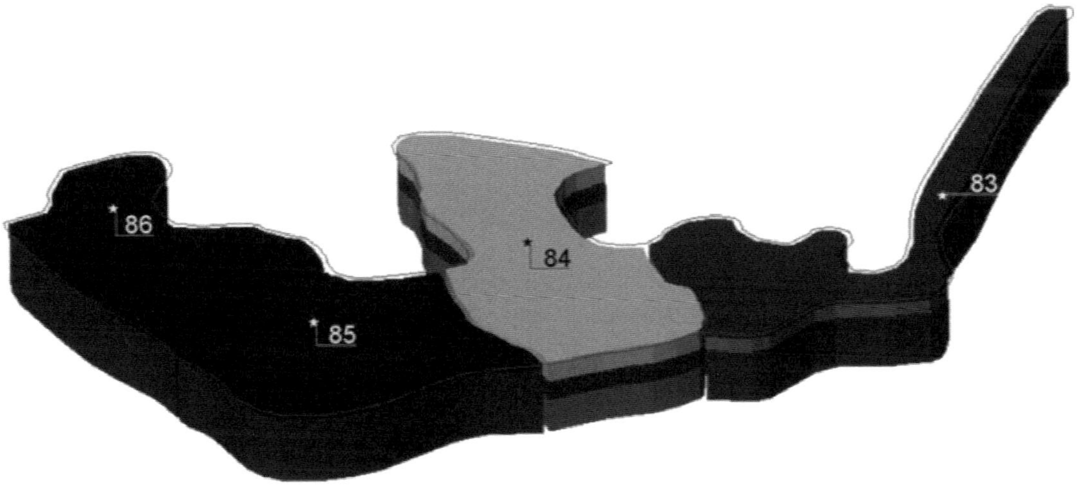

Figure 2. 3D-model of chloride concentration of a land plot X-Y-ZZZZ

The acidity or pH of the soil is a biochemical indicator that characterizes its ability to exhibit (neutralize) the properties of acids. During the exchange of hydrogen ions with soil minerals and organic substances, acids and bases (alkalis) are formed in the fertile layer. PH indicates their balance in the soil solution, denote it by numbers from 1 to 14. The lower the pH value, the more acidic the environment.

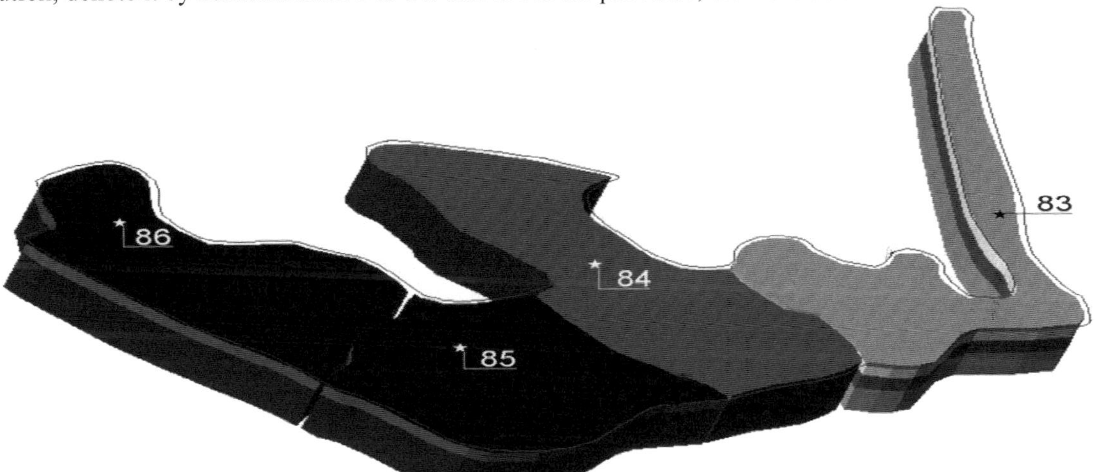

Figure 3. 3D-model of pH concentration of a land plot X-Y-ZZZZ

6. Distribution of plots and their parts into reclamation groups:
a) areas or parts thereof where biological remediation is required;
b) the plots or their parts, which do not require the use of biological remediation;
c) areas or parts of them where it is currently impossible to achieve the required results during biological remediation due to the lack of necessary or insufficient existing technologies;
d) areas or parts of them where biological remediation is temporarily impractical due to the high probability of their repeated contamination due to insufficient operational reliability of pipelines [1] (table 3).

Table 3. Volume of oil product residues, m^3

Location	Volume of oil product residues, m^3	Total, m^3

	parcel «a» (0,1m)	parcel «b» (0,2m)	parcel «c» (0,3m)	parcel «d» (0,5m)	
83	N	3,17 (0,012)	N	N	3,17
84	13,64 (0,078)	4,06 (0,012)	5,72 (0,011)	N	23,42
85	10,79 (0,072)	10,7 (0,04)	N	N	21,49
86	10,96 (0,078)	5,25 (0,019)	N	N	16,21
Total	**35,39**	**23,18**	**5,72**	N	**64,29**

7. Create QR codes and parcel codes to improve the process of accounting and systematization of oil-contaminated sites.

This encoding will speed up, objectify, and unify the acquisition and use of information (Fig.4).

Figure 4. QR code of a three-dimensional model of an oil -contaminated land plot.

A parcel is the smallest part of an oil-contaminated area that is allocated at the intersection of the point zone and the layer. Each parcel has a number of characteristics and parameters. They can be encoded. For example, X-Y-ZZZZ/84b/350/g. in this case, the cipher indicates: 1) the number of the site ;the parcel in the area of the point 84 of layer b; the volume is 350 m3; the excess of the content of petroleum products over the permissible level in the range from 31 to 40%. Options of forms and types of coding you can choose many, the main thing is to be consistent. Parcel coding precedes QR coding.

8. Conclusions.

The proposed method allows us to present the parameters of an oil-contaminated site in three-dimensional and tabular form for a more accurate assessment of the state of oil-contaminated sites. Accurate assessment helps to implement effective reclamation, which allows you to hand over land to environmental services. The 3D model in color reflects the pH, chlorides, and petroleum products in each layer of soil. The use of QR codes will speed up the collection and processing of information, simplify the certification of oil-contaminated sites, and combine a variety of databases from various industries.

References

[1] Mukhametshina E and Aitov I 3D-MODELS AS A METHOD FOR REFLECTING VOLUME AND SPATIAL PARAMETERS OF OIL-CONTAMINED SITES 2019 E3S Web of Conferences **140**, 09003 EECE-2019

IOP Conf. Series: Materials Science and Engineering **921** (2020) 012018 doi:10.1088/1757-899X/921/1/012018

Ecological State of Technogeneous Saline Soil of Oil-Contaminated Alluvial Ecosystems and Their Remediation Techniques

Mariya Vladimirovna Nosova[1,2], **Valentina Petrovna Seredina**[2] and **Alexander Sergeevich Rybin**[3]

[1]Joint Stock Company Tomsk Research and Design Institute of Oil and Gas, JSC TomskNIPIneft, 634027, 72 Mira ave, Tomsk, Tomsk Region, the Russian Federation.
[2]National Research Tomsk Polytechnic University, TSU, Biological Institute, Department of Soil Science and Soil Ecology, 634050, 36 Lenin ave, Tomsk, Tomsk Region, the Russian Federation.
[3]National Research Tomsk Polytechnic University, TPU, 634050, 30 Lenin ave, Tomsk, Tomsk Region, the Russian Federation.

e–mail: NosovaMV@tomsknipi.ru, seredina_v@mail.ru.

Abstract. The article discloses the trends of technogenic salts distribution in alluvial soils contaminated with oil and petroleum products. The salt distribution was examined in the lateral direction – impact zone, oil contamination boundary, as well as radial salts distribution within the soil profile (to 100 cm depth). Specifics and chemistry of technogenic halogenesis process (freely soluble including phytotoxic salts content, qualitative composition, migration and distribution trends) has been identified for soils of different contaminated zones. It was revealed that technogenic salinization of alluvial soils (that is the end body of the cascaded geochemical systems) is an essential geoecological factor contributing to formation within oil-contaminated areas of new technogenic soil-geochemical successions – chemical-contaminated soils, showing the features of salinity not typical for humid climate. The technique of technogeneous saline soil remediation was suggested based on indigenous halophyte plants phytomelioration sowing.

1. Introduction

The Western Siberian superaqual landscapes are the end point of organic pollutants accumulation, where the intense technogenic stream flow of petroleum hydrocarbons typical for oil pipeline accidents produces maximum impact to soil landscape [1– 3], which indirectly affects the state of the hydrological network. At shallow groundwater occurrence, the presence of freely soluble salts may result in soil salinization, and no proven salinized soil remediation technique is currently available. In the conditions of growing technogenic environmental load, its ecological status issues are of high priority. According to multiple studies [3–6], oil contamination impact on natural ecological systems is a function of geochemical and physical loads. Soils are a natural barrier and a center of accumulation of all pollutants and various chemicals. In this respect, the soil landscape investigation in the technogenic zone is essential, as the contaminants upset the balance of natural processes resulting in a change in soil system parameters [6–9].

IOP Conf. Series: Materials Science and Engineering **921** (2020) 012018 doi:10.1088/1757-899X/921/1/012018

The issues of radial and lateral salts distribution within the arid zone soil were the subject of numerous surveys [1–11]. In areas of humid soil genesis, salts migration during accidental oil spills as well as their secondary (post-technogenic) redistribution has not been actually studied yet. Hence, for the zone of excessive humidification, the technogeneous saline soils may provide the models for prediction of geochemical changes related to oil contamination; this makes the study of the above issues relevant and essential.

Therefore, this study is focused on identification of specifics of halogeochemical phenomena in alluvial ecosystems of Western Siberia under circumstances of soil local contamination with oil and petroleum products, and selection of the most effective salinized soil remediation technique.

2. Material and methods

Actual data originate from field studies and laboratory tests fulfilled by this article authors in Aleksandrovsky and Kargasoksky districts of the Tomsk Region, and in Nizhnevartovsk District of the Khanty-Mansi Autonomous Okrug – Yugra. To reveal soil morphological transformations, a routing method was applied with subsequent comparative/geographic and profile/genetic analysis of reference and contaminated soil. The trends of radial and lateral salts and oil distribution were revealed with the help of reference sections (2 full-scale sections) and trenching (26 samples) in different technogenic impact zones. The contamination source area is the zone affected with the heaviest oil contamination and closest to the technogenic impact source (a field pipeline). The impact zone is the area of oil contamination impact from the contamination source area contour (abrupt drop of petroleum products content) to contamination diffuse halo (bitumen cake thickness reduction). The morphological boundary of the contamination diffuse halo was assumed the lack of visible petroleum products (bitumen cake) on soil surface, and the vegetation condition (signs of oppression or disturbance).

According to soil classification of Russia [12], common alluvial gley heavy loamy soil formed in the central part of the flood plain land was used for reference. Oil-contaminated soils are defined as oil-contaminated chemical-contaminated soils occurring within the areas of gray typical/gley medium-shallow soil. To specify vertical soil characteristics, main genetic horizons were sampled, and to assess salts lateral migration, the envelope technique was used as per national standard [13]. Petroleum products content in soil was determined by the fluorimetric method using a Fluorat-02 fluorometer. Acidic reaction (aqueous extract pH), as well as analytical studies of readily soluble salts content and composition were fulfilled as per national standard [14]. Statistica 6.0 software package was used for data mathematical processing.

3. Results

Due to specific genesis conditions, the West Siberian soils show low ecological/geochemical immunity from technogenic impacts and low potential of autopurification and self-regeneration [4]. Hence, oil can migrate laterally from eluvial and downslope landscape locations, as shown in the diagram suggested by V. P. Seredina and M. E. Sadykov (Fig. 1). Immediately after an accident, oil is accumulated in soil near the area of accident within the contamination halo center.

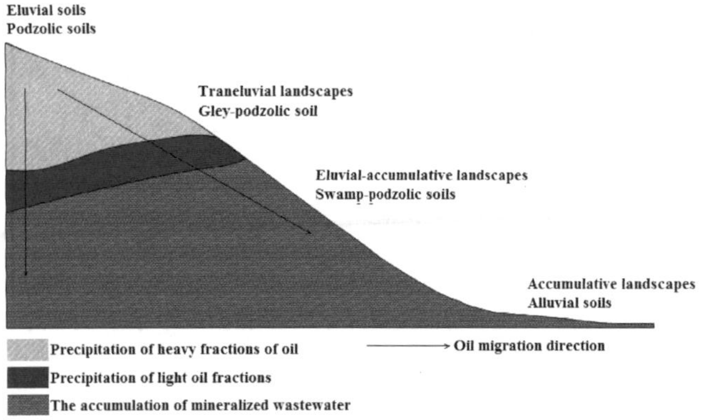

Figure 1. Oil pollutant migration pattern in geochemically related landscapes of the study area [4].

IOP Conf. Series: Materials Science and Engineering **921** (2020) 012018 doi:10.1088/1757-899X/921/1/012018

However, pollutants content growth is observed towards the contamination halo periphery and in deeper soil horizons that is indicative of oil seeping in the contamination central area and its lateral migration via migration paths (Table 1). Biogenic accumulative chemical-contaminated soil horizons (0-10 cm) of the entire contamination halo typically show higher content of petroleum products fixed with the soil humus matrix forming a toxic mixture of soil organic matter and hydrocarbon pollutant ($NaCl$, Na_2SO_4, $MgCl_2$).

Table 1. Oil content in various pollution zones. Oil products content in g/100 g of soil.

	Epicenter of pollution		*Impact zone*		*Pollution boundary*	
	0–10 см	10–30 см	0–10 см	10–30 см	0–10 см	10–30 см
Min	16.23	11.82	11.53	7.06	4.43	3.98
Max	41.37	17.48	26.56	12.24	13.74	7.08
Mean	36.78	10.15	22.42	6.54	13.3	7.69

Contamination heterogeneity is a function of pollutants accumulations in microdepressions. Such oil products diffusion creates spotty contamination pattern. Therefore, morphological features of contamination are the sensitive indicator of physical/chemical changes in soil: soil horizons gleying as a result of anaerobic conditions establishing; bluish shades of cutan; ferromanganesian nodulous neoformations; humus/aluminum nodules; oil strong smell; generation of a compact bituminous cake on soil surface.

Any accidental crude spill is accompanied with high-mineralized salts discharges to the ecological system; these cause soil technogenic salinization and create bituminous varieties of soils. Statistical analysis data are indicative of a direct correlation between the content of freely soluble salts and petroleum products in contaminated soils: correlation factor (r) is 0,87 at p=0,91 (at 0–10 cm depth) and 0,83 at p=0,076 (at 10–30 cm depth).

Maximum discharged salts quantity is accumulated in the root layer (Table 2).

Technogenic liquids migration inside the soils is associated with ion composition fractionation. In a contamination source zone salts distribution shows prismoidal configuration: essential quantities of freely soluble salts are found within the entire soil profile showing the highest accumulation level in biogenic accumulative soil horizon.

Table 1. Oil content in various pollution zones. Oil products content in g/100 g of soil.

	Dense residue, %	The amount of toxic salts,%	pH	Salinity	Anionic salinity type	Type of salinity according to cationic composition
			Epicenter of pollution (0–10 cm)			
Min	0.35	0.28	7.7			
Max	1.12	0.33	8.5	medium	sulfate	sodium chloride
Mean	0.54	0.31	7.6			
			Epicenter of pollution (10–30 cm)			
Min	0.45	0.18	6.9			
Max	0.65	0.29	8.5	low	sulfate	sodium
Mean	0.48	0.26	7.3			
			Impact zone (0–10 cm)			
Min	0.45	0.19	6.3			
Max	0.55	0.25	7.5	low	sulfate	sodium
Mean	0.48	0.22	6.9			

Impact zone (10–30 cm)

IOP Conf. Series: Materials Science and Engineering **921** (2020) 012018 doi:10.1088/1757-899X/921/1/012018

Min	0.15	0.07	5.9			
Max	0.29	0.13	6.7	low	sulfate	sodium
Mean	0.24	0.09	6.2			
			Pollution boundary (0–10 cm)			
Min	0.29	0.14	5.9			
Max	0.6	0.17	6.7	low	sulfate	sodium
Mean	0.32	0.16	6.2			
			Pollution boundary (10–30 cm)			
Min	0.15	0.07	5.8			
Max	0.29	0.13	6.5	non–saline	sulfate	sodium
Mean	0.24	0.09	6.1			

The trend of high-mineralized stream flows accumulation is observed in soil top horizons and in contamination impact zone; however, another zone of salts accumulation is observed in the heavy grain size horizons at their general low content in the remaining soil thickness (Fig. 2).

As the total salinization impact is caused with different salts and ions toxicity for plants, qualitative salts composition is essential for salinized soils agronomical value. After pollutants batch-like discharge, the most mobile ion Cl^- prevails in aqueous extracts from soil bottom horizons within the spill source and contamination zone periphery; note that total less mobile ions Ca^{2+} gradually decrease from top to bottom soil layers. Such ions distribution within the soil stratum causes high cation Na^+ and anion SO_4^{2-} content. Therefore, salinization typically shows sulfate and chloride/sulfate nature. Despite of low soil salinization, total toxic salts content is assumed the main diagnostic feature of the salts environmental impact. In the soils under consideration, technogenic salts are represented by compounds of the toxic salts NaCl, Na_2SO_4, $MgCl_2$. These salts inhibit plants growth that is an essential factor during phytomelioration sowing of grass at a biological remediation phase, since humid areas vegetation, unlike that of arid areas, is not adapted for periodic soil salinization.

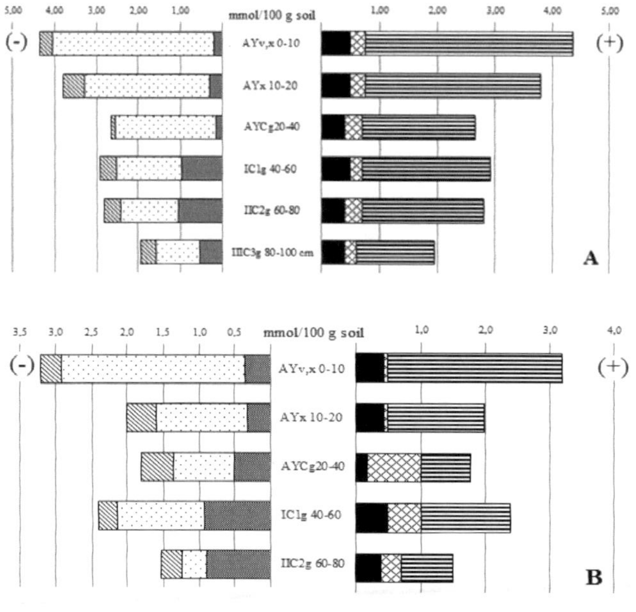

Figure 2. Salt soil profile of oil–contaminated alluvial soils in the epicenter (A) and impact zone (B) of pollution.

One of the critical diagnostic features of oil contamination is soil acidic reaction. In reference soils, pH values vary from 4.6 to 5.2 that is indicative of acidic and low acidic reaction. Unlike the reference samples, the oil-contaminated soils show low alkaline reaction. Hence, in the contamination source area, two months after the oil spill, the top soil layer shows pH equal to 7.82 that is indicative of the maximum oil and petroleum products content in the layer, as well as of the fact that crude has a certain content of freely soluble salts. So, acidic/alkaline properties are sensitive to oil contamination. Changes in absorbing soil complex, such as H^+ and Al^{3+} ion exchange for Na^+ resulting from soil contamination with oil (especially with crude showing high salinity) cause the shift in alkaline/acidic conditions: alkalization of originally acidic and low-acidic soils is observed.

Hence, primary geochemical structure of infiltration bodies is formed: gradual reduction of the contamination core salinization towards its periphery, and simultaneous growth of chlorides percentage in salts. This differentiation of salts at contaminants single injection discharge to soil is indicative of progressive salinization, aggravating soil salinization in line with main phenomena forming the alluvial soil profile (turfy, alluvial, gleying).

The landscape features of the Western Siberian alluvial soils, such as poor drainage and shallow groundwater table contribute to forming the high-contrast and stable salt accumulations creating toxic environment for plants growth and development. Besides, shallow occurrence and subsequent rising of groundwater table may result in soil resalinization in presence of freely soluble salts. In this respect, alluvial soil technogenic salinization is an essential geoecological factor, because these soils accumulate extra amount of different chemical compounds including different pollutants incoming from the elevated sites and contributing to forming new technogenic soil-geochemical successions (chemical-contaminated soils) in oil-contaminated areas.

Over the past ten years, different remediation technologies for technogeneous saline areas of the oil fields were put forward more than once. Contaminated lands irrigation with fresh water and its subsequent pumping-off and disposal at landfills may be a promising technique. This technique has been proven in practice, and is typically use for agricultural lands [14]. However, technical procedures involving irrigation of technogeneous saline areas have certain specifics: the irrigation rate of 5 l/m^2 is not capable of significant reduction of toxic water-soluble salts content in top soil, and is comparable with natural rainfalls. Contaminated irrigation water shall be accumulated to special tank trucks and removed; this activity involves extra costs and makes the technique extremely expensive and labor consuming.

Toxic salts of the root layers may be removed by means saline soil cutting and transportation from the area under consideration. This land remediation technique requires the use of special machinery, such as excavation equipment and trucks; this approach is extremely expensive and ineffective for organogenic soils and large areas. Besides, salinized soil landfill shall be authorized by law. Removed soil shall be replaces with pure soil (peat, sand). Such an approach is also ineffective due to salt accumulation in the underlying soil horizons, which may instigate resalinization in future.

The technique of acceleration of freely soluble salts surface carry-over with thaw water [15] is known; it may be rather effective for contaminated land remediation, as the major part of precipitations is accumulated in a form of snow cover in humid climate of Western Siberia. To intensify snow thawing, top soil mulching with peat or ash may be used. This technique may have a dual effect: despite of the fact that snow water washes the top soil horizons and salt concentration is somehow reduced, in general the technique does not actually influence the surface water composition and results in forming the strips of various concentration of water-soluble salts in surface horizons. Mobile drainage systems may be effective. In bogged and waterlogged areas, where near-surface water table is close to the day light, salinized groundwater inflow may be increased to natural or artificial microdepressions [16].

Efficient remediation of salinized areas may be achieved by means of combination of a number of techniques with consideration of environmental conditions, such as salinization level, area dimensions, topography, soils physical and mechanical properties, and watering conditions. Summarizing the lessons learned, it may be concluded that the methods combined with the techniques used for oil-contaminated land remediation may be effective. Based on field observations and qualitative chemical analysis data on freely soluble salts content and distribution the systems of drain ditches may be recommended with their subsequent washout with thaw water; and indigenous halophyte species seeds shall be used for phytomelioration. The use of indigenous (local) plant species or those adapted for a

IOP Conf. Series: Materials Science and Engineering **921** (2020) 012018 doi:10.1088/1757-899X/921/1/012018

specific climate and environment is an essential feature of remediation in humid climate conditions. The plants of *Carex, Taraxacum, Artemisia, Ranunculus, Poa, Plantago* families are capable of freely soluble salts accumulation in the Western Siberian environment. The technique provides for subsequent phytomass removal, and perennial grasses resowing till freely soluble salts content in soil achieves permissible values. The technique is also based on the landscape geochemical potential involvement into soil remediation making it essentially relevant at soil resalinization, when the groundwater table rises. Freely soluble salts will be carried-over with water to biogenic accumulative soil horizons and then will be accumulated in phytomass.

4. Conclusion

In reducing conditions of the middle forest of Western Siberia, in the event of accidental oil spills, the technogenic halogenesis causes highly mineralized flows, wherein sulfates and sodium are significant. Salt compounds create phytotoxic environment during phytomelioration at biological phase of land remediation . Chloride/sodium type of salinization and toxic salts content in soil profile governs the salinization that is not typical for soils of alluvial genesis type, and is indicative of the initial stage of technogenic halogenesis. Therefore, remediation activities shall involve the mechanisms of natural soil demineralization. The factors affecting chloride ions migration in soil are: amount of precipitations, area topography, air temperature mode, depth of soil seasonal freezing and its filtration properties. The soils under consideration require additional irrigation with fresh water, which will ensure salts removal from top and bottom soil horizons.

Besides, alluvial ecosystem soils accumulate an extra amount of various chemical compounds including various pollutants arriving from elevated sites. Groundwater may cause soil recontamination, because groundwater table rising will initiate post-technogenic soil transformation, that is migration of freely soluble salts from the underlying horizons to biogenic accumulative ones. Hence, alluvial soils are of high danger, and even after remediation activities monitoring and inspection is a must. Based on the study, it is recommended to compile new standard maps for technogeneous saline soil remediation, and it is necessary to include a section on studying the processes of technogenic soil salinization in the monitoring programs for oil-contaminated areas.

References

[1] Solntseva N P 2002 *Soil evolutionary trends in the technogenesis zone (Soil Science* vol 1*)* (Moscow: Moscow State University) pp 9–20
[2] Seredina V P and Kolesnikova E V 2017 *Features of the effect of oil pollution on soils of the middle taiga of Western Siberia (Oil industry* vol 5*)* ed V A Kondykov, A I Nepotrebnyy and S A Ognev pp 9–20
[3] Seredina V P and Sadykov M E 2011 *Soils of oil fields in the middle taiga of Western Siberia and a predictive assessment of the danger of pollution by organic pollutants (Siberian Journal of Ecology* vol 5*)* chapter 18 pp 617–623
[4] Seredina V P and Sadykov M E 2011 *The soils of West Siberia middle taiga oil deposits and a predictive estimate of contamination hazard with organic pollutants (Contemporary Problems of Ecology* vol 4*)* chapter 5 pp 457–461
[5] Seredina V P and Nepotrebnyy A I 2010 *Property change pattern of the soils in the oily ecosystems in the humid soil–formation conditions (Vestnik KrasGAU* vol 10*)* ed M Sadykov pp 49–54
[6] Seredina V P and Andreeva T A 2006 *Oil–contaminated soil: properties and remediation* ed T P Alekseeva, T I Burmistrova and N N Tereshchenko (Tomsk: TPU Publishing) p 270
[7] Chen K and Kao C 2010 *Control of petroleum hydrocarbon contaminated groundwater by intrinsic and enhanced bioremediation (J. Environ. Sci.* vol 22*)* ed C Chen, R Surampalli and M Lee pp 864–871
[8] Belabbas A and Boutoutaou D 2016 *Soil contamination by petroleum products. Southern Algerian case Proceedings AIP Conference* ed S Segaï and L Segni p 1758
[9] DeLaune R and Wright A 2011 *Projected impact of the Deepwater Horizon oil spill on U.S. Gulf Coast Wetlands (Soil Science Society of* America vol 5*)* chapter 75 pp 1602–1612

[10] Pinedo J and Ibáñez R 2014 Human risk assessment of contaminated soils by oil products: total TPH content versus fraction approach (Hum. Ecol. Risk Assess vol 20) chapter 5 ed P.A Lijzen and A Irabien pp 1231–1248
[11] Shuguang W and Yan Xu 2017 The harm of petroleum–polluted soil and its remediation research Proceedings AIP Conference ed L Zhaofeng, J Zhang, N Norbu and Wei Liu p 1864
[12] Shishov L 2004 Classification and diagnostics of Russian soils *(Smolensk: Oikumena)* p 342
[13] State Standart 26423–85 The soil: Methods for determining the electrical conductivity, pH and the solid residue of the aqueous extract
[14] State Standart 28168–89 The soil: Sample selection
[15] Harris T and Tapp J 2005 *Remediation drainage system and hay (Environmental Geosciences* vol 12*)* ed K Sublette pp 101–113
[16] Moralwar A and Sablet L 2005 *Remediation of a Spill of Grude Oil and Brine without Gypsum* vol 1 *(Environmental Geosciences* vol 12*)* ed K Ford, Duncan, G Thomas and J Brokaw pp 115–125

IOP Conf. Series: Materials Science and Engineering **921** (2020) 012019 doi:10.1088/1757-899X/921/1/012019

Thermal properties of polyethylene compositions with micro- and nanocellulose

A N Perova[1], S V Usachev[2], E V Koverzanova[2], A V Khvatov[1], S M Lomakin[1]

[1] N M Emanuel Institute of Biochemical Physics, Russian Academy of Science, Moscow 119334, Russia
2 N N Semenov Institute of Chemical Physics, Russian Academy of Science, Moscow 199334, Russia

E-mail: aleksandra.perova1994@mail.ru

Abstract. The effect of micro- and nanocellulose addition on the thermal properties of polyethylene compositions has been studied by means of thermogravimetric analysis and pyrolysis-gas chromatography-mass spectrometry. The main temperature characteristics of the thermo- and thermo-oxidative degradation of the compositions are determined. The influence of micro- and nanocellulose additives on change of the distribution of hydrocarbon pyrolysis products of composite materials was established.

1. Introduction

Cellulose is the most common natural biopolymer with a low density, high specific strength and specific elastic modulus [1-4]. Various forms of cellulose can be considered as promising fillers for thermoplastic and thermosetting polymers [5-6] due to the possibility of creating materials with improved physical and mechanical characteristics due to the reinforcement of the polymer matrix. The resulting materials can be widely used in medicine, textile industry, construction, etc.

Despite the advantages of such materials, their properties are limited by the characteristics of cellulose fillers, such as: relatively low interfacial adhesion, low thermal stability, and hydrophilicity [7-8]. In the process of preparing polyolefin composites containing cellulose, problems arise in their thermodynamic incompatibility, which makes it very difficult to achieve uniform distribution of cellulose, especially nanocellulose (NC) in the polymer matrix. To solve these problems, methods of physical [9-10] and chemical [11-12] cellulose modification are used, as well as the use of functionalized polymers [13], which allows to increase the compatibility of the polar cellulose filler and non-polar polymer matrix.

To increase the degree of distribution of nanocellulose (NC) in polymer composites, various methods are proposed based on the combined use of solution and melt technology in their preparation. So in [13] the problems of increasing the dispersion of crystalline NC in the hydrophobic matrix of polypropylene (PP) were solved by a preliminary step of preparing a suspension of NC at 130 °C in toluene, followed by the addition of PP powder until it was completely dissolved. The resulting suspension of the PP/NC composite was dried to obtain a solid powder of the PP/NC composition, which was then extruded and molded under pressure.

The main goal of the present work was an assessment of the effect of micro- and nanocellulose on the thermal properties of polyethylene (PE) compositions.

2. Experimental

2.1. Materials

For the synthesis of compositions, high-density polyethylene HDPE 273-83 (Stavrolen, LLC, Russia), low-density polyethylene LDPE 15803-020 (Ufaorgsintez, open society, Russia), microcrystalline cellulose (Lachema, Czech Republic) and nanocellulose gel (0.5 %), obtained by the method described in [14].

Compositions HDPE 273-83 with microcrystalline cellulose (MC) were obtained by mixing the components in a two-rotor laboratory mixer with a chamber volume of 25 cm^3 (N N Semenov Institute of Chemical Physics, RAS) at a temperature of 170 °C and a rotor speed of 30 rpm for 5 minutes.

Compositions based on LDPE 15803-020 and nanocellulose (NC) were prepared as follows. LDPE 15803-020 was dissolved by heating in o-xylene to 115-120 °C, after which an aqueous solution (gel) of NC was added with vigorous stirring. Heating and stirring were carried out until the azeotropic mixture water-o-xylene was completely withdrawn. When the boiling point of o-xylene (144-145 °C) was reached, stirring was stopped. The resulting clear solution was evaporated to form a solid film.

2.2. Thermogravimetric analysis(TGA)

TGA of PE, MC, NC, PE/MC and PE/NC composites was made by Netzsch TG 209 F1 thermal balance at the heating rate of 20 °C/min in inert atmosphere of Ar and the samples weight of about 5±0.5 mg.

2.3. Pyrolysis

Pyrolysis of sample compositions was carried out in a flow pyrolytic cell with a heating zone of 50 mm. A tube with a diameter of 3 mm with the test sample was placed in a cell preheated to 500±2 °C. Argon was purged for 15 minutes through a tube at a rate of 15 ml/min. The pyrolysis products were condensed in 1.5 ml n-hexane cooled to 4 °C.

2.4. Gas chromatography-mass spectrometry(GC-MS)

GC-MS measurements were performed on a gas chromatograph Trace-1310 and Single Quadrupole MS ISQ Detector. The pyrolysis products were injected into the GC capillary quartz column 15 m × 320 μm i.d. and 0.25 μm film (TG-SQC, Thermo Scientific Phase) under the temperature-programmed conditions in the range of 40 – 290 °C with gradient 15 °C/min. MS-spectra of degradation components were identified and interpreted by matching the results with NIST Mass Spectral library and literary data [15—16].

The yields of all degradation products were standardized by relative peak area of individual product (RPAi) according to equation

$$RPAi = \frac{PAi}{\Sigma PAi} \times 100\%$$

PAi - peak area of individual product, ∑PAi – sum of peak areas of all individual product.

Relative peak area for hydrocarbon fraction (RPAf) presents a sum of RPAi for corresponded fraction.

3. Results and discussion

3.1. TGA of PE/MC and PE/NC compositions

The results of TGA of PE/MC and PE/NC compositions are shoved in Figure 1 and Table 1. The studies showed that with an increase in the concentration of MC (Figure 1a) in the composition in argon, the decomposition onset temperature decreases compared to the initial sample, and thermal degradation proceeds in two stages, which is associated with the decomposition of the cellulose additive. The decomposition temperature of PE (the second stage of thermal decomposition of the compositions in Table 1) is practically independent of the content of MC, which indicates the absence of the influence of MC on the thermal decomposition of PE (Figure 1 a, Table 1). You can also see the absence of the influence of NC additives in a concentration of 0.5 to 5 % for thermal degradation of PE (Figure 1 c and Table 1).

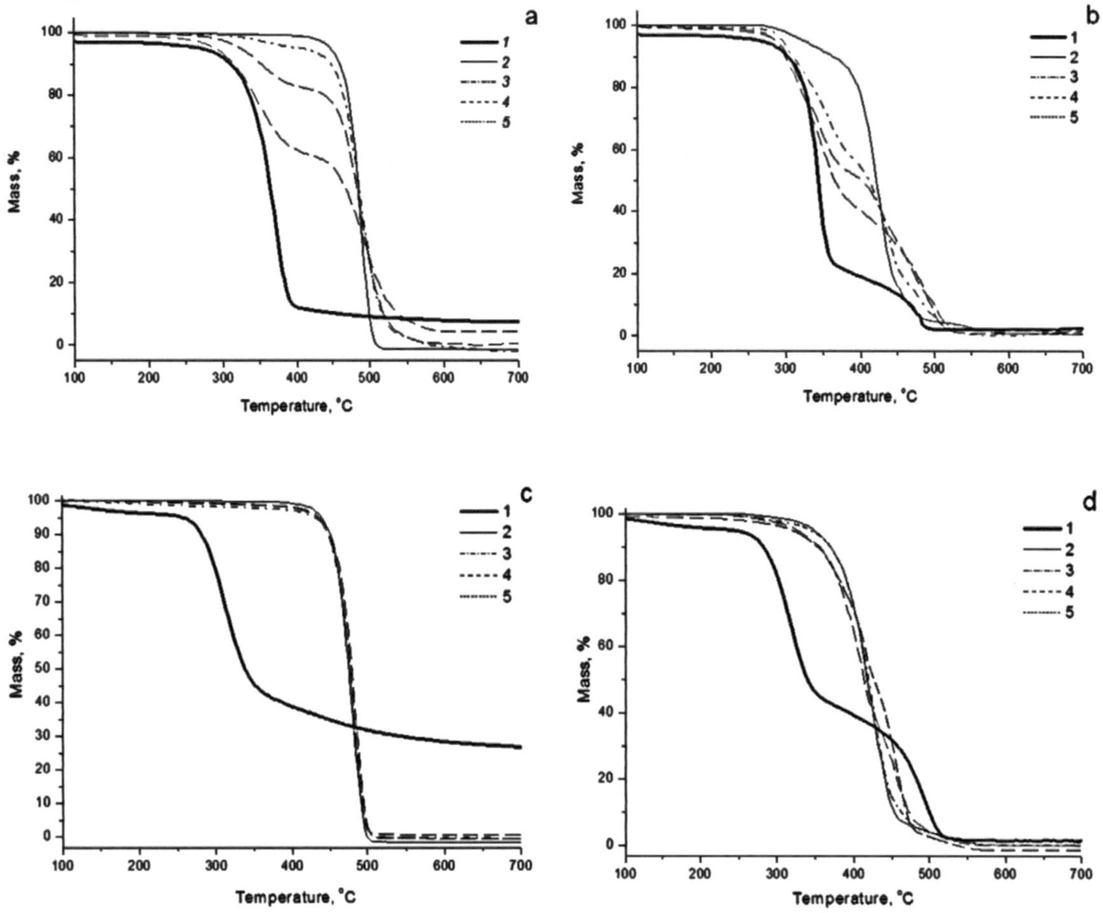

Figure 1. a - TG curves of PE/MC composites heated in Ar flow (1 - MC, 2 - PE, 3 - PE/MC - 5 %, 4 - PE/MC - 20 %, 5 - PE/MC - 50 %), b - TG-curves of PE/MC composites heated in air flow (1 - MC, 2 - PE, 3 - PE/MC - 5 %, 4 - PE/MC - 20 %, 5 - PE/MC - 50 %), c - TG curves of PE/NC composites heated in Ar flow (1 - NC, 2 - PE, 3 - PE/MC - 0.5 %, 4 - PE/MC – 3 %, 5 - PE/MC - 5 %), d - TG-curves of PE/NC composites heated in air flow (1 - NC, 2 - PE, 3 - PE/MC - 0.5 %, 4 - PE/MC – 3 %, 5 - PE/MC - 5 %).

Under the conditions of thermo oxidative degradation (Figure 1b, d), a tendency is observed for the temperature of the onset of reactions similar to that observed in an inert medium. The appearance of the third stage of decomposition with the introduction of the MC filler in the composition with PE is associated with the burnout of the carbonized cellulose residue.

A distinctive feature of the influence of MC and NC fillers on the thermal-oxidative degradation of compositions with PE is the increase in the temperature of the maximum decomposition rate at the second stage of thermooxidative degradation (Table 1, Figure 1 b, d). In our opinion, this effect of PE stabilization is due to inhibition of the radical chain process of composition destruction due to the recombination of macroradicals of PE and cellulose coke, leading to chain termination.

Table 1. Key TGA features of PE, PE/MC and PE/NC composites

PE matrix (brand)	Filler	The filler content, %	T_d[1], °C Ar/Air	T_{md}[2], °C Ar/Air		
				Stage 1	Stage 2	Stage 3
LDPE 273-83	MC	0	420 / 285	- / 323	488/ 424	- / -
		5	340 / 280	363 / 342	490 / 439	- / 481
		20	300 / 280	360 / 346	490 / 438	- / 487
		50	270 / 257	353 / 346	488 / 446	- / 490
		100	285 / 263	370 / 344	- / -	- / 482
HDPE 15803-020	NC	0	413 / 312	- / -	478 / 417	- / -
		0.5	407 / 315	- / -	483 / 415	- / -
		3	415 / 312	- / -	484 / 455	- / -
		5	405 / 308	- / -	483 / 453	- / -
		100	245 / 235	312 / 316	- / -	- / 492

[1] - decomposition onset temperature, °C.
[2] - temperature of the maximum decomposition rate, °C.

3.2. Pyrolytic chromatography-mass spectrometry of PE, PE/MC and PE/NC compositions

In Figure 2 shows chromatograms of volatile products of pyrolysis of HDPE 273-83 and compositions of HDPE 273-83 containing 5, 20 and 50 % MC, and in Figure 3 - LDPE 15803-020 and composition LDPE 15803-020 containing 5 % NC. The main products of pyrolysis consist of unsaturated (alkenes) and saturated (alkanes) hydrocarbons of various structures, as well as dienes. In addition, the chromatograms of the HDPE 273-83/MC compositions revealed the products of the pyrolysis of cellulose additives: a - furfural, b - 5-methyl-2-furfural, and c - levoglucosenone (Figure 2).

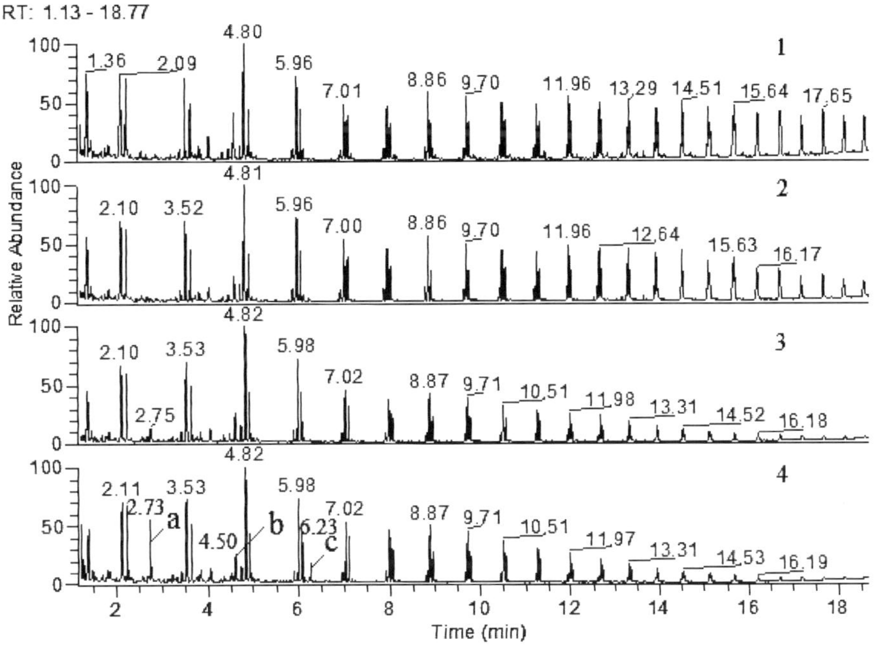

Figure 2. Chromatograms of hexane solutions of pyrolysis products of PE (1), PE/MC - 5 % (2), PE/MC - 20 % (3), PE/MC - 50 % (4) compositions, a - furfural, b - 5-methyl-2-furfural, and c - levoglucosenone

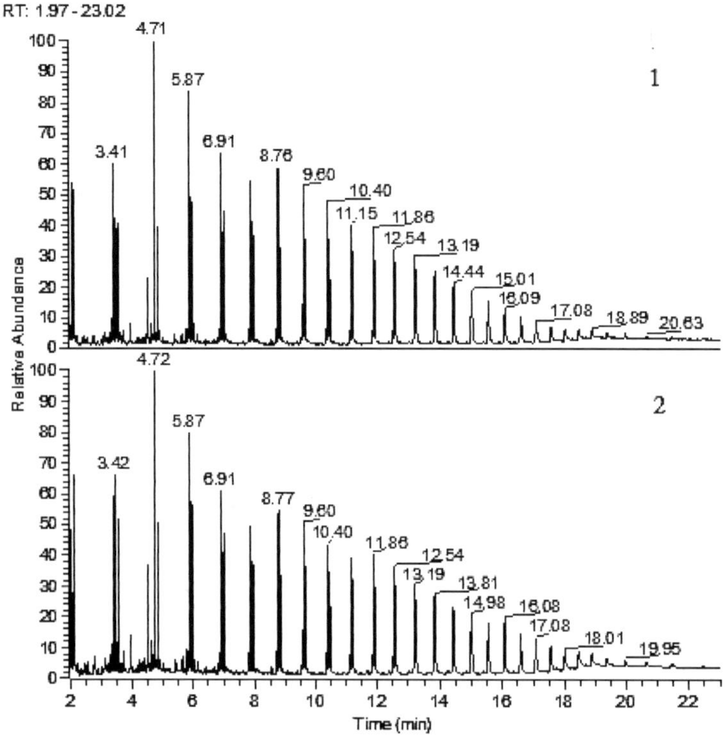

Figure 3. Chromatograms of hexane solutions of pyrolysis products of PE (1), PE/NC - 5 % (2)

Table 2 presents data reflecting the distribution of hydrocarbon products of the pyrolysis of PE and its compositions with MC and SC at different concentrations of cellulose fillers. From the data provided in Table 2 one can easily see that with the rise of concentration of MC and NC the content of lighter gaseous products of C8-C17 pyrolysis in PE/MC and PE/NC compositions increases as compared with unfilled PE.

Table 2. The distribution by molecular weight of hydrocarbon pyrolysis products of PE and PE compositions with different concentration of MC and NC.

PE matrix (brand)	Filler	The filler content, %	Content of components in the sample, %	
			C_8-C_{17}	C_{18}-C_{34}
LDPE 273-83	MC	0	48.25	51.75
		5	56.20	43.80
		20	78.58	21.42
		50	82.25	17.75
HDPE 15803-020	NC	0	62.34	37.66
		5	62.70	37.30

In works [17–18] on the study of the joint pyrolysis of PE compositions and cellulose-containing derivatives at a temperature of 700 °C, it was found that the products of thermal degradation of PE compositions containing up to 50 % cellulose there was a significant increase in the amount of low molecular weight aliphatic hydrocarbons (C_5-C_{15}), compared with the original PE. The authors explained this fact by the acceleration of hydrogen abstraction reaction in polyethylene macromolecules by the radical-acceptor products of cellulose pyrolysis (carbonaseous char residue).

In Figure 3 one can note the characteristic products of cellulose pyrolysis: furfural, 5-methyl-2-furfural, and levoglucosenone, which we found on chromatograms of hexane solutions of products of PE / MC compositions. According to the mechanism of cellulose pyrolysis, along with volatile degradation products, the main product of aromatization and condensation is formed - carbonaceous coke.

Thus the shift in the distribution of hydrocarbon pyrolysis products observed in our experiments towards the formation of lighter gaseous C_8-C_{17} products for PE compositions with MC and SC is due to an increase in the rate of initiation of the radical chain process of PE degradation by the coke residue of cellulose additives at temperatures above 400 °C.

4. Conclusions

Studies of TGA and pyrolytic chromatography-mass spectrometry of the PE/MC and PE/NC compositions showed the effect of cellulose additives at the stage of thermal and thermal-oxidative degradation of polyethylene. This is explained by the radical acceptor properties of cellulose coke, which affect the elementary stages of initiation and termination during the destruction of PE compositions.

The work was carried out within the framework of budget financing on the theme of state assignment No. of state registration 01201253305.

References

[1] Elsabbagh A, Steuernagel L and Ziegmann G 2009 Processing and modelling of the mechanical behaviour of natural fiber thermoplastic composite: flax/polypropylene *Polym. Compos.* **30(4)** 510–519

[2] Bledzki A K, Reihmane S and Gassan J 1996 Properties and modification methods for vegetable fibers for natural fiber composites *J. Appl. Polym. Sci.* **59(8)** 1329–1336

[3] Hornsby P R, Hinrichsen E and Tarverdi K 1997 Preparation and propeties of polypropylene composites reinforced with wheat and flax straw fibres: Part II analysis of composite microstructure and mechanical properties *J. Mater. Sci.* **32(4)** 1009–1015

[4] Khalil H P S A, Bhat A H and Yusra A F I 2012 Green composites from sustainable cellulose nanofibrils: A review. *Carbohydr. Polym.* **87** 963–979

[5] Herrera-Franco P and Aguilar-Vega M 1997 Effect of fiber treatment on the mechanical properties of LDPE – cellulose fiber composites *Appl. Polym. Sci.* **65(1)** 197–207

[6] Sdrobis A, Darie R N, Totolin, Cazacu G and Vasile C 2012 Low density polyethylene composites containing cellulose pulp fibers *Composites: Part B* **43(4)** 1873–1880

[7] Saheb D N, Jog J P 1999 Natural fiber polymer composites: a review *Advan. Polym. Tech.* **18(4)** 351–363

[8] Georgopoulos S T, Tarantili P A, Avgerinos E, Andreopoulos A G and Koukios E G 2005 Thermoplastic polymers reinforced with fibrous agricultural residues *Polym. Degrad. Stab.* **90(2)** 303–312

[9] Low I M, McGrath M, Lawrence D, Schmidt P, Lane J, Latella B A and Sim K S 2007 Mechanical and fracture properties of cellulose-fibre-reinforced epoxy laminates *Composites: Part A.* **38(2)** 963–974

[10] Alamri H, Low I M 2012). Mechanical properties and water absorption behavior of recycled cellulose fibre reinforced epoxy composites. Polym. Test. 31(5),620–628.

[11] Khalil A, Bhat I U H, Jawaid M, Zaidon A, Hermawan D and Hadi Y S 2012 Bamboo fibre reinforced biocomposites: a review *Mater. Des.* **42(2)** 353–368

[12] Mohamed H G, Mostafa A E, Kazuya O and Toru F 2010 Effect of microfibrillated cellulose on mechanical properties of plain-woven CFRP reinforced epoxy *Compos. Struct.* **92(5)** 1999–2006

[13] Gwon J G, Cho H D, Lee D, Choi D H, Lee S, Wu Q and Lee S Y 2018 Physicochemical and mechanical properties of polypropylene-cellulose nanocrystal nanocomposites: effects of manufacturing process and chemical grafting *BioResources* **13(1)** 1619–1636

[14] Voskoboynikov I V, Konstantinova S A, Korotkov A N, Mikhailov A I and Nikolsky S N 2010

Preparation of hydrogels of nanocrystalline cellulose from plant raw materials *Forestry Bulletin* **6** 151–154

[15] Poutsma M L 2003 Reaxamination of the pyrolysis of polyethylene: data needs, freeradical mechanistic considerations, and thermochemical kinetic simulation of initial productforming parhways *Macromolecules* **36** 8931–8957

[16] Sojak L, Kubinec R, Jurdakova H, Hajekova and Bajus M 2006 GC-MS of polyethylene and polyprophylene thermal cracking products Petroleum & Coal **48(2)** 1–14

[17] Xue Y, Kelkar A and Bai X 2016 Catalytic co-pyrolysis of biomass and polyethylene in a tandem micropyrolyzer *Fuel* **166** 227–236

[18] Matsuzawa Y, Ayabe M and Nishino J 2001 Acceleration of cellulose co-pyrolysis with polymer *Polym. Degrad. Stab.* **71(3)** 435–444

IOP Conf. Series: Materials Science and Engineering **921** (2020) 012020 doi:10.1088/1757-899X/921/1/012020

Interference analysis of wells in oil and gas reservoir with bottom water

LL Ryzhova, R D Kanevskaya

Institute of Geology and Development of Fossil Fuels

E-mail: l_ryzhova@igirgi.su, r_kanevskaya@igirgi.su

Abstract. A mathematical model of the inflow and algorithm for calculating the pressure field for interfering wells in a complex oil and gas reservoir with bottom water were developed in this paper. The interference model explores well interaction and when the operating mode of one of them is changing it allows to evaluate the change of bottom hole pressure in the reacting wells. While using the developed model it is possible to calculate a pressure field for a group of wells, to simulate the process of pulse-code well testing and also to specify the boundary conditions for more detailed inflow models. All the results were successfully tested on real data from one of the Eastern Siberia fields.

1. Introduction

Recently, there is an increasing tendency to complicate a process of hydrocarbon production, associated both with the properties of reservoir rocks and hydrocarbons. Modeling of reservoir with a complex structure of pore space is characterized by a high indeterminacy of the input parameters, due to both the insufficiency of initial information and the excessive complexity of the physical processes described in the models. In such cases it is important to create the methods for quick calculations adapted to the specified characteristics of the hydrocarbon system and based on simplified analytical or numerical models.

The main objectives of this research are creating of the mathematical model of inflow to a group of interfering wells in a complex reservoir with bottom water and a gas cap and its testing on real data for one of the Eastern Siberia oil fields. It should be noted that by the interference model it is possible not only to evaluate the pressure behavior of piezometric wells and to build a pressure field for the simultaneous operation of a group of wells, but also to simulate the process of pulse-code well testing and correct the boundary conditions for more detailed inflow models.

2. Mathematical model of well interference

The well interference consists in fact that while starting, stopping or changing the operating modes of one of the wells, the behavior of other wells in the same reservoir changes too [2].

The problem of calculating the pressure field during the simultaneously operation of a group of wells in a massive oil and gas reservoir with bottom water is considered. To solve it, we will use a model of inflow to a point sink, which was developed by the authors earlier. The linearity of this model makes it possible to use the principle of superposition to account for well interference.

A schematic model of the problem under consideration is characterized by the following features: it is believed that oil is contained in limited areas with sufficiently high porosity, which allows to ignore capillary forces. Hence, we can assume that there is a gravitational fluid segregation. The densities of oil, water and gas ρ_o, ρ_w, ρ_g and porosity m are assumed to be linearly dependent on pressure [2, 3]:

$$\rho_o = \rho_o^0\left(1 + c_o(p_o - p_o^0)\right), \rho_w = \rho_w^0\left(1 + c_w(p_o - p_o^0)\right),$$
$$\rho_g = \rho_g^0\left(1 + c_g(p_o - p_o^0)\right), \ m = m^0\left(1 + c_r(p_o - p_o^0)\right). \tag{1}$$

Here c_o, c_w, c_g are the compressibility of phases, p_o– oil pressure, upper index means parameters value at the initial conditions.

The mathematical model of the inflow can be described by the system of mass balance equations under conditions of gravitational equilibrium. These equations can be solve for oil pressure at a given level (for example, horizontal wellbore depth) $p_o(x, y, t)$ and the shapes of the interface - oil-water (WOC) $z_{ow}(x, y, t)$ and gas-oil contacts (GOC) $z_{og}(x, y, t)$. Also, it is important to study the water and gas coning and determine the conditions for the stability of contacts [4, 5]. However, nonequilibrium effects in the reservoir, including the breakthrough of cones, occur only in the vicinity of the wells. While studying the interaction of wells, we are considering a large-scale approximation, for which such effects are insignificant. Therefore, while modeling interference, the use of the equilibrium model is quite reasonable.

The shape of the contacts in large-scale calculation can be determined by the following equations:

$$z_{ow} = \alpha_w(p_o - p_o^0) + z_{ow}^0,$$
$$z_{og} = \alpha_g(p_o - p_o^0) + z_{og}^0.$$
$$\alpha_g = \begin{cases} -H_g(c_g + c_r), \text{compressibility effect only} \\ \dfrac{1}{\gamma_g - \gamma_o}, \text{ with active gas inflow} \end{cases} \tag{2}$$
$$\alpha_w = \begin{cases} H_w(c_w + c_r), \text{ compressibility effect only} \\ \frac{1}{\gamma_w - \gamma_o}, \text{with active water inflow.} \end{cases}$$

Here γ_o,γ_w ,γ_g – specific gravities of oil, water and gas, α_w, α_g– WOC and GOC "velocities", H_w, H_g – thickness of water and gas parts respectively.

Substitution of these expressions in the equation of oil balance gives the following nonlinear equation:

$$div\left[\frac{\rho_o^0 k}{\mu_o}\left(1 + c_o(p_o - p_o^0)\right)\nabla p_o\left(z_{ow} - z_{og}\right)\right]$$
$$= \frac{\partial}{\partial t}\left[m^0\rho_o^0\left(1 + (c_o + c_r)(p_o - p_o^0)\right)(z_{ow} - z_{og})\right], \tag{3}$$

Subsequently this equation can be linearized [1, 2].

The solution of equation (3) for the point sink and the principle of superposition which is valid for linear equations have been giving us a possibility to estimate the well interference and calculate the pressure field for simultaneously operations of wells group. It is advisable to note that in a large-scale approximation any well will behave as a point sink. In this regard, the following problem is the basis of the proposed well interference estimation algorithm.

Consider a group of wells in the plane Oxy. It is assumed that these wells are located inside the rectangle area with impermeable borders. Its dimensions can be set arbitrarily, but in such way that all wells have to be inside this area and its center coincided with geometric center of analyzed well group (Figure1).

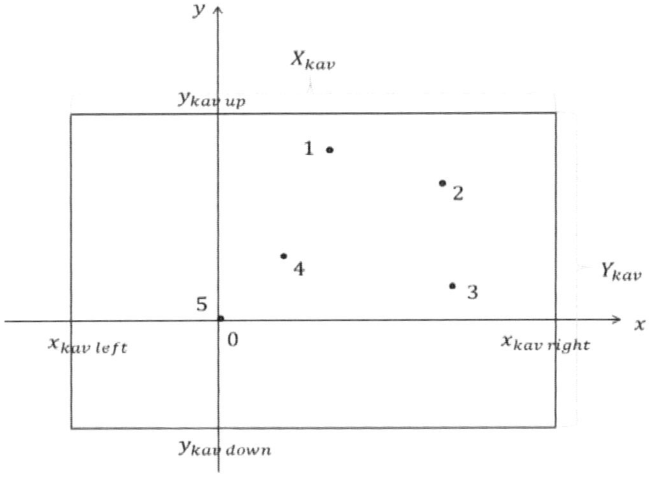

Figure 1. A schematic representation of the well location in the inflow area.

The impermeability of the area boundaries is simulated by multiple reflection of all sinks, relative to each area boundary. The accuracy of the solution will depend on the selected number of reflections. Ultimately, the use of reflection and superposition methods [2] gives the following expression for calculating the pressure field for interfering wells:

$$\begin{aligned} p_o(x,y,t) = P^0 &+ \sum_{i=1}^{n} Q_i'(t)Ei\left(-\frac{(x-x_i)^2+(y-y_i)^2}{4\chi t}\right) \\ &+ \sum_{j=1}^{\infty}\sum_{i=1}^{n} Q_i'(t)Ei\left(-\frac{\left(x-x_{ijLR}\right)^2+(y-y_i)^2}{4\chi t}\right) \\ &+ \sum_{j=1}^{\infty}\sum_{i=1}^{n} Q_i'(t)Ei\left(-\frac{(x-x_i)^2+\left(y-y_{ijUD}\right)^2}{4\chi t}\right) \\ &+ \sum_{j=1}^{\infty}\sum_{k=1}^{\infty}\sum_{i=1}^{n} Q_i'(t)Ei\left(-\frac{(x-x_{ikLR})^2+\left(y-y_{ijUD}\right)^2}{4\chi t}\right) \end{aligned} \quad (4)$$

Here values of P^0 and $Q_i'(t)$ are determined by the initial pressure and flow rates of the wells, n - number of wells in the considered area, x_i, y_i – its coordinates, x_{ikLR} and y_{ijUD} – coordinates of imaginary wells reflected from the left and right and, respectively, upper and lower boundaries of the considered area, j, k are the reflection numbers.

The proposed interference model is quite useful in practice. First, the calculated pressure distribution allows to correct the boundary conditions for a single well fine-scale model. Secondly, the model can be used to estimate the reservoir parameters and to evaluate the response of each well disturbances in the simultaneous operations of several wells. Third, while using this interference model, it is possible to successfully solve the problems of well testing.

3. Simulation results

The model presented in this paper was successfully tested on real data of one of the Eastern Siberia oil fields. Adjusting the parameters of the flow area, a relatively small deviation value from the actual data was achieved. The correct matching of reservoir characteristics allowed to repeat the trend of wells pressure behavior despite many assumptions of the presented model.

IOP Conf. Series: Materials Science and Engineering **921** (2020) 012020 doi:10.1088/1757-899X/921/1/012020

One example of piezometry process is demonstrated. The process of pulse-code piezometry includes series of short-term and small changes in the flow rate (pulses), which lead to changes in the measured pressures. Since the pulses are short-lived, pressure changes are small too.

Figure 2 shows the simulation results of piezometry modeling for one reacting and four production wells. Green color in the figure shows the actual measured pressure change in the reacting well R1 when starting and stopping production wells P1, P2, P3, P4 (other curves). The changes obtained by interference model calculation are represented in black color in the figure. The graph demonstrates how the pressure in the observation well responds to the start and stop of the production wells. It can be seen that the interference model made it possible to successfully reproduce the reaction to all disturbances and repeat the actual trend with enough quality.

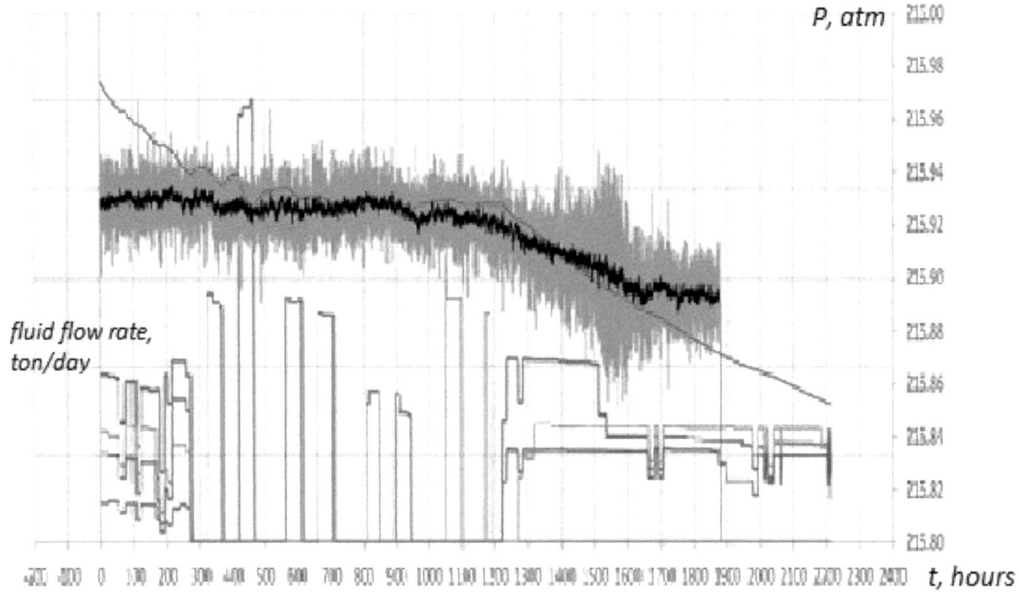

Figure 2. Adaptation of the interference model for the reacting well R1 (black color - calculation with noise filtering with a moving average, green color – actual data) when changing the flow rates of production wells P1 (blue), P2 (red), P3 (green), P4 (purple).

4. Conclusion

The mathematical model of well interference in complex oil and gas reservoir with bottom water was constructed in this research. All developed algorithms were implemented as a program in FORTRAN. During programming, all the computational complexity of the resulting algorithms was taken into account and methods that speed up the calculation, including parallel, were applied. The model has been successfully tested and adapted to real data of one of the Siberian oil fields.

IOP Conf. Series: Materials Science and Engineering **921** (2020) 012020 doi:10.1088/1757-899X/921/1/012020

5. References

[1] Barenblatt G I, Entov V M and Ryzhik V M 1984 *Theory of fluid flow through natural rocks* (M.: Nedra) pp 38–51

[2] Basniev K S, Dmitriev N M, Kanevskaya R D and Maksimov V M 2005 *Subsurface hydromechanics* (M.: Izhevsk: Institute of computer research) pp 1–496

[3] Kanevskaya R D 2002 *Mathematical modeling of hydrodynamic processes of hydrocarbon field development* (M.: Izhevsk: Institute of computer research) pp 23, 86

[4] Charny I A1963 *Reservoir hydrodynamics* (L.: Gostoptekhizdat) pp 202–218

[5] Efros D A 1963 *Research of a filtration of non-uniform systems* (L.: Gostoptekhizdat). pp 37–59

IOP Conf. Series: Materials Science and Engineering **921** (2020) 012021 doi:10.1088/1757-899X/921/1/012021

Mineral gels with a condensed dispersed phase

P A Sakharov[1], A V Khvatov[1] and E V Belenko[2]

[1]Emanuel Institute of Biochemical Physics, Russian Academy of Sciences, Kosygin str., 4, Moscow, 119334 Russia
[2]MIREA - Russian Technological University, 78 Vernadsky Avenue, Moscow, 119454 Russia

E-mail: hvatovanatoliy@gmail.com

Abstract. In the modern oil industry, water-polymer gels are widely used as drilling fluids. The main structural and mechanical units of these gels are macromolecules that participate in the construction of a single spatial structural grid due to donor-acceptor or ion bonds. The purpose of this work is to study the possibility of obtaining mineral hydrogels that combine high structural and mechanical characteristics characteristic of polymer hydrogels, but differ in the pseudoplastic flow model in combination with stability in the harsh thermobaric conditions of oil wells. Condensed hydrogels can potentially solve the most important problem in the production of liquid and gaseous minerals, which is to preserve the natural ecological balance and ensure the normal mode of biological self-purification of biogeocenoses in the areas of oil field development and disposal of drilling waste.

An example of an ionic mechanism for the formation of a polymer hydrogel is the coordination of functional groups of polyelectrolytes (carboxymethylcellulose (CMC), polyacrylamide (PAA), and others) around polyvalent metal cations [1] (Figure 1). The strength of such gels increases with an increase in the molar fraction of the complexing cation. In work [2], the mechanism of formation of spatially crosslinked polymer structures is presented due to substitutional interactions between functional groups of polyanions (CMC) and polyester chains of nonionic surfactants (polyoxyethylene (POE)).

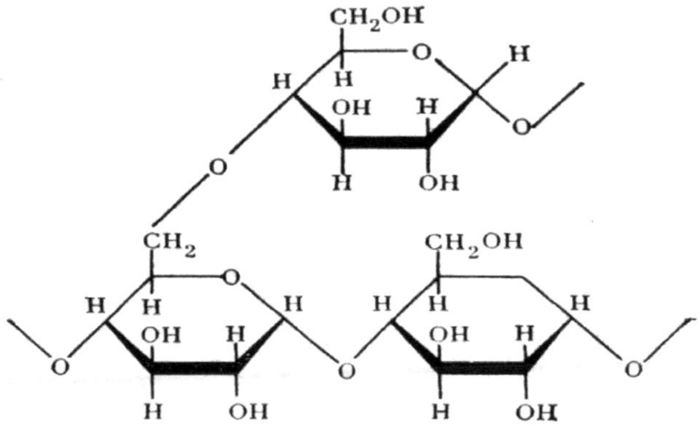

Figure 1. Cross-linked polymer hydrogel.

In this case, polyoxyethylene groups of surfactants form an associated polycation with counterions (Na^+), which is capable of forming a spatial structure due to interpolymer complexation with macromolecules of an anionic stabilizer. On the other hand, in aqueous solutions of xanthan biopolymer, hydrogen bonds are formed between neighboring macromolecules, characterized by regular alternation of branches of oligosaccharides (Figure 2).

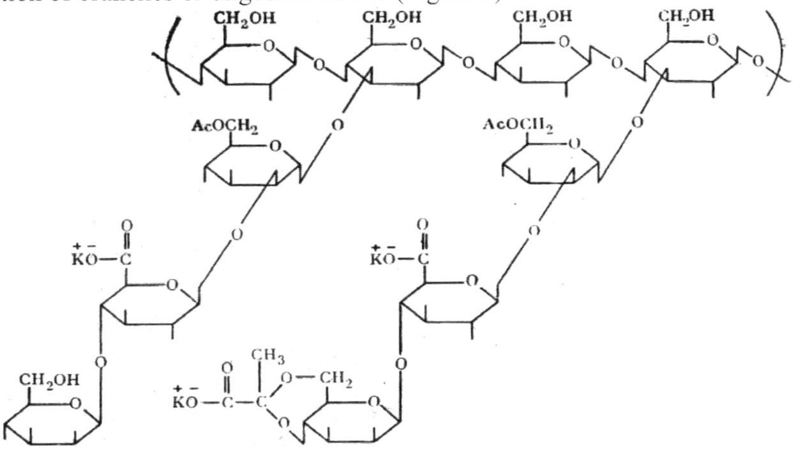

Figure 2. Hydrogel with a branched structure of the base polymer forming a system of hydrogen bonds.

This ensures an optimal ratio of the structural and rheological properties of biopolymer gels, which, however, are highly unstable to thermobaric effects [3]. Polymer-dispersed gels formed by non-ionic polymers (hydroxyethylcellulose (HEC), PoE) and salts of transition and amphoteric elements (Fe, Ni, Al) have increased strength of the structure [4, 5].

Emulsion hydrogels based on concentrated surfactant solutions also have acceptable structure formation. In such solutions, macromycellar aggregates interact at a close distance by forming hydrogen, dispersion, or ionic bonds, forming a cellular structure. To ensure the necessary level of kinetic stability of supramicellar structures, a combination of three specially selected components is necessary: polyanion, an oxyethylated nonionic surfactant, and a relatively polar component, which uses acids from vegetable oils. In [6], the principal possibility of obtaining structured hydrogel systems by stabilizing solubilization microemulsions with polysaccharide derivatives that carry oxy – ethyl and carboxymethyl groups, and as a surfactant, an oil-soluble copolymer of ethylene and propylene oxides is shown.

For almost all types of washing liquids (suspensions and emulsions), the height of the electrostatic barrier that prevents the convergence and coagulation of dispersed particles is quite high. In other words, flushing fluids are aggregatively stable systems. If the depth of the far potential minimum is large enough, then, regardless of the height of the electrostatic barrier, there is a long-range interaction of two particles, fixed at a distance of about 10^3 A, corresponding to the secondary potential minimum. In this case, 2 particles can neither disperse nor come close to each other and exist as a pair participating together in Brownian motion. When coagulating in the far minimum, the particles retain their individuality, and the system retains its dispersion and specific surface area, while when coagulating in the near minimum, the particles coalesce, which leads to a decrease in the specific surface area and surface energy of the system.

For concentrated washing liquids (weighted polymer-clay systems or hydrocarbon emulsions), the depth of the secondary potential minimum is more than 10 times higher than the energy of the thermal movement of molecules, so aggregates of dispersed particles formed by long-range interaction can only disintegrate if the kinetic energy of the particles is sufficiently large: the "Sol ↔ Aggregate" equilibrium is established. With a sufficient concentration of Sol (in conditions of high concentration of the clay phase or low-soluble salt), the fixation of dispersed particles in the far potential minimum

IOP Conf. Series: Materials Science and Engineering **921** (2020) 012021 doi:10.1088/1757-899X/921/1/012021

leads to the transformation of the Sol into a structured system. When the number of particles in the aggregate increases, the depth of the far minimum increases, which leads to the formation of "periodic colloidal structures" that have a far order and have a quasi-crystalline structure.

In the process of flushing flat and horizontal wells, pseudo-plastic flushing fluids have many technological advantages, since they reduce hydraulic losses when the solution circulates in the annular space, and also contribute to high-quality cleaning of the wall sections of the borehole from the drilled rock. For the transition from the bingamian to the pseudo-plastic rheological model, an adsorption-solvate barrier must exist between the dispersed phase particles, which provides the system with high viscosity, elasticity, and shear resistance. These properties prevent the interpenetration of the solvation layers at the convergence of particles. Strong hardening of solvate layers is achieved by adsorption on the surface of dispersed particles of long-chain surfactants or high-molecular compounds (HMC). In this case, adsorption-solvate layers are formed, covering even the far potential pit, and the washing liquid acquires the properties of a lyophilic system. The adsorption layers of the HMC have shear resistance, high viscosity and do not have time to "squeeze out" during the collision of particles, forming a structural and mechanical barrier that prevents coagulation.

Surfactants and HMC that create a structural-mechanical barrier are stabilizers of washing liquids. In contrast to surfactants, the cohesive strength of polymer adsorption layers increases over time (it reaches its maximum a few hours after the introduction of the polymer stabilizer into the system as a result of slow diffusion and orientation of macromolecules). Increasing the charge due to dissociation of ionogenic groups of macromolecules in the adsorption layer, which increases hydration, increases the repulsion forces between the macromolecules, reducing the lateral cohesion, and, consequently, the strength of the adsorption layers. Thus, you can control the stability of the washing liquid by changing the pH level, electrolyte concentration, and using stabilizers (surfactants and HMC).

To create an adsorption-solvate barrier on the surface of dispersed particles, which leads to the implementation of a pseudo-plastic model of the flow of the washing liquid, mainly non-ionic water-soluble polymers of the polysaccharide series and oxyethylated surfactants are used. Thus, the authors [7] link the appearance of pseudoplastic characteristics of a polymer-clay solution with the use of water-soluble polyethylene oxide oligomers in its composition. In [8], the implementation of pseudo-plastic properties of the emulsion system is associated with the use of an oil-soluble copolymer of ethylene and propylene oxides with an ethylene oxide content of no more than 15 % mol. as an emulsifier, and as a hydrocolloid filler - derivatives of cellulose and starch. In polysaccharide reagents (oxyethylcellulose, carboxymethylcellulose, etc.) are used to create pseudo-plastic properties of heat-resistant polymer washing systems, which are lined with aluminum and iron (III) salts.

The success of using low-charge and non-ionic polymers (starch, xanthan resin, hydroxyethylcellulose) in washing liquids is associated with ensuring the aggregate stability of the system, as well as with the formation of a three-dimensional structural grid. Stabilization of the dispersed system and structure formation are interrelated processes in which the polymer provides the implementation of structural-mechanical, sorption-solvate and electrostatic stability factors of the system. There are several mechanisms of structure formation involving a nonionic polysaccharide:

- in aqueous solutions of xanthan resin, there is a very strong tendency to macrostructure as a result of the formation of supramolecular spatial grids consisting of spiral structural units that contact by means of hydrogen and dispersion bonds. Macromolecules of xanthan resin, in addition to polar functional groups, also carry anionic carboxylate groups, the source of which is the structural links of pyruvic and glucuronic acids, located on the side branches inside the main (d) glucose helix. It is this screening of charged sections of the xanthan macromolecule that seems to explain the unique reversibility and shear strength of hydrated macromolecular structures that provide a solid, pseudoplastic character of the rheological behavior of xanthan flushing fluids. This is, in particular, confirmed by the fact that the maximum solubility of xanthan resin is observed at pH=4-6, while the solubility region of conventional polyanion stabilizers lies at pH>8 ;
- nonionic polymers (hydroxyethylcellulose, starch) form spatially structured systems when interacting with dispersed particles of high-grade bentonite as part of polymer-bentonite washing

IOP Conf. Series: Materials Science and Engineering **921** (2020) 012021 doi:10.1088/1757-899X/921/1/012021

fluids. An important property of such solutions is their pseudoplasticity, which is characterized by a pronounced dependence of the effective viscosity on the shear rate;

- polymer structured washing fluids are prepared by complexation using polyvalent metal ions characterized by a large charge and a small ionic radius. Metals whose ions have a large hydrate capacity include Cr, Al, Ca, Zr, and Ti. To enhance the thixotropic properties of xanthan resin solutions, trivalent chromium salts are introduced into them, the effect of which is due to the ability to form multi-core complex compounds, for example, aquacomplexes, in the inner sphere of which water molecules and hydroxyl groups are contained. The coordinating ability of cations depends on their ion potential (the ratio of the ion charge to its radius), according to the value of which the complexing agents are located in the following row: Al>Fe>Cu>Ni>Mn. The thixotropic properties and viscosity of polymer solutions containing a particular complexing agent are directly proportional to its ionic potential. Thus, Al^{3+} cations are able to coordinate very tightly the hydroxyl and carboxyl groups of polysaccharide derivatives in the pH range < 6, while a three-dimensional hydrogel structure can be formed on the basis of a complex of iron (III) salts and hydroxyethylcellulose.

The accumulated data on the physical, chemical and technological properties of polymer hydrogels with different spatial structure formation nature allow us to assert that these systems combine high structural and visco-plastic properties, as well as significant inhibitory and lubricating characteristics characteristic of saturated emulsion systems. However, polymeric hydrogels have several significant disadvantages associated with the peculiarities of their colloid-chemical structure. Thus, the lifetime of the polymer hydrogel is limited by the resistance of the base polymer to bio-and thermal degradation. In addition, purified varieties of carboxymethylated polysaccharides are required to produce polymer hydrogels, the market value of which is constantly growing.

The purpose of this work is to study the possibility of obtaining mineral hydrogels that combine high structural and mechanical characteristics characteristic of polymer hydrogels, but differ in the pseudoplastic flow model in combination with stability in the harsh thermobaric conditions of oil wells.

Mineral hydrogels are significantly cheaper than polymer analogues, since they are made from large-capacity products of processing and enrichment of rocks. The most commercially popular mineral hydrogel is a clay suspension, which is structured to achieve the effect of swelling and dispersion of smectite-type layered aluminosilicate in an aqueous medium. This kind of mineral hydrogels can be attributed to systems with a peptised dispersed phase (Figure 3).

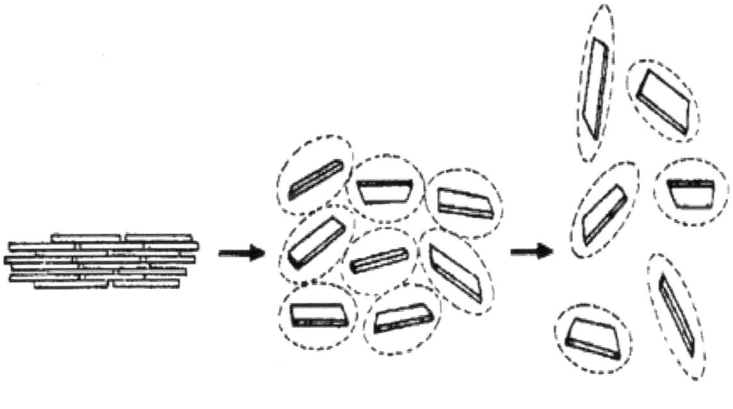

Figure 3. Coagulation-type Hydrosols with a peptised dispersed phase (clay suspensions).

When condensed hydrogel is obtained in an aqueous solution supersaturated with polyvalent metal salts (a phosphate-carbonate complex of iron, magnesium, and aluminum), conditions are created for the formation of new phase microcrystals that grow to a certain limit. When the necessary size of the condensed phase particles is reached, which is sufficient to create a strong hydrogel macrostructure, a crystal growth inhibitor is introduced into the system – a surface-active substance that is adsorbed on the active centers of the dispersed phase particles and prevents their further enlargement.

IOP Conf. Series: Materials Science and Engineering **921** (2020) 012021 doi:10.1088/1757-899X/921/1/012021

Condensed hydrogels, in contrast to polymer structured systems, are characterized by high pseudoplastic characteristics (Figure 4, Table 1).

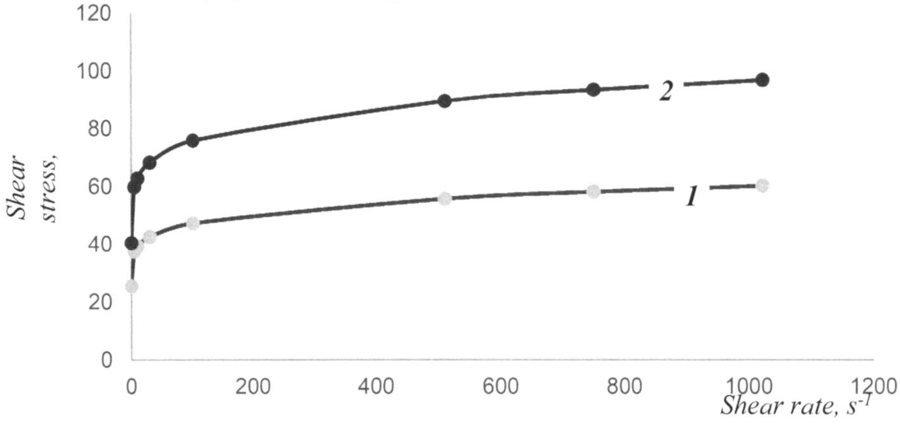

Figure 4. Flow curves of hydrogels based on a phosphate-carbonate (PhC) complex of magnesium and aluminum salts with a condensed microphase content of 3.7% wt. (curve 1) and 4.5% wt. (curve 2).

Table 1. Structural and mechanical parameters of condensed hydrogels.

Rheological parameters	2,2 % wt. PhC	3,7 % wt. PhC	4,5% wt. PhC
Yield strength (γ), dPa	25	40	75
Plastic viscosity, mPa·s	8	9	14
Exponent *n* in the model of Ostwald-Weyl	0,25	0,2	0,2

As a result of the research, the main features of the physical and chemical structure and technological properties of condensed hydrogels were formulated:
- hydrogels with a condensed solid phase form a structure due to the electrostatic interaction between the micron nuclei of condensation;
- the flow of the hydrogel is described by a pseudoplastic rheological model that provides optimal properties of the washing liquid;
- the structural and mechanical characteristics of the hydrogel are formed without the participation of organic polymers, which ensures the stability of the system to biodegradation;
- hydrogel has high coagulation resistance in the presence of acids and mineral salts;
- hydrogel is safe for the environment; gelling salt complex is used as a common mineral fertilizer.

Condensed hydrogels can potentially solve the most important problem in the production of liquid and gaseous minerals, which is to preserve the natural ecological balance and ensure the normal mode of biological self-purification of biogeocenoses in the areas of oil field development and disposal of drilling waste. The most consistent solution to this problem is the use of effective resource-saving methods for influencing the productive formation, which ensure maximum well flow rate with minimal initial opening time. This makes it possible to achieve a relatively low level of contamination of the drilled rock and drilling waste water, both with reservoir fluid and reagents in the composition of washing fluids. Hydrogel technology is designed to improve the productivity of oil reservoirs by upgrading the dispersion medium of washing liquids as a result of the use of environmentally friendly reagents that prevent colmatation and ensure the growth of oil recovery of the exposed layers.

References

[1] Antonov K V, Lukmanov R R and Sharipov A U 2000 Resource-Saving technology for opening productive layers during drilling *Abstracts of the international scientific and practical conference "Geo-ecology and modern geodynamics of oil and gas regions"* pp 45-55

[2] Penkov A I, Vakhrushev L P, Belenko E V, Ostryagin A I and Dzhioev T E 2000 Influence of polyanion electrolytes on the state of adsorption layers of polyalkylene glycols in bentonite suspensions News Of Universities North Caucasus region Natural science № 1 pp 90-94

[3] Penkov A I, Koshelev V N and Shishkov S N 1998 Compositions of drilling fluids used in drilling horizontal wells and assessment of their properties compliance with the requirements of drilling horizontal wells Issues of flushing wells with horizontal sections pp 21-37

[4] US patent 4021355, MK E 21 B 43/26, 1977

[5] Penkov A I and Fedosov R I 1998 Development of environmentally compatible drilling fluids for drilling and completion of wells on the Arctic shelf Issues of flushing wells with horizontal sections pp 49-53

[6] Belenko E V, Vakhrushev L P, Koshelev V N and Penkov A I 2001 Pseudoplastic hydrogels based on surfactant emulsions stabilized by polysaccharide derivatives Oil and gas technologies № 2 pp 18-21

[7] Ostryagin A I, Stamps A I and Rastegaev B A 2003 RU № 2139315

[8] Penkov A I, Vakhrushev L P and Koshelev V N 2003 RU № 2212429

Study of hydraulic resistance in collapsible pipelines

L V Seoev, D A Drozdov, K M Plotnikova, A V Elkin

[1] Federal Autonomous Establishment «The 25-th State Research Institute of Himmotology, Ministry of Defence of Russian Federation», 10 Molodogvargeyskaya Street, Moscow, Russia

E-mail: nio21-25gosnii@yandex.ru

Abstract. The features of conducting research on hydraulic resistances of collapsible pipelines with a bell and spigot joint are shown. The importance of taking into account the influence of pipe joints on the flow of the pumped liquid, as well as deviations from straightness due to angular mobility, is noted. Data on the determination of pressure losses in collapsible pipelines PMTP-100 and TSR-MK-100 based on the results of tests carried out on sections of 1200 m long when pumping water are presented. The values of Darcy friction factor are also set. As a result of processing experimental data using mathematical statistics, the values of the coefficient of equivalent roughness of the inner surface of pipes of prefabricated pipelines PMTP-100 and TSR-MK-100 were determined. Using the methodological apparatus of hydraulic modeling, calculations were made to simulate the pumping of various groups and brands fuels through collapsible pipelines. In the analysis of the data revealed that by reducing the coefficient of equivalent roughness of the inner surface of pipes, supply pipeline TSR-MK-100 can be higher than by pipeline PMTP-100 on average, 5-13%.

Introduction

Actuality due to the need for public testing of pipeline TSR-MK-100 for the integrated assessment of conformity with the technical and performance requirements of the Ministry of Defence of the Russian Federation. The aim of the study is to assess the actual performance of the system is collapsible pipeline TSR-MK-100.

The main objectives of the study:

– to analyze the methods of determining pressure losses and performance pressure in pipelines;

– to develop a method of determining pressure losses and performance pressure in collapsible pipelines;

– on the basis of data processing of experimental researches to determine the values of the Darcy friction factor and the coefficient of equivalent roughness of the inner surface of the pipe;

– to evaluate the efficiency of the pipeline TSR-MK-100 for pumping fuels from various groups and grades.

One of the main objectives when performing hydraulic calculations collapsible pipelines is the determination of the magnitude of the head losses due to friction [1]. They are usually expressed depending on the square of the average speed of the pumping according to Darcy-Weisbach equation:

$$h_{\mathrm{f}} = \frac{\lambda l w^2}{d 2g}, \tag{1}$$

where: h_{f} – friction head loss; λ – Darcy friction factor; l – estimated pipeline length; d – pipeline inner diameter; w – average fluid velocity; g – acceleration of gravity.

The Darcy friction factor depends on flow regime [2]. In the existing guidance documents manual for collapsible pipelines [1] applied Altshul equation [3]:

$$\lambda = 0.11\left(\frac{68}{Re} + \varepsilon\right)^{1/4}, \tag{2}$$

where: Re – Reynolds number; ε – coefficient of relative roughness of the inner surface of the pipe.

In the equation (2), the coefficient that takes into account the roughness of the inner surface of the pipe ε accepted for pipes with enameled coating equals 0.0001 for pipes with zinc coating – 0.00011 for all types of collapsible pipelines increased capacity [1]. It should be noted that the value of this ratio was obtained when testing on the plot of the collapsible pipelines, a length of less than 100 m, while the study was conducted only with collapsible pipeline PMTP-150, collapsible pipeline PMTP-100 was not investigated.

The coefficient of relative roughness as a dimensionless criterion of hydrodynamic similarity, defined as the ratio of the coefficient of equivalent roughness to the internal diameter of the pipeline:

$$\varepsilon = \frac{k_{\mathrm{e}}}{d}, \tag{3}$$

where: k_{e} – the coefficient of equivalent roughness of the inner surface of the pipe.

Obviously, for pipes with different diameter and type of the inner surface of this coefficient will change. This statement formed the basis of a new method for determining head losses and collapsible pipelines capacity.

Results

When assessing the hydraulic resistance of collapsible pipelines, it is also necessary to consider the influence on fluid flow the joints connections as well as deviations from straightness due to the angular movement of the pipes. This requires that the length of the experimental section of the pipeline must not be less than 1000 m, and the length of the entire collapsible pipelines deployed on the ground should not be less than 5000 m.

Experimental research, according to the method of determining the pressure loss in the collapsible pipelines was conducted at the test site. With this purpose was developed the technological scheme of the pipeline, which included two experimental parts:

– the first consisted of a pipeline PMTP-100 for checking and confirmation or refutation of the existing value of the coefficient of equivalent roughness;

– the second pipeline TSR-MK-100 for the determination of the equivalent roughness of the pipeline of a new generation.

Each experimental plot consisted of 200 pipes with a total length 1200 m. Both the pipeline section lying parallel to each other for eliminating the difference in elevations of the initial and final areas, and other terrain features. As a means of transfer of used fuel transfer pump station PSG-240 and a mobile pumping unit PNU-100/200K.

If you set the mode of transfer using flow meters liquid at the same time the following quantities were measured: the flow of the pipeline, the pressure at the beginning and end of test parts. The measurements were carried out by pumping water. Figure 1 presents a graph of pressure loss from the flow for two types of pipeline. It is obvious that the pressure loss when pumping through a new pipeline TSR-MK-100 less than that PMTP-100.

IOP Conf. Series: Materials Science and Engineering **921** (2020) 012022 doi:10.1088/1757-899X/921/1/012022

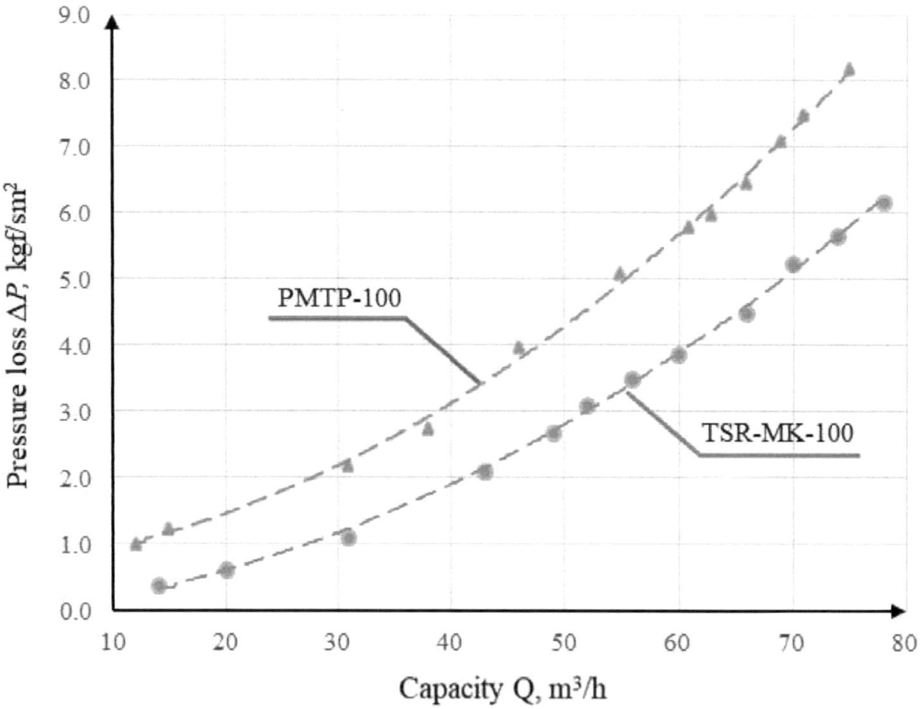

Figure 1. Dependence of the pressure loss in each experimental parts from the capacity of collapsible pipelines PMTP-100 and TSR-MK-100

According to the known value of the pressure loss and pipeline capacity was determined by Darcy friction factor by the equation:

$$\lambda = \frac{\Delta P \pi^2 d^5}{Q^2 8 l \rho}, \tag{4}$$

where: ΔP – pressure loss; Q – pipeline capacity; ρ – fluid density.

The equivalent coefficient of roughness of pipe inner surface was determined by the equation:

$$k_e = \left(6860\lambda^4 - \frac{68}{Re}\right)d. \tag{5}$$

Assuming that the distribution law of the measured values closes to the normal law, for the assessment of errors you can define the boundaries of the confidence intervals, for a given value of confidence [4].

The average value of the equivalent coefficient of roughness is determined by the equation:

$$\tilde{k}_e = \frac{1}{n} \cdot \sum_{i=1}^{n} k_{ei}, \tag{6}$$

where: n – number of experiments.

The standard deviation is determined by the equation:

$$\tilde{\sigma}_{ke} = k_n \left[\frac{1}{n-1} \sum_{i=1}^{n} \left(k_{ei} - \tilde{k}_e\right)^2\right]^{1/2}, \tag{7}$$

where: k_n – coefficient value which is selected in accordance with the number of experiments [5].

The value of the confidence interval is determined from the equation [6]:

$$\alpha = F\left(\frac{\theta n^{1/2}}{2\tilde{\sigma}_{ke}}\right), \qquad (8)$$

where: α – a confidence probability to the mathematical expectation in the case of known standard deviation; θ – confidence interval; $F(z)$ – is the Laplace function.

The value of the confidence interval at Student's distribution is determined from the equation [6]:

$$\theta = \frac{t_\alpha \tilde{\sigma}_{kэ}}{n^{1/2}}, \qquad (9)$$

where: t_α – the amount determined by the table of Student's distribution.

Processing of results of measurements methods of mathematical statistics have shown that the value confidence probability, the values of the coefficients of the equivalent roughness of the inner surface of the pipe amount to 0.005 mm and 0.035 mm, respectively for pipelines TSR-MK-100 and PMTP-100.

Also the values of the standard deviations by the formula (7), when the number of experiments $n = 22$; ratio $k_n = 1.0134$, confidence probability to the mathematical expectation in the case of known standard deviation $\alpha = 0.95$; the amount determined by the table of student's distribution $t_\alpha = 2.103$; the value of the standard deviation of 0.0036 and 0.0086 for pipelines TSR-MK-100 and PMTP-100, respectively.

The magnitude of the confidence intervals is determined by the formula (8) and amounted to 0.0016 inch 0.0043 and for pipelines TSR-MK-100 and PMTP-100, respectively.

The dependence of Darcy friction factor from Reynolds number for TSR-MK-100 and PMTP-100 subject to the equivalent roughness coefficients for each type of pipeline are shown on figures 2, 3.

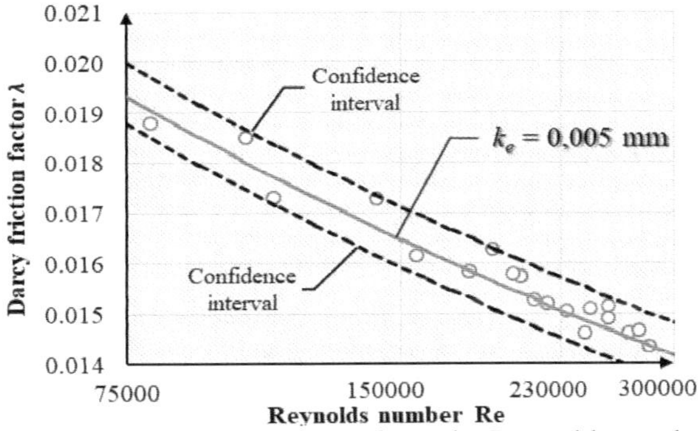

Figure 2. The dependence of the Darcy friction factor from the Reynolds number for composite pipeline TSR-MK-100

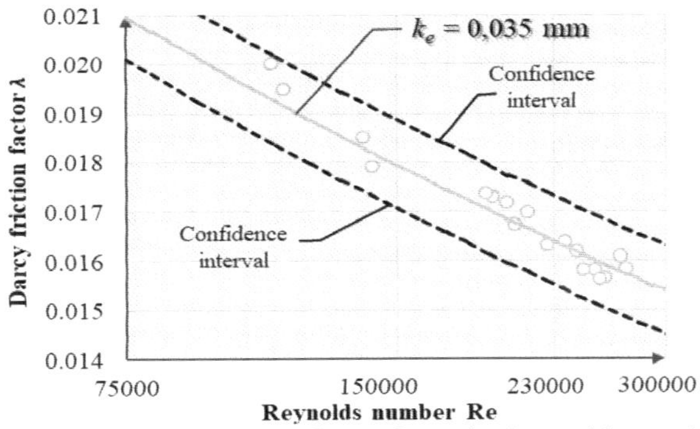

Figure 3. The dependence of the Darcy friction factor from the Reynolds number for steel pipeline PMTP-100

Using the methodological apparatus of the hydraulic calculations were plots of differential pressure performance pump through pipelines TSR-MK-100 and PMTP-100 of the following types of fuels: motor gasoline Regular 92, jet fuel TS-1, diesel fuel Z-minus 35 and diesel fuel EURO class 4 (figure 4).

In the analysis of the data revealed that by reducing the coefficient of equivalent roughness of the inner surface of pipes, supply pipeline TSR-MK-100 can be higher than by pipeline PMTP-100 when pumping:

– diesel fuel for 3-7% (average 5%);
– jet fuel 6-9% (average 8%);
– motor gasoline at 11-15% (average 13%).

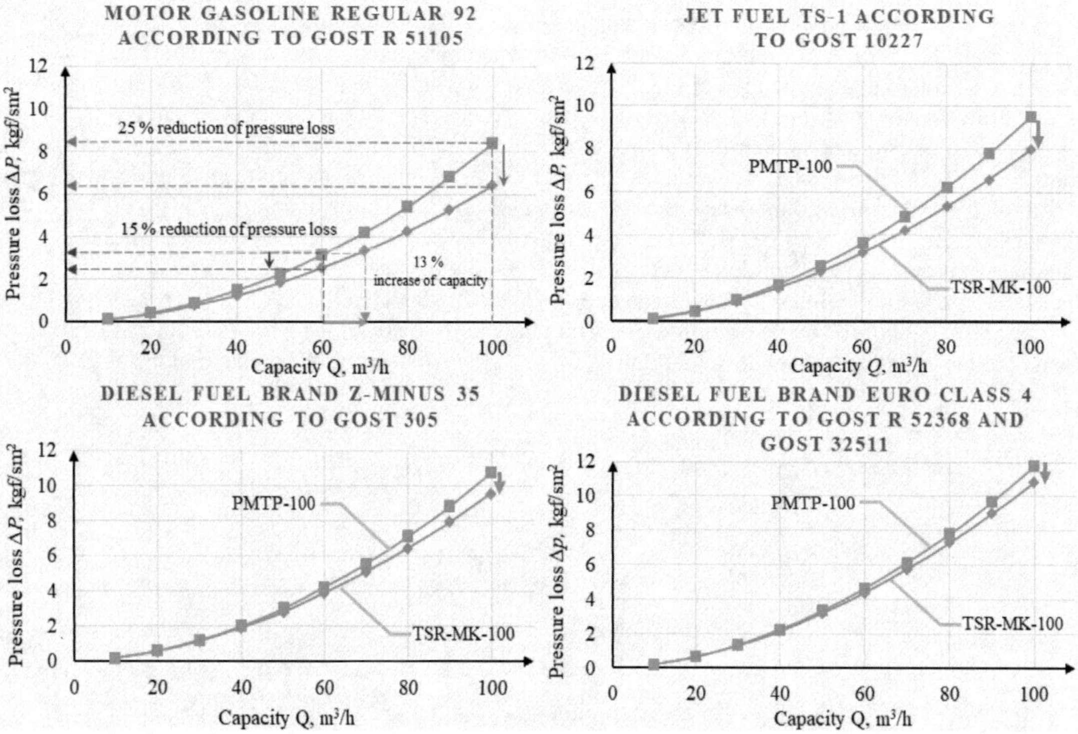

Figure 4. The dependence of pressure loss ΔP from capacity Q at composite pipeline TSR-MK-100 and PMTP-100 for pumping of various groups and brands fuels

Specific value increase flow through the pipeline depends on the physico-chemical properties of the product transfer (density and kinematic viscosity), as well as the number and type of means of pumping.

The study evaluated the effectiveness of the pipeline TSR-MK-100. Estimated cost daily fuel consumption mobile pumping unit (PNU-100/200K) was determined by the equation:

$$S_{\text{mpu}} = Q_{\text{mpu}} t q_{\text{f}} , \tag{10}$$

where: S_{mpu} – estimated cost daily fuel consumption mobile pumping unit; Q_{mpu} – the rate of fuel consumption, l/hour; t – working time will mobile pumping unit in the day, h; q_{f} – the price for 1 liter of fuel, RUB/l.

Was calculated the estimated cost of daily fuel consumption engine performance mobile pumping unit PNU-100/200K. The study established that under the existing organization the application of the set of pipeline length 150 km, the economic effect from reduction of operating costs to conduct pumping can be up to 160 thousand rubles per day.

Conclusion

The study was carried out to estimate actual productivity pressure in the collapsible pipelines TSR-MK-100. When it was obtained the following research results:

– the analysis of the current methodology for determining the head losses and capacity for collapsible pipelines, the results showed that the existing method of determining the pressure losses in collapsible pipelines has a significant drawback, since it does not take into account the impact of the inner surface and the pipe diameter on the Darcy friction factor;

– the technique of determining pressure losses and capacity in collapsible pipelines with regard to the influence of pipes joints as well as deviations from straightness due to the angular movement of the pipes;

– using experimentally obtained data were used to determine the coefficients of the equivalent roughness of the inner surface of the pipe for pipelines PMTP-100 and TSR-MK-100, the value of which is made up of 0.035 and 0.005, respectively;

– the evaluation of the effectiveness of pipeline TSR-MK-100 showed that by reducing the coefficient of equivalent roughness of the inner surface of pipes, supply pipeline TSR-MK-100 can be higher than by pipeline PMTP-100 by pumping in an average of 5-13%.

An evaluation of the effectiveness of the implementation of the results of the study showed that the economic effect could be up to 160 thousand rubles per day.

References

[1] Drozdov D A and Plotnikova K M 2019 Determination of Hydraulic Resisistanse Coefficient of Demountable Field Pipelines at Turbulent Flow Regime *Industrial Servies* **2** (71) pp 22–27

[2] Idelchik I E 1992 *Handbook of hydraulic resistances* (Moscow: Mashinostroyeniye Press) p 672

[3] Altschul A D 1982 *Hydraulic resistance* (Moscow: Nedra Press) p 224

[4] Novitsky P V and Zograf I A *Assessment of Measurement Result Errors* (Leningrad: Energoatomizdat Press p 248

[5] Abezgauz G G, Tron A P, Kopeykin U N, Korovina I A 1970 *Probability Calculation Reference Book* (Moscow: Voenizdat Press p 536

[6] Abchuk V A, Matveychuk F A, Tomashevsky L P 1979 *Reference Book of Operations Research* (Moscow: Voenizdat Press p 368

IOP Conf. Series: Materials Science and Engineering **921** (2020) 012023 doi:10.1088/1757-899X/921/1/012023

Evaluation of organomineral fertilizer for restoration under the conditions of the Far North of disturbed illuvious-ferruginous podzol

S M Sevostyanov, N F Deeva, D Yu Aladin, D V Demin

142290, Russia, Pushchino, Moscow Region, 2 Institutskaya st.
Institute of Basic Biological Problems RAS.

E-mail: sevost2000@rambler.ru

Abstract. The results of a model experiment on the use of organomineral fertilizer (OMF) based on food waste for the sowing of three mixtures of perennial cereals on a sample of the soil type of the Far North: an illuvial -ferruginous podzol with the aim of developing a technology for the restoration of disturbed soil cover are presented. It was revealed that when OMF is added in the amount of 25 and 50%, the physicochemical properties of the soil improve; there is an increase in the green mass of plants relative to control by an average of 3.3 times. When 50% of this fertilizer is applied, signs of plant inhibition are observed on day 18, which indicates an overabundance of nutrients in the soil. The mass of plants at this concentration is close to the values obtained at 25% OMF content. Organomineral fertilizer based on food waste with the obtained characteristics can be used in the restoration of the illuvial-ferruginous podzol as a reclamation fertilizer.
Keywords: organic-mineral fertilizer, perennial cereals, physicochemical properties of soils.

1. Introduction

In the Far North, the most vulnerable is the land cover - the main structure-forming element of natural landscapes. The presence of permafrost at shallow depth sharply limits the thickness of the biologically active layer, in which nutrients, plant roots, soil microorganisms and animals are concentrated. Destruction, scalping of the surface soil layer leads to the destruction of the vegetation cover, access to the surface of the mineral part of the profile, and, as a result, the development of solifluction processes, erosion, and karst. As a result of the development of mineral deposits, the laying of gas pipelines, roads and other technological disturbances in the Yamalo-Nenets Autonomous Okrug, the area of completely disturbed lands amounts to 377557 hectares, according to other sources, land losses are estimated at 6 million hectares or 8% of the territory [9]. Therefore, there is an acute question of reclamation of disturbed soil cover taking into account bioclimatic conditions. Often, enterprises exploiting natural resources carry out only technical restoration of disturbed territories, considering it acceptable to leave the land for self-growth. The processes of natural restoration of vegetation of technogenic landscapes are complex and time-consuming (up to 30 years) and are complicated by random factors [8].

Bioremediation is now recognized as possibly the most attractive clean-up approach for Polar Regions. Numerous contaminated sites in Polar Regions have legacy contamination that is often

IOP Conf. Series: Materials Science and Engineering **921** (2020) 012023 doi:10.1088/1757-899X/921/1/012023

associated with weathered contaminants that tend to be recalcitrant to bioremediation. There is growing interest in cold-region landfarming because it is a relatively inexpensive and effective method for dealing with contaminated soils in inaccessible areas. This technique has proven to be effective in the Arctic. Reports of effective landfarming of higher-molecular weight compounds in cold climatesare limited and concerns remain about the effectiveness of this technology [15].

Polar soil environments differ from other soil environments in a number of ways that might affect hydrocarbon biodegradation. Polar soils have unique periglacial features, including permafrost and numerous types of patterns primarily due to freeze-thaw effects. In the Arctic tundra, an active soil zone, above the permafrost, thaws for a period of typically 1±2 months in the summer. Summer temperatures in the active zone are highly dynamic and vary greatly, from near freezing at the permafrost interface to occasionally above 20°C at the surface. Permafrost restricts water movement and sometimes results in a saturated active zone. The active zone is where most hydrocarbon contaminants exist and is presumably the site of most biological activity. Little is known about the relationships between permafrost and hydrocarbon contaminants. Many Arctic tundra soils are low in organic content, a characteristic which can be expected to directly affect sorption of hydrocarbons to the soil and to indirectly affect biodegradation of hydrocarbons. Arctic tundra soil microbial communities have not been well characterized. There is presently no convincing evidence that the composition or structure of these communities is unique to Arctic tundra; although, the biomass may be low and the organisms more cold-adapted, relative to soils in temperate regions. Organic decomposition is slow in Arctic tundra soils, largely due to low temperatures. Some studies indicate the presence of hydrocarbon-degrading microorganisms in Arctic tundra soils and groundwater. Similarly, hydrocarbon-degrading microorganisms have been reported in Antarctic soils. Physical, chemical and biological factors have complex effects on hydrocarbon biodegradation in soil. For this reason, experts frequently recommend that soil bioremediation projects begin with treatability studies to empirically test the biodegradability of the hydrocarbon contaminants and to optimize treatment conditions. On the other hand, it is possible that the expense of such treatability studies could be avoided or minimized, if certain soil characteristics could be measured and used to predict the potential for bioremediation of a site, the kinetics of hydrocarbon removal or the optimal values for certain controllable treatment conditions. For example, certain co-contaminants such as heavy metals might preclude hydrocarbon bioremediation. Or, soil particle size distribution might partly dictate the potential rate and extent of hydrocarbon removal [16].

Measures to restore the disturbed lands of the Far North should be aimed at creating a fertile layer and selecting affordable mixtures of perennial grasses adapted to specific environmental conditions that can form a stable grass and turf in a short time [3, 6, 7]. Creating a stable turf cover is difficult without additional application of organomineral substrates - peat, land, manure, mineral fertilizers [4]. The best results of restoration of tundra soil are given by the combined use of mineral and organic fertilizers [11]. In the case of reclamation of sandy soils, the best result also comes from the integrated use of mineral fertilizers with peat [12].The introduction of peat is complicated by the fact that it is necessary to develop it, and this is a violation of the soil cover with a subsequent set of measures for its restoration. As an alternative, technologies are proposed for the use of potassium humate preparations, which can be used as a stimulator of the growth and development of perennial grasses sown on disturbed tundra soils for the purpose of their reclamation [13].

In addition to man-made disruptions of natural ecosystems, the amount of human waste around human settlements is growing. Their disposal is an important problem that must be solved in parallel with the restoration of technogenic landscapes. For various remediation purposes, various types of waste can be used. In proceedings [14], the possibility of using organic waste (excess activated sludge, sedimentary brewer's yeast, chitosan) from the Murmansk region for the restoration of oil-contaminated podzol soils. Comparison of indicators of the content of oil products in soil samples with indicators of phytotesting on plant seedlings allowed us to conclude that it is advisable to use the tested waste as reclamants of oil-contaminated soils.

IOP Conf. Series: Materials Science and Engineering **921** (2020) 012023 doi:10.1088/1757-899X/921/1/012023

One of the promising directions for solving the problem of organic fertilizer and at the same time solving the problem of waste from settlements is the processing of food waste into composts. An approach to soil disinfection and restoration using plants, mushrooms and composts, without the use of synthetic fertilizers, is promising for reclamation in facilities with a cold climate [15].

In the Far North, composting in an open way is inefficient, due to low temperatures, a short period of positive temperatures. At the same time, if there is an isolated base and a greenhouse structure, it becomes possible to protect the substrate from climatic factors. A technology has been developed to produce environmentally friendly organic-mineral fertilizer based on food waste directly within the places of their formation and accumulation.

The use of this technology will help to improve the environmental situation by eliminating stored food waste and reducing the cost of biological reclamation of land in adjacent territories.

The aim of this work is to evaluate the effectiveness of the application of organic fertilizer based on food waste when sowing various mixtures of perennial cereals to restore disturbed soil cover in the Far North.

2. Objects and methods

The experiment was conducted in laboratory conditions with soil sampled in the Novy Urengoy district of the Yamalo-Nenets Autonomous Okrug. The soil was used from a depth of 5-30 cm, as these horizons go to the day surface because of an intense anthropogenic load.

The soil is an illuvial-ferruginous podzol, and horizon A_2 is a grayish-whitish fine-grained sand. [5].

Organomineral fertilizer (OMF), obtained by solid-phase fermentation from food waste from the village of Yubileiny, Yamalo-Nenets Autonomous Okrug. In the process of fermentation, lime was introduced into the composition to form the fertilizer structure and deoxidize the substrate.

The physicochemical analysis showed that OMF has an alkaline reaction (Table 1), according to the grouping of soils by nitrifying ability, the content of mobile phosphorus, exchange potassium and humus belong to a very high class of security. Moreover, all of the above indicators (except for humus, the excess of which within the limits of this class) are exceeded several times [1], which confirms its high agrochemical effectiveness.

The soil is characterized by a neutral aqueous pH, a very low content of organic matter, and a low content of alphitite (Table 1).

Table 1. Chemical and physical properties of soils, soil mixtures and organomineral fertilizers (OMF)

Measured indicators	Podzol illuvial-ferruginous	Soil mix with organomineral fertilizer (25%)	Organomineral fertilizer
pH, aqueous extract	7,0	8,7	8,7
pH, salt extract	5,4	8,3	8,2
Organic matter, %	0,2	2,7	10,6
$N\text{-}NH_4$, mg/kg	11,9	342,5	1977,8
$N\text{-}NO_3$, mg/kg	1,7	86,2	112,8
P_2O_5, mg/kg	12	55	430
K_2O, mg/kg	3	1963	8062
Alphitite content (<0,01 mm), %	6,9, (cohesive sand)	7,8 (cohesive sand)	19,8 (clayed sand)

A mixture of soil with organic fertilizer was prepared in a ratio of 3: 1 (75: 25%) and 1: 1 (50:50%) by dry weight. The physicochemical properties of the obtained substrate were determined.

For analysis, samples were prepared with one dose of application - 25% by dry weight, which allows you to recalculate the result of application for any dose.

IOP Conf. Series: Materials Science and Engineering **921** (2020) 012023 doi:10.1088/1757-899X/921/1/012023

After the addition of 25% OMF, the chemical characteristics of the soil mixture improved according to the content of nutrients (NPK) and organic matter (Table 1).

To conduct a model experiment, soil samples of the illusionally-glandular soil podzol and soil-mixture with a OMF content of 25 and 50% were placed in plastic vessels, according to the principle of Mitcherlich's vessels, with openings on the bottom and trays for collecting water. The surface area of the vessel is 95 cm2. After filling the vessels, the humidity of the mixtures in all cases was brought to the ultimate field water capacity and maintained at this level during the growing of plants. The experiment was carried out in triplicate.

In all vessels with organic soil-samples carried crop of 0.5 g three gramineous grass mixtures.

Mix No. 1: perennial ryegrass (*Lolium perenne*) - 25%; red fescue (*Festuca rubra*) - 47%; meadow bluegrass (*Poa praténsis*) - 13%; creeping bent (*Agróstis stolonífera*) - 15%.

Mix No. 2: red fescue altered (*Festuca rubra commutata*) - 15%; red fescue (*Festuca rubra*) - 35%; meadow bluegrass (*Poa praténsis*) - 10%; sheep fescue (*Festuca ovina*) - 5%; perennial ryegrass (*Lolium perenne*) - 35%.

Mix No. 3: couch grass (*Elytrígia répens*) - 20%; meadow fescue (*Festuca pratensis*) - 60%; perennial ryegrass (*Lolium perenne*) - 20%.

The selection of these perennial cereals is based on literature data [2,9]. Basically, the mixtures are represented by grassroots perennial cereals that are most stable in the Far North and are capable of reproduction.

Similar cereal mixtures were seeded into control soil samples. Tanks with crops were covered with glass and placed in a greenhouse.

Thus, for each soil sample, nine experimental variants were obtained: three - with sowing of three mixtures of herbs in control samples; three - a mixture with the introduction of 25% OMF and sowing herbs, as well as with the introduction of 50% OMF.

The experimental conditions are the greenhouse of the greenhouse, which eliminates accidental effects on the growth and development of plants.

Duration of illumination is 12 hours a day, 12 hours a night, illumination 25000 lm/m^2, temperature + 20 ± 2°C, air humidity 60 ± 5%.

After emergence (on the third day), the coverslips were removed. Irrigation was carried out every three days with distilled water with a pH of 6.3, at the same time the height of the plants (maximum, average, minimum) was measured in all variants of the experiment.

This is due to the fact that mixtures of herbs were sown, the seeds of which have different sizes, different rates of germination and differ in productivity (mass of one plant).

The time of the vegetation experiment is 21 days after germination. Further, in all variants of the experiment, the weight of the green mass in the wet and dried state was evaluated, and a comparison was made relative to the control.

3. Results and discussion

During the experiment, it was noted that during the entire time in the control samples of the lluvial-ferruginous podzol, there was a lag in the formation of the green mass of plants in comparison with all variants of the experiment, which is associated with the physicochemical properties of this soil. In the variant containing 50% organic fertilizer, more intensive plant growth was observed, but on the 18th day the tips of the leaves began to dry and darken, which indicates an excessive content of nutrients when using such a dose of fertilizer, which leads to growth inhibition.

Table No. 2 presents data on the average weight (in three replicates) of the top mass of plants according to 9 experimental variants. In the control, the moisture content of plant samples was 72-73%, with the addition of 25 and 50% OMF, the humidity increased to 86%. Regarding control, the weight of the top parts of plants in the wet state increased more than three times, in the dried state - more than 1.5 times. For a grass mixture consisting of *Lolium perenne, Festuca rubra, Poa praténsis; Agróstis stolonífera*, the yield increase was maximum in both wet and dried state. With the introduction of 50% OMF into the soil, the weight of the green mass of plants slightly decreased for

IOP Conf. Series: Materials Science and Engineering **921** (2020) 012023 doi:10.1088/1757-899X/921/1/012023

all variants with respect to the fertilizer dose of 25%, which confirms the fact of inhibition of plant development due to an excess of nutrients.

Table 2. The mass of plants grown in illuvial-ferruginous podzol

Variant	Grass mixture	Green mass weight (g)	The increase in control, %	The weight of the dried green mass (g)	The increase in control, %
Control	1	0,52±0,02		0,140± 0,002	
Control	2	0,51±0,02		0,139± 0,002	
Control	3	0,49±0,02		0,138± 0,002	
Soil + 25% OMF	1	1,85± 0,25	256	0,258± 0,002	84
Soil + 50% OMF	1	1,80±0,02	246	0,250 ± 0,002	79
Soil + 25% OMF	2	1,65±0,05	224	0,230± 0,005	65
Soil + 50% OMF	2	1,58± 0,05	210	0,230±0,005	65
Soil + 25% OMF	3	1,70± 0,05	247	0,240± 0,005	74
Soil + 50% OMF	3	1,63± 0,05	233	0,238±0,005	72

4. Conclusions

During the experiment, it was found that organomineral fertilizer, based on food waste, is characterized by a high content of nutrients NPK, which confirms its high agrochemical effectiveness.

OMF improves the physicochemical properties of the illuvial-ferruginous podzol. No drying of the soil surface is observed, and optimal conditions are created for plants to consume available moisture and nutrients.

At 50% content of OMF by external signs, inhibition of plants is observed on the 18th day of the experiment, which indicates an excessive content of nutrients. Therefore, the use of this dose is not recommended.

The introduction of OMF in the amount of 25 and 50% leads to an increase in the yield of the presented mixtures of herbs relative to the control by an average of 3.3 times.

For illuvial-ferruginous podzol in control and variants OMF best results are obtained for the second mixture of cereals: perennial ryegrass (*Lolium perenne*) - 25%; red fescue (*Festuca rubra*) - 47%; meadow bluegrass (*Poa praténsis*) - 13%; creeping bent (*Agróstis stolonífera*) - 15%.

Organomineral composition based on food waste, with the obtained characteristics, is suitable for use in the restoration of the disturbed illuvial-ferruginous podzol as a complex organic fertilizer and ameliorant in the Far North.

References

[1] Agrochemical examination and monitoring of soil fertility: a training manual. Stavropol: AGRUS, 2012, 352 p.

[2] Biological reclamation of disturbed lands in the Yamal Peninsula: Recommendations. - Novosibirsk, 1994, - 48 p.

IOP Conf. Series: Materials Science and Engineering **921** (2020) 012023 doi:10.1088/1757-899X/921/1/012023

[3] Likhanov A. I., Archegova I. B. Development of theoretical and practical aspects of the restoration process of disturbed lands in the north of the Komi Republic // Theoretical and Applied Ecology. 2014. No3. pp. 79-85
[4] Nazarova G.V., Ivanov V.V., Gavrilyeva L.D., Mironova S.I. On the issue of using waste in the restoration of disturbed lands // Successes in modern natural science. - 2012.-No 11. pp.135-136.
[5] National Atlas of Soils of the Russian Federation / Under the General Ed. corresponding member RAS S.A.Shoba. Faculty of Soil Science, Moscow State University M.V. Lomonosov. - M.: Astrel Publishing House, 2011, pp. 500-502.
[6] Popov A.I., Kapelkina L.P. Restoration of vegetation on disturbed lands in the Nenets Autonomous Okrug // Plant Resources, - 2012.- V. 4. No. 2.- pp. 278-287.
[7] Pystina N. B., Baranov A.V., Listov E.L., Budnikov B.O. Improvement of technologies for the restoration of disturbed and contaminated lands in hydrocarbon deposits of the Far North. // Scientific Bulletin of the Yamal-Nenets Autonomous Okrug. - 2016.- No: 2 (91). pp. 4-8.
[8] Sumina O.I., Koptseva E.M. On the classification of technogenic vegetation in the Far North of Russia // Collection of the scientific works of GNSS. 2016. V.143. pp. 224-230.
[9] Tikhanovsky A.N. Problems and methods of biological reclamation of technologically disturbed lands of the Far North. // Successes of modern science. - 2017. - No. 2 pp. 43-47.
[10] Korneykova M.V., Myazin, V.A., Fokina, N.V. Restoration of Oil-Contaminated Soils in Mountain Tundra (Murmansk Region, Russia) // Springer Geography. 2020, Pages 187-198
[11] Iglovikov A.V., Motorin A.S. Emerging Technologies for Recultivation of Disturbed Sandy Soil after Anthropogenic Disturbances in the Industrial Development of the Far North // IOP Conference Series: Earth and Environmental Science Volume 194, Issue 9, 15 November 2018, Номер статьи 092009
[12] Galiulin R.V., Bashkin V.N., Galiulina R.A., Alekseev A.K. Potassium humate for the restoration of disturbed tundra soils // Business journal Neftegaz.ru. 2018. - No. 3 (75) - p. 104-106.
[13] Vasilieva Zh.V., Gorbovskaya T.D., Pavlov A.V. Reclamation of oil-contaminated soils using organic waste in the Kola North // Bulletin of the Murmansk State Technical University. 2019.Vol. 22. No.1. pp. 72-82.
[14] Robichaud K., Stewart K., Labrecque M., Hijri M.. Cherewyk, J., Amyot, M. An ecological microsystem to treat waste oil contaminated soil: Using phytoremediation assisted by fungi and local compost, on a mixed-contaminant site, in a cold climate // Science of the Total Environment Volume 672, 1 July 2019, Pages 732-742
[15] Sanscartier D., Laing T., Reimer K., Zeeb B. 2009. Bioremediation of weathered petroleum hydrocarbon soil contamination in the Canadian High Arctic: Laboratory and field studies. Chemosphere 77:1121-1126.
[16] Mohn, W.W., Stewart, G.R., 2000. Limiting factors for hydrocarbon biodegradation at low temperature in Arctic soils. Soil Biol. Biochem. 32, 1161-1172.

IOP Conf. Series: Materials Science and Engineering **921** (2020) 012024 doi:10.1088/1757-899X/921/1/012024

Laser ablation method for the generation of chromium, iron, manganese, nickel, scandium, titanium and vanadium, nanoparticles: control of size and properties

Alexander V Simakin[1,*], Ilya V Baimler[1,2], Oleg V Uvarov[1], Ignat I Rakov[1] and Sergey V Gudkov[1]

[1] Prokhorov General Physics Institute of the Russian Academy of Sciences, Vavilove St. 38, Moscow, 119991 Russian Federation
[2] Moscow Institute of Physics and Technology, Institutskaya St. 9, Dolgoprudny, Moscow Region 141700 Russian Federation

E-mail: avsimakin@gmail.com

Abstract. Using method of laser ablation of massive targets in liquids, nanoparticles (NP) of chromium, iron, manganese, nickel, scandium, titanium and vanadium with controlled characteristics were generated. To generate NP with the desired characteristics, both ablation of macro-sized butt and fragmentation are consumed. Laser fragmentation is the process of melting large NP into small ones using laser radiation. The influence on the distribution of NP by diameter (size) of such key parameters of laser irradiation as the density of energy of beam of laser in the medium and power of the laser is determined. To obtain NP devoid of surface oxide films, organic solvents or water saturated with molecular hydrogen were used as a working fluid. The proposed approaches allow one to obtain NP that meet the following criteria: 1). the type of NP tends to be spherical; 2). 95% of NP have a certain radius; 3). NP consist of a single chemical element, the quantity of impurities is minimal; 4). the crystal lattice of NP is metallic. If necessary, it is possible to generate NP with a surface layer consisting from oxides, as well as fully oxidized NP.

Keywords: nanoparticles, laser fragmentation, laser ablation, transition metals

1. Introduction

Currently, the processes accompanying laser breakdown in colloidal solutions of various nanoparticles, processes associated with the generation of nanoparticles and their characterization are being actively studied [1-3]. It is known that during optical breakdown of aqueous solutions containing nanoparticles, fragmentation and oxidation of these nanoparticles and their surfaces occurs. During optical breakdown, generation of molecular hydrogen [4], molecular oxygen [5], ozone, hydrogen peroxide and other active molecules and atoms is observed [6]. When ablating pieces of minerals, metals or polymers, the above products are also formed. These products are formed by laser plasma. The formation of plasma on macroscopic pieces or nanoparticles proceeds as follows. Laser radiation is absorbed. Rapid heating of the material occurs. The material is melting. A vapor-gas shell forms on the surface of the material. When the shell collapses, molten particles fall into the medium.

Acoustic and shock waves accompany the process, plus optical and ultraviolet radiation is observed. Physicochemical processes occurring during optical or ablation are affected by the material with which laser radiation interacts [7]. It is fundamentally difficult to obtain nanoparticles of the same size. If we are talking about obtaining nanoparticles with a similar surface topology, oxide-free, with the same properties, then the task is complicated many times [8-11]. Of course, there are exceptions to this rule [12]. The aim of this work is to obtain nanoparticles of the same size, topology, deprived of oxide layers or, conversely, coated layers of the desired size, up to complete oxidation.

2. Methods

2.1. Generation of nanoparticles

Metallic nanoparticles were obtained by laser ablation in a liquid. The experimental setup for nanoparticle generation by laser ablation is shown in Fig. 1. We used a ytterbium-doped fiber laser with a variable pulse duration of 4-200 ns, a pulsed wave of 1064 nm, an energy of 1 mJ per pulse, a pulse repetition rate of 20 kHz, and an average power of about 20 W. As a working medium, deionized Milli-K water was used in an amount of 10-15 ml. Polished metal pieces were used as a primary target for laser ablation. The target of the corresponding material was immersed in the medium so that the liquid layer above the target surface was about 1-2 mm. The exposure times of the target were from 5 to 20 minutes. For laser fragmentation of nanoparticles in a medium, an Nd: YAG laser with a fixed pulse duration of 10 ns, λ = 1064 nm, pulse energy of 2 mJ, pulse repetition rate of 10 kHz, and an average power of about 20 W was used. Laser radiation was introduced into the experimental cuvette from below, to prevent defocusing of the laser radiation on the resulting air bubbles. The times of radiation exposure on nanoparticles during their fragmentation ranged from 15 to 30 minutes.

2.2. Methods used in the study of nanoparticles

The average size of NP was studied using DC 24000 analytical disk centrifuge (CBC Instruments). Centrifugation on an analytical disk is a fast, reliable and sensitive method for analyzing medium-sized nanoparticles that are contained in a liquid. To determine the morphology, confirm the presence of an oxide layer or impurities on the surface of the nanoparticles, and additionally determine the size of the nanoparticles, we used a 200FE (CarlZeiss) transmission electron microscope with a spectroscopic attachment to measure the energy loss of electrons. The methods we use, the necessary settings, and the subtleties of the experiment are described earlier in our articles [13-15].

Figure 1. Experimental installation for laser ablation of a volumetric target in a liquid medium.

3. Results

Laser ablation yielded scandium, titanium, vanadium, chromium, manganese, iron and nickel nanoparticles. If deionized water is used as working media, then in the ablation, large metal NP can be obtained. Such NP mainly consist of metal, although its may contain oxide surface layers or its parts. As a representative example, Figure 2A shows micrographs of a NP generated from scandium. After the complete fragmentation of such particles in aqueous solutions or pure water, the manifestation of small scandium nanoparticles is observed. It is shown that during fragmentation in aqueous solutions, the formation of fully oxidized NP is observed. In rare cases, the formation of NP with a surface oxide layer and a metal core is observed. A typical microphotography of the scandium preparation after partial fragmentation in water media is shown in figure 2B. It Should be noted that when changing aqueous solutions to aliphatic liquids, oxide layers are practically not formed.

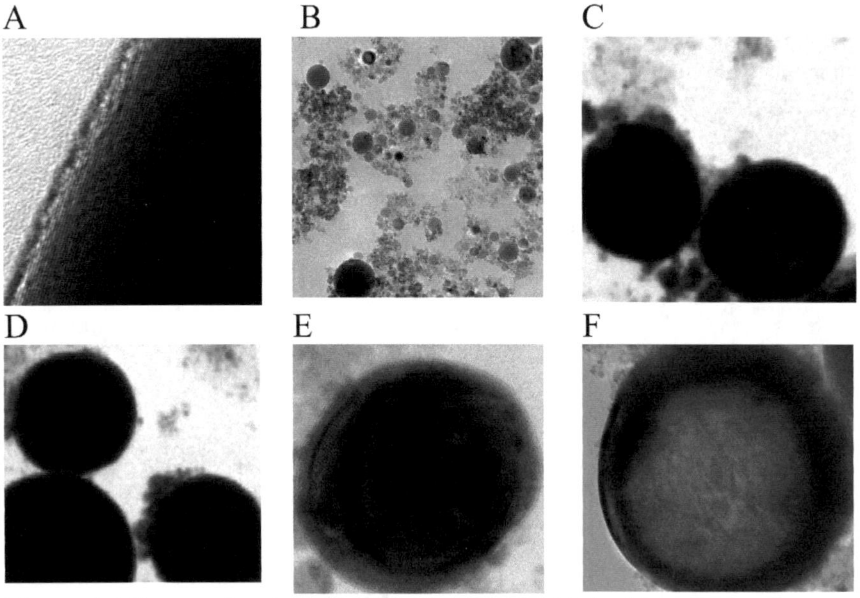

Figure 2. Micrographs of nanoparticles generated obtained by using transmission electron microscopy.
A - Microphotography of a part of a nanoparticle consisting of scandium about 100 nm in size.
B - Micrograph of scandium preparation obtained after partial fragmentation of large nanoparticles. The non-fragmented black nanoparticle (lower left corner) has a characteristic diameter of 100 nm.
C — Micrograph of nickel nanoparticles made by laser fragmentation in isopropanol.
D - Micrograph of nickel nanoparticles generated by laser fragmentation in isopropanol.
E - Micrograph of an iron nanoparticle made by laser fragmentation in pure alcohol.
F - Micrograph of a nickel nanoparticle made by laser fragmentation in pure alcohol.

It is known that during laser ablation oxidation of NP is observed. This is due to the generation of molecular oxygen and its reactive species during ablation. It is obvious that the fewer oxygen atoms enter the ablation medium, the less molecular oxygen and its active forms are formed. However, there is another part of the problem. When using media that do not contain oxygen atoms, the oxidation of nanoparticles is not observed. Nanoparticles are generated entirely metal. however, such nanoparticles are difficult to transfer to water environments. Such nanoparticles aggregate, precipitate, and a large amount of aliphatic substances is transferred to the water with them. In addition, under certain boundary conditions, nanoparticles are observed to form a shell from polymerized aliphatic media. To solve the problem, we used amphiphilic liquids, mainly low-atom alcohols. The use of different alcohols as ablation and fragmentation media makes it possible (depending on the boundary conditions) to generate as nanoparticles devoid of a surface oxide layer (completely metallic) (Fig. 2C, 2D), and nanoparticles containing an oxide layer on the surface (a metal core surrounded by an oxide layer) (Fig. 2E, 2F). Alcohols must be saturated with hydrogen before use. This measure allows us to obtain the purest nanoparticles. Thus, employing laser fragmentation, NP of various specified diameters were generated. Figure 3, as an example, shows a series of diameters of nanoparticles. In this case, we obtained nanoparticles from various metals with the same average diameter of about 20 nm.

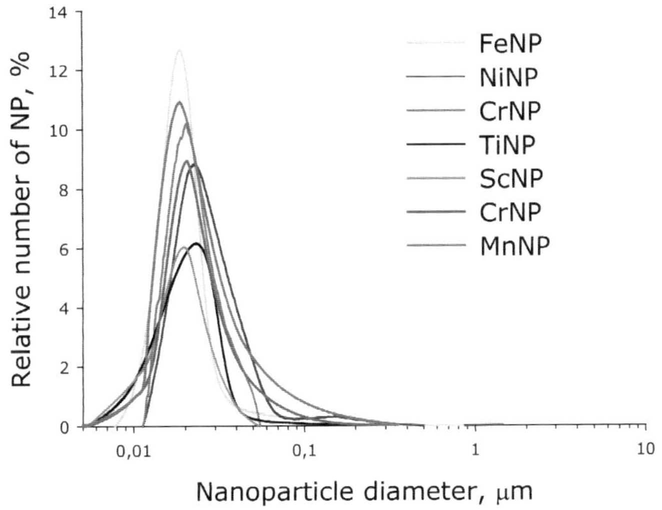

Figure 3. Size distribution of nanoparticles. Nanoparticles of iron (FeNP), chromium (CrNP), scandium (CsNP), manganese (MnNP), nickel (NiNP), titanium (TiNP) and vanadium (VNP).

4. Conclusions

Thus, it has been shown that using laser ablation of metal pieces in liquid media it is possible to generate chromium, iron, manganese, titanium, scandium, nickel and vanadium NP with a clearly defined diameter and characteristics. Laboratory regulations for the generation of metal NP are developed.

5. Acknowledgments

Experiments were supported by Russian Foundation for Basic Research, grant No 19-02-00061-a. The authors are grateful to the Center for Collective Use of the GPI RAS for the equipment provided.

References

[1] Barcikowski S., Compagnini G. 2013 *Phys. Chem. Chem. Phys.* **15** 3022-3026.
[2] Kalus M.R., Lanyumba R., Lorenzo-Parodi N., Jochmann M.A., Kerpen K., et al. 2019 *Phys. Chem. Chem. Phys.* **21** 18636-18651.
[3] Frias Batista L.M., Meader V.K., Romero K., Kunzler K., Kabir F., Bullock A., Tibbetts K.M., 2019 *J. Phys. Chem. B* **123** 7204-7213.
[4] Baymler I.V., Barmina E.V., Simakin A.V., Shafeev G.A., 2018 *Quantum Electron.* **48** 738
[5] Barmina E.V., Simakin A.V., Shafeev G.A., 2017 *Chem. Phys. Lett.* **678** 192-195.

IOP Conf. Series: Materials Science and Engineering **921** (2020) 012024 doi:10.1088/1757-899X/921/1/012024

[6] Simakin A.V., Astashev M.E., Baimler I.V., Uvarov O.V., Voronov V.V. et al., 2019 *J. Phys. Chem. B* **123** 1869-1880.

[7] Vogel A., Venugopalan V., 2003 *Chem. Rev.* **103** 577-644.

[8] Menéndez-Manjón A., Barcikowski S., 2011 *Appl. Surf. Sci.* **257** 4285-4290.

[9] Amendola V., Meneghetti M., 2007 *J. Mater. Chem.* **17** 4705-4710.

[10] Callegari A., Tonti D., Chergui M., 2003 *Nano Lett.* **3** 1565-1568.

[11] Zayed M.F., Eisa W.H., El-kousy S.M., Mleha W.K., Kamal N., 2019 *Spectrochimica Acta A* **214** 496-512.

[12] Ayyyzhy KO, Voronov VV, Gudkov SV, Rakov II, Simakin AV, Shafeev GA. 2019 *Phys. Wave Phenom.* **27** 113–118.

[13] Dolgaev SI, Simakin AV, Voronov VV, Shafeev GA, Bozon-Verduraz F. 2002 *Appl. Surf. Sc.* **186** 546-551

[14] Gudkov S.V., Guryev E.L., Gapeyev A.B., Sharapov M.G., Bunkin M.F., et al., 2019 *Nanomedicine* **15** 37-46.

[15] Sharapov M.G., Novoselov V.I., Penkov N.V., Fesenko E.E., Vedunova M.V. et. al., 2019 *Free Radic. Biol. Med.* **134** 76-86.

IOP Conf. Series: Materials Science and Engineering **921** (2020) 012025 doi:10.1088/1757-899X/921/1/012025

Modeling of the development of the Vuktyl field hydrocarbon system in the depletion mode using a bench installation

Skibitskaya N.A.[1], Volkov A.N.[2], Latishev A.A.[2], Indrupsky I.M.[1], Popov A.A.[2], Bolshakov M.N.[1], Kuzmin V.A.[1], Marutyan O.O.[1]

[1]Oil and Gas Research Institute of the Russian Academy of Sciences (OGRI RAS), Moscow, Russia
[2]OOO «Gazprom VNIIGAZ» Ukhta, Russia

E-mail: rgu2006@mail.ru

Abstract. The work substantiates the location of liquid hydrocarbons (LH – oil and retrograde condensate) in the dynamic, filtering part of the capacitive volumes of gas-saturated productive deposits of gas condensate and oil and gas condensate fields (GCF and OGCF). The necessity of the development and bench modeling of the production of LH from gas-saturated deposits of GCF and OGCF in the late stages of field development is shown due to the high geological reserves of LH in them and the need to develop them with the aim of component recovery increasing and extending the life of the fields. The experience of bench modeling on the experimental installation of the life cycle of the oil and gas condensate system of the Vuktyl OGCF for the period of its development on the "depletion" mode is described. On the basis of a highly permeable core model, a simulation of the initial fluid saturation was successfully held which fits the initial state of the reservoir system of the Vuktyl OGCF, including the LC (reservoir oil), and also the current thermobaric state and fluid saturation of the reservoir system were simulated. The total mass of LH (reservoir oil and retrograde condensate) in the reservoir model is determined which can be taken as a basis for calculating the LH extraction rate while carrying out further research on bench modeling of LH production technologies in the late stages of Vuktil field development on the core model created. As a result of the works described above, an oil and gas condensate system was created in a highly permeable bench core model which is the model of depleted to maximum condensation pressure reservoir of Vuktyl OGCF.

1. Introduction

It is considered that the production of hydrodynamically immobile reservoir liquid hydrocarbons (LH) from gas-saturated productive OGCF deposits is technologically impossible. In this case, a comparison is made with the oil recovery factor (ORF) from productive deposits of oil fields, which differ from oil and gas condensate fields by 100% filling of effective pore volumes with oil.

If in the oil-saturated reservoir the central (structurally pinched for filtration) part of the effective pore volumes is filled with dynamically immobile oil, then in the gas-saturated collector it is filled with structurally-pinched reservoir gas (see Figure 1). At higher gas saturation, the dynamic (continuously connected through channels, filtering) part of the effective pore volume and the central part of the channels connecting the pores begin to fill with gas (see Figure 1) [1].

IOP Conf. Series: Materials Science and Engineering **921** (2020) 012025 doi:10.1088/1757-899X/921/1/012025

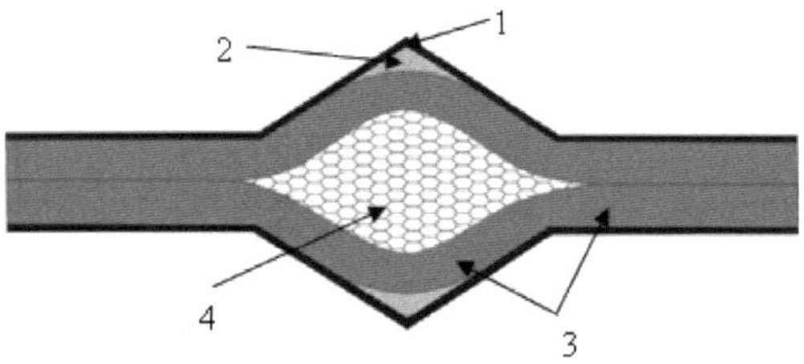

Figure 1. Structural characteristic of reservoir fluids in the capacitive volumes of the gas-saturated part of gas-condensate and oil and gas condensate fields: 1 - film water (the rock is more philic to water) or film oil (the rock is more philic to oil); 2 – pore corners water (capillary pinched, meniscus) or pore corners oil (the rock is more philic to oil); 3 - dinamical (continuously connected) drained oil (the rock is more philic to water) or drained water (the rock is more philic to oil); in the central part of the pore channels - dynamic drained gas; 4 - structurally pinched gas (when flooding during the development process or with a high concentration of liquid fluids – oil, water).

The dynamic part of the gas is filtered in the reservoir during its development. Structurally pinched in the pores gas is gradually discharged into the dynamic part of the pore volumes during the development process (through its expansion while creating depression on the bed) and only then begins to participate in filtering.

With a high content of LH and full filling of dynamic volumes with it, the gas in the central part of the pore volumes turns out to be structurally pinched, isolated

At gas saturation values exceeding the structurally pinched volumes, and high initial reservoir pressure, the free gas energy may be enough to break the continuity of the liquid hydrocarbon phase in the filtration channels, that ensures single-phase gas filtration in the reservoir even at high saturation of dynamic volumes of gas-saturated rocks with liquid hydrocarbons. As the reservoir pressure decreases during the development the energy of the remaining gas is insufficient to break the continuity of the liquid phase and the two-phase filtration process begins (gas - LH) or if free reservoir water is available the three-phase filtration of reservoir fluids (gas – LH - water) has place.

Such a pattern of filtration of reservoir fluids in a gas-saturated reservoir during the development process, as well as an increase in the saturation of the pore volumes with liquid hydrocarbons due to precipitation of retrograde condensate in the process of reducing reservoir pressure, suggests that the most favorable period of production of LH (oil and retrograde condensate) is the final (late) stage of development of oil and gas condensate fields.

The presence in the gas part of depleted oil and gas condensate deposits of Vuktil OGCF (more than 200 million tons), Orenburg OGCF (1.406 billions tons only in pore-type rocks) and other OGCF of significant volumes of reservoir oil and retrograde condensate, which, however, do not reach the threshold of hydrodynamic mobility, makes it urgent to search for possibilities of their directional production [2].

Maximum, close to limiting, accumulation of LH due to precipitation of retrograde condensate in reservoir oil and significantly reduced reservoir pressures in the reservoir at the final stages of development before developing the LH production technologies need the preliminary bench modeling of the whole period of oil-gas-condensate system development on the "depletion" mode.

2. Bench modeling of the life cycle of the Vuktyl field's oil and gas condensate system

Bench modeling according to the developed program was carried out on a bench experimental installation (Figure 2) incuding: reservoir model, hydraulic rock overburden press, pressure transmitters "Metran-150", gas meter, chromatograph to determine the composition of the gas and liquid phases at the outlet of the reservoir model, measuring separator-meter, low-temperature cryostat to maintain a negative temperature in the separator, piston recombinators with working agents, pumps to maintain a constant pressure in the recombinators.

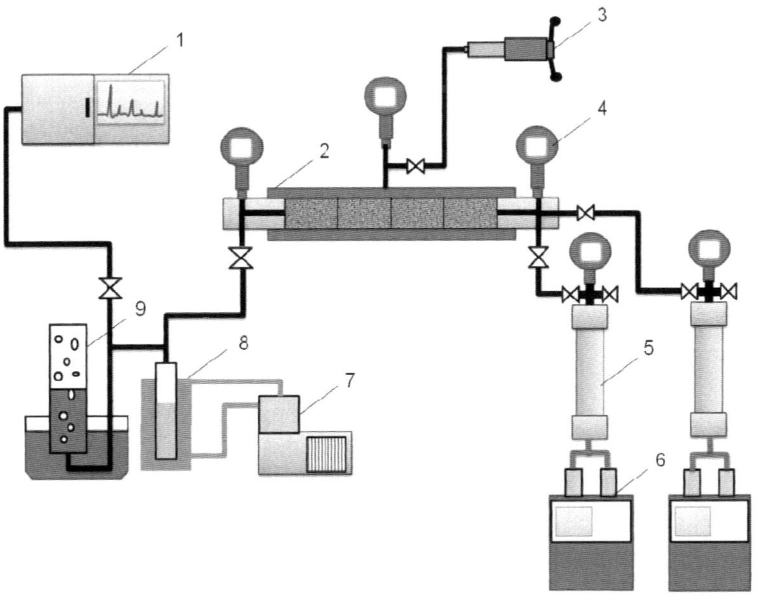

Figure 2. Experimental installation scheme.
1 – chromatograph; 2 – reservoir model; 3 – press for rock oveburden; 4 – pressure transmitter; 5 – separation piston recombinator; 6 – pump; 7 – cryostat; 8 – measuring separator; 9 – gasmeter

Before starting the assembly of the installation and the experimental studies, all measuring and auxiliary equipment was checked and calibrated. Before performing chromatographic analyzes, chromatographs were calibrated using certified calibration mixtures. For the analysis of the composition of the gas phase, a chromatograph of the type "Chromatek-Crystal 5000.1" was used, directly connected to the experimental setup. The composition of the liquid phase was studied also on a chromatograph of the type "Chromatek-Crystal 5000.1". The analysis was carried out on the temperature ramp mode (from 32 to 320 °C) using a high-temperature thin-film capillary WCOT column 30–50 m long and a flame ionization detector. The studies were conducted on a high-permeable core model from rock samples of the Vuktyl OGCF. The high-permeable reservoir model (HRM) is shown on Figure 3 in the form of a scheme created on the basis of electron microscopic cathodoluminescent images (electronic zoom x100) [3].

Bench modeling of the life cycle of the field development includes next key stages:

- simulation of the initial fluid saturation fitting the initial state of the reservoir system of the Vuktyl OGCF, including LH (reservoir oil),
- simulation of the current thermobaric state and fluid saturation of the Vuktyl OGCF reservoir system.

IOP Conf. Series: Materials Science and Engineering **921** (2020) 012025 doi:10.1088/1757-899X/921/1/012025

№№	Pore space structure parameters	8/19B	8/20B	8/15	8/13	8/20A	8/19A
1	Pores and channels quantity	2090	1900	3299	3299	1914	2427
2	Sum perimeter of pores and channels, um	332761	326170	368335	368335	381789	376114
3	Sum square of pores and channels, um^2	4144983	4116041	3580490	3580490	5199678	3440836
4	Av square of pores and channels, um^2	1983,25	2166,34	1085,33	1085,33	2716,66	1417,74
5	Av equivalent diameter of pores and channels, um	50,27	52,54	37,19	37,19	58,83	42,5
6	Av perimeter of pores and channels, um	159,22	171,67	111,66	111,66	199,48	154,98
7	Coordinate number, N_{ch}/N_p	4,49	4,34	4,74	4,74	4,35	4,39
8	Av channel diameter, um	6,47	6,32	4,84	4,84	7,25	6,1
9	Av pore diameter, um	88,27	89,67	69,52	69,52	105,09	72,5
10	Av ratio d_p/d_{ch}	13,6	14,2	14,4	14,4	14,5	11,9
11	Porosity (measured), %	19,4	17,62	18,76	17,42	17,8	16,86
12	Permeability (measured) mD	269,18	226,78	195,94	168,95	168,11	122,0

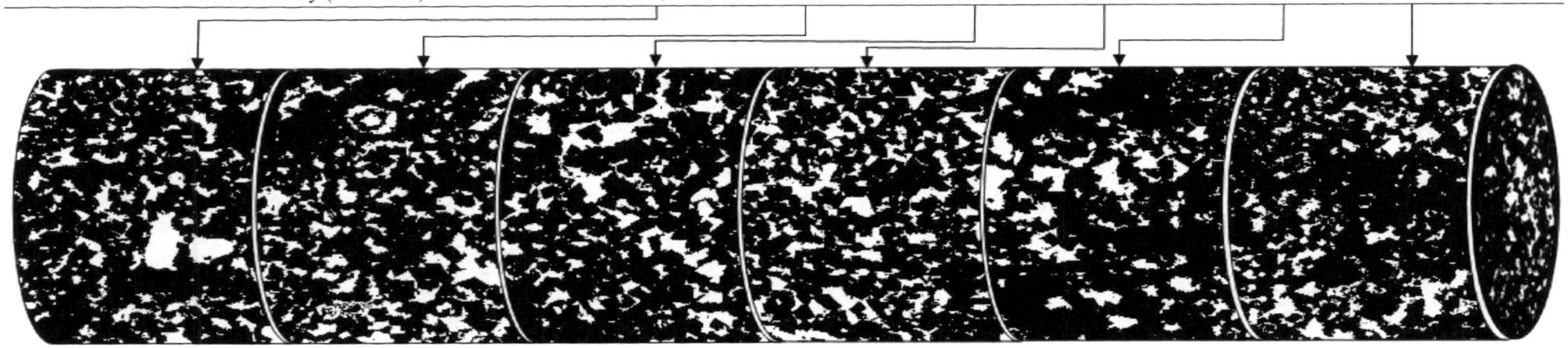

Figure 3. High-permeable bench model - scheme based on electron microscopic cathodoluminescent images, electronic zoom x100, (pores are white colored).

3. Simulation of the initial state of the reservoir system of the Vuktyl OGCF, including reservoir oil

For simulation of the initial state of the reservoir system, including reservoir oil, the next operation consequence is hold:

- The prepared reservoir model, after measuring porosity and permeability, is exhausted and saturated with the reservoir model water (salinity 230 g/l). The model is drained with reservoir water until the air stops outgoing.
- Residual water saturation is created by pumping modeled degassed reservoir oil through a 100% water-saturated reservoir model until the water phase stops outgoing. Saturation is controlled by matter balance.

The process of residual water saturation modeling in an experimental reservoir model is shown in Figure 4.

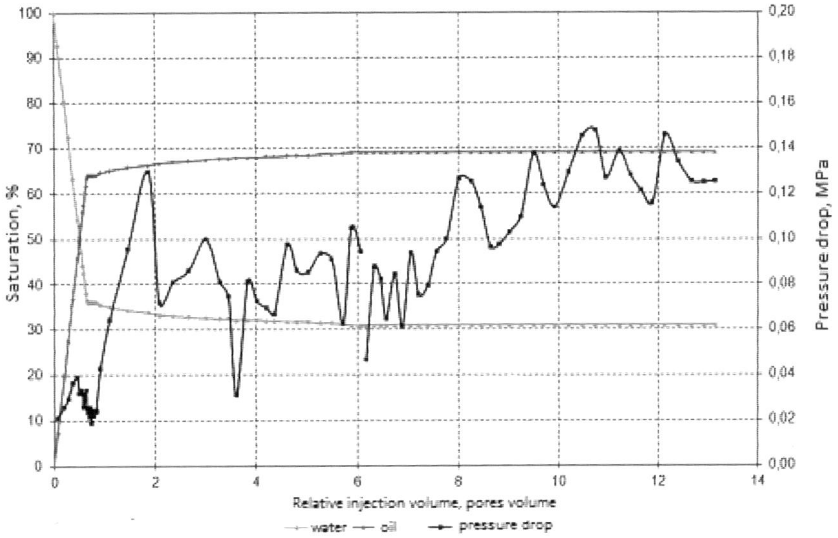

Figure 4. Simulation of residual water saturation in high permeable reservoir model (HRM).

After pumping about 5.8 of pore volumes of oil, the removal of water stopped. The value of residual water saturation was 30.91%.

As a model of degassed oil, a mixture of distillate after 110°C of degassed oil from well 279 of the Severo-Vuktylskoye field with n-pentane and n-octane in such a proportion to fit the initial density of the degassed oil of well 279 (823.4 kg/m^3). The choice of this model of oil is due to two factors:

- estimated proximity of the composition and properties of reservoir oil in the main gas condensate reservoir of the Vuktyl OGCF to the reservoir oil of the Severo-Vuktylskoye field,
- the need to separate the production of LH and the model solvent - hexane (with the planned production of LH) from the reservoir model according to chromatographic analysis.

The accepted way to the preparation of the reservoir oil model made it possible to almost completely eliminate the content of hexane in it while maintaining the proximity of properties to the degassed reservoir oil of the Vuktyl OGCF.

Residual oil saturation is created by pumping methane through the model of the reservoir until the outflow of oil from the model stops under thermobaric conditions close to the average initial pressure (35 MPa) and temperature (63 °C) of the reservoir system of the Vuktyl OGCF. Saturation is

IOP Conf. Series: Materials Science and Engineering **921** (2020) 012025 doi:10.1088/1757-899X/921/1/012025

controlled by matter balance. Modeling of water-, oil- and gas-saturation by methane pumping to the HRM is shown in Figure 5.

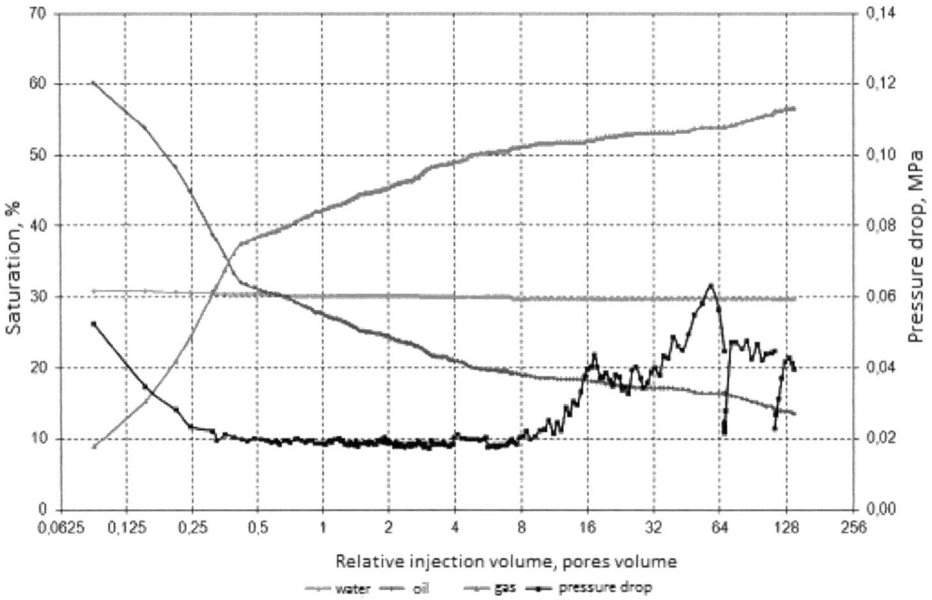

Figure 5. Simulation of water, oil and gas saturation in the process of methane injection to high permeable reservoir model (HRM).

While maintaining pressure and temperature, methane is replaced by a model of a reservoir gas-condensate mixture (GCM) from a recombinator until the composition of the outgoing product is stabilized. The composition is controlled according to the chromatographic analysis of gas and liquid samples.

The model of the reservoir gas-condensate mixture (GCM) is prepared by recombining the gas separator samples of the well 68 from Yuguidsky OGCF and pure hydrocarbon components to achieve the initial hydrocarbon content (HC) of the C_{5+} group at 360 g/m^3, which is typical for average initial conditions of Vuktyl OGCF [4].

In order to be able to correctly interpret the data of the bench modeling, the experiment of differential condensation with the same GCM model is carried out previously in the PVT vessel and the relative volume of the liquid phase is determined at different stages of pressure reduction. The experiment is carried out in two versions - only with the GCM model and with the GCM model in a mixture with oil and water models in proportions corresponding to the initial phase saturations in the reservoir model.

Thus, in the group hydrocarbon composition of the gasoline fraction (IBP - 125 °C) of the liquid phase of HRM production in the process of replacing GCM methane over time, there is a slight increase in the proportion of alkane compounds with a parallel decrease in naphthenic and aromatic. The change in the group hydrocarbon composition of the gasoline fraction (IBP – 125 °C) of the liquid phase of the HRM production in the process of replacing GCM methane is clearly shown in Figure 6.

Changes are also recorded in the content of normal and isoprenoid alkanes (fraction above 125 °C) of the liquid phase of HRM production. At the initial stage of pumping, isoprenoid hydrocarbons are found in the composition of the liquid phase, even in small concentrations. In the future (after pumping 2.7 pore volumes of the GCM), they are no longer registered as part of the studied fraction. The change in the content of normal alkanes in the liquid phase of HRM production in the process of replacing GCM methane is shown in Figure 7. From Figure 7 it can be seen that the composition of normal alkanes changes according to the observed tendency to lightening of liquid production — at the beginning of pumping, the length of the concentration range of normal alkanes is traced down to C_{40}, then it is shortened to C_{19}.

In general, according to gas-liquid chromatography (GLC), in the process of replacing GCM methane, the relief of the liquid phase of products was noted, associated with expeling the residual oil and increasing the amount of gas condensate hydrocarbons in the liquid phase of HRM production.

During observing the process of replacing GCM methane, it was found by changing of the condensate-gas factor (CGF) and the content of components in the production gas and liquid phases at the outlet of the reservoir model that their stabilization started when approximately seven pore volumes were injected. The pumping up of one more pore volume ensured the constancy of the composition of the mixture within the error of the chromatographic analysis. Then the reservoir model was held for two days to establish sorption equilibrium and uniform distribution of the GCM in the entire pore volume system. At this stage, the preparation of the initial reservoir oil and gas condensate system in the HRM was completed. The residual water saturarion was 29,7 %. The value of residual oil saturation at the same time decreased to 5.88%.

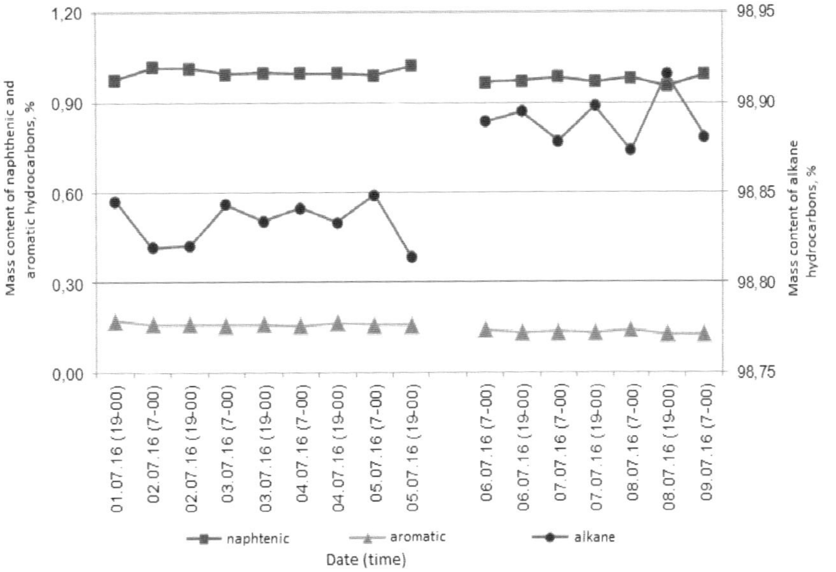

Figure 6. Changes in the group hydrocarbon composition of the gasoline fraction (IBP – 125 °C) of the production liquid phase of HRM in the process of replacing GCM methane.

IOP Conf. Series: Materials Science and Engineering **921** (2020) 012025 doi:10.1088/1757-899X/921/1/012025

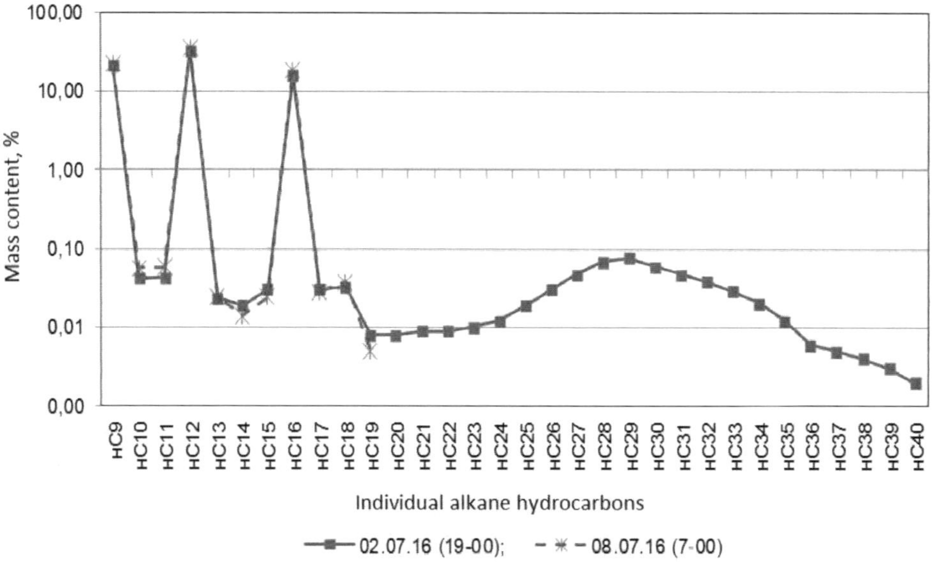

Figure 7. The change in the mass content of normal alkane hydrocarbons in the liquid phase of the production of HRM in the process of replacing GCM methane.

4. Simulation of the current state (depleated) of the Vuktyl OGCF reservoir system

At present, the thermobaric conditions and the composition of the reservoir fluids vary greatly inside the main reservoir of the Vuktyl OGCF. Nevertheless, it can be affirmed that in the most volume of the deposits the reservoir pressure has already reached the value below the pressure of maximum condensation of the reservoir GCM. As for the injection of dry ("Tyumen") gas, which has been implemented since 1992, in general, in relative terms, it did not make a fundamental contribution to the change in the current C_{5+} content in the liquid and gas phases [4]. In particular, this applies to the selected pilot sites, where the condition of maintaining of sufficient potential for HC group C_{5+} was accepted.

Based on these assumptions, the modeling of the conditional current state of the reservoir system was carried out as follows.

- While maintaining the set temperature of 63 °C, the model was depleted to the pressure of maximum condensation of the reservoir GCM of Vuktyl OGCF - 5 MPa. According to the results of these studies, calculations of the content of hydrocarbon components in the production of the reservoir model were performed.
- The depletion rate was about 0.1 - 0.2 MPa/h to ensure the conditions of condensation of the liquid phase in the reservoir model and prevent carrying out its drops with the gas phase.

Figure 8 shows the dynamics of pressure change over time in the process of reservoir depletion modeling on the reservoir model. The duration of the process of the reservoir model depletion was 290 hours. Experimental studies were carried out at reservoir conditions in around-the-clock mode.

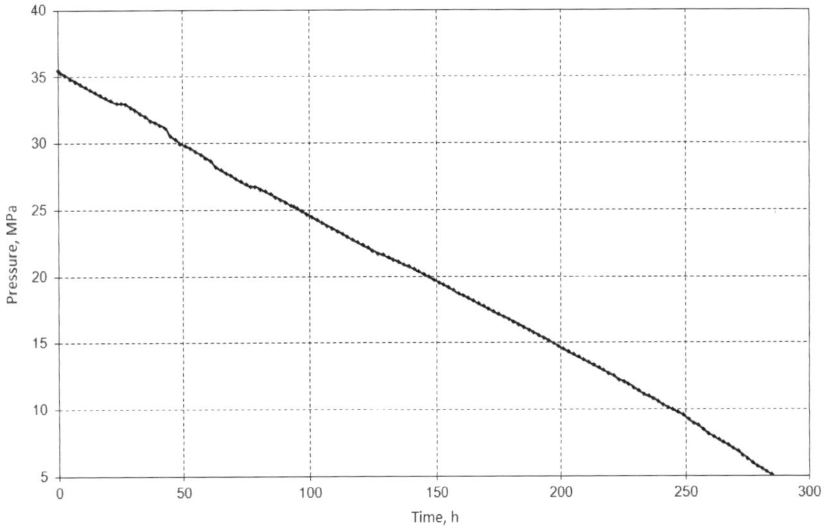

Figure 8. Dynamics of pressure change over time.

During the experimental studies, the volume of production, the mass of the liquid phase (after the separator box), the composition of the phases by chromatographic analysis of samples were monitored. The final saturation of the reservoir model with water, LHC and gas were estimated.

Chromatographic studies of separation gases obtained by simulating the depletion of an oil and gas condensate reservoir on HRM showed that trends in the content of all components of the studied gas phase are similar to the character of changes in the composition of the produced gas reservoir observed in the actual development of Vuktyl OGCF during natural depletion.

Chromatographic studies of the liquid phase obtained in the simulation of the depletion process of the Vuktyl oil and gas condensate reservoir on HRM showed that as the pressure decreases, the resulting liquid products become more light (the molar mass of the condensate decreases, the mole fraction of the heaviest components of the condensate significantly decreases, starting from the sum of the pseudo C_{11} and up to the sum of the pseudo C_{20+}). The revealed dependences are typical for the liquid hydrocarbons, produced in the process of developing a gas-condensate field in the mode of natural depletion of reservoir energy. Experimental studies were carried out on the pVT facility to determine the LHC reserves (condensate + oil) in the reservoir model after reducing the pressure to 5 MPa. In the pVT-cell, the system was recombined (GCM + water + oil) in the ratios obtained in the reservoir model after pumping the GCM.

At the end of the differential condensation test, the degassing of the saturated liquid phase was performed. The mass of LHC (reservoir oil and retrograde condensate) according to the reservoir model was 4.8348 g. This value can be taken as a basis for calculating the LHC extraction coefficient when conducting further research on the development of technologies for extracting hydrodynamically immobile LHC (reservoir oil and retrograde condensate) in the later stages of the development of the Vuktyl field. It should be noted that the additional extraction of LHC entails an increase in the final product content of high-boiling paraffins, resins and asphaltenes, reducing its quality, as well as a decrease in the content of C_3-C_4 components, which must be investigated using additional, for example, optical methods [5-7].

5. Conclusion

Thus, according to the results of depletion in the gas-saturated reservoir model:

- the saturation of LHC was formed, including the original reservoir oil and retrograde condensate from the GCM,

- the saturation of the liquid hydrocarbons with gas and intermediate components to an equilibrium state at the current pressure and temperature took place.

As a result of the work described above, a system was created in the HRM that simulates the reservoir depleted to maximum oil-gas-condensate reservoir condensation pressure. The developed methodical approach can be used to imitate realistic reservoir conditions for bench modeling to estimate the possibility of extracting of hydrodynamically immobile LHC (reservoir oil and retrograde condensate) from the gas-saturated part of the fields at the final stages of depletion.

Acknowledgements
The results presented in the paper were obtained under the OGRI RAS state research contract, № AAAA-A19-119030690047-6 "Forecast of the resource base state of the Russia oil and gas complex based on system studies of the prospects for the oil and gas potential of natural reservoirs in carbonate, terrigenous and shale formations".

References

[1] Skibitskaya N A, Bolshakov M N, Burkhanova I O, Kuzmin V A, Surnachev D V Tight Oil in Oil-And-Gas Source Carbonate Deposits' Gas Saturation Zones of Gas-Condensate and Oil-Gas Condensate Fields **2016** *SPE Russian Petroleum Technology Conference and Exhibition, 24-26 October, Moscow, Russia.* SPE-182076-RU

[2] Gafarov N A, Karnaukhov S M, Rezunenko V I, Skibitskaya N A and others Matrix Oil of the Orenburg oil and gas condensate field - prospects for the development of the Orenburg gas chemical complex *Gazovaya promyshlennost* **2012** №9(680) pp 56-60

[3] Kuzmin V A Cathodoluminescence Investigation of Pore Space of Rocksin Scanning Electron Microscope *Poverkhnost. Rentgenovskiye, sinkhrotronnyye i neytronnyye issledovaniya* **2007** №11. pp 76-80

[4] Dolgushin N V Methodology for studying the gas condensate characteristics of oil and gas condensate fields with a high condensate saturation and a large gas column *Dissertation*, Ukhta SEVERNIPIGAZ. **2007**, 400 p

[5] Kuryakov V N, De Sanctis Lucentini P G, Ivanova D D Tricosane ($C_{23}H_{48}$) and octacosane ($C_{28}H_{58}$) mixture phase transition insight via light scattering techniques *IOP Conference Series: Materials Science and Engineering "3rd International Youth Scientific Forum with International Participation "New Materials""* **2018** p 012034

[6] Belyakov M Y, Voronov V P, Gorodetskii E E, Grigorev B A, Deshabo V A, Kosov V I, Kulikov V D, Kuryakov V N, Yudin I K, Yudin D I Thermodynamics of multicomponent mixtures near the liquid-vapor critical point *High Temperature* **2012** T 50. № 4 pp 479-487

[7] Belyakov M Y, Gorodetskii E E, Kulikov V D, Kuryakov V N, Yudin I K Light-scattering anomaly in the vicinity of liquid-vapor critical point of multicomponent mixtures *Chemical Physics* **2011** T 379. № 1-3 pp 123-127

IOP Conf. Series: Materials Science and Engineering **921** (2020) 012026 doi:10.1088/1757-899X/921/1/012026

Impact of environmental factors on agrofibers based on "green" polymers

Yu V Tertyshnaya[1,2], A A Olkhov[1], N S Levina[2], I A Bidey[2], M N Moskovskiy[2]

[1] Emanuel Institute of Biochemical Physics, Russian Academy of Sciences, 4 Kosygina str., Moscow, 119991, Russia

[2] Federal Research Agro-Engineering Center VIM, 1st Institutskiy proezd, 5, Moscow, 109428, Russia

E-mail: moraxella@bk.ru

Abstract. Influence of different environmental factors on the degradation of fibrous samples based on polylactic acid and poly-3-hydroxybuturate was studied in this work. The article presents data obtained by differential scanning calorimetry (DSC), scanning electron microscopy (SEM). The hydrolytic degradation of nonwoven fibers, degradation in soil and under UV light were investigated. It was found that the degree of degradation of the material obtained from the 5% PHB solution was higher than from 7 and 9 %. It was shown that after 60 days of soil exposure, the samples of nonwoven fibers PLA and PLA-PHB collapsed into fragments and melting temperature of both samples was reduced by 11-12 °C. The degree of crystallinity also significantly decreased.

Introduction

World consumption of polymers has increased exponentially since the nineteen fifties. According to numerous forecasts, the crude oil resources will become appreciably exhausted after 2050; therefore, the use of alternative materials will become inevitable. Already now the development and production of biodegradable "green" polymers of vegetable origin is a topical problem for many developed countries. Interest in natural polymers such as poly-3-hydroxybutyrate (PHB), polylactide (PLA), starch, chitin, cellulose, etc., increased in the past decade [1-3]. Items made of these polymers are safe for the environment and humans. Under the influence of ultraviolet, heat, air and soil microorganisms they degrade to low molecular compounds and easily assimilate in soil, being included in closed ecological circuit [4].

PLA is a biodegradable thermoplastic polyester derived from biomass such as sugar, corn, beet, which possesses excellent physical and mechanical properties combined with biocompatibility and biodegradability properties [5]. Due to its initial production costs, the starting applications of PLA have been focused on high value products such as medical devices [6]. Because of the high market demand of biodegradable based products in recent years, PLA based products are manufactured commercially in many countries. Their applications include film, food packaging, textiles, disposal bottle and tableware [7].

Another promising, biodegradable polymer, which will be the subject of this work, is poly-3-hydroxybutyrate. Poly-3-hydroxybutyrate (PHB), for example, is synthesized inside cells and is degraded to CO_2 and water when disposed in biological systems [8]. The main advantage of PHBs

IOP Conf. Series: Materials Science and Engineering **921** (2020) 012026 doi:10.1088/1757-899X/921/1/012026

over other types of biodegradable polymers is its biodegradability under both aerobic and anaerobic conditions, such as buried in soil, immersed in activated sludge, and submerged in sea water.

In recent years, studies of PLA combined with PHB have shown that the mechanical properties of the blend are intermediate between those of the individual components. Specifically, PLA-PHB blends exhibit greater flexibility and hydrolytic biodegradation than PLA or PHB alone [9].

Polymer-based nanoscale composites can be produced by different fabrication methods. In this work the fibrous materials of PLA-PHB were prepared by electrospinning. Electrospinning is one of the most prominently used techniques for producing micro- to nanometer fibers because it is a simple, cheap and versatile process.

Experimental

Preparation of Fibrous Material Poly(3-hydroxybutyrate) (Biomer, Germany) was used in the form of powder obtained by microbiological synthesis with a molecular mass of 2.5×10^5 g/mol, a melting point (T_m) of 175–180°C, and a density of 1.25 g/cm^3. To prepare the spinning solutions, chloroform of reagent grade was used. Polylactide (Nature works 4032D, USA) with a molecular mass of 1.9×10^5 g/mol and a melting point (T_m) of 163–165°C, and a density of 1.24 g/cm^3 was used. The fibers were obtained by electrospinning.

DSC The degree of crystallinity and melting temperature of the samples were studied with differential scanning calorimeter DSC 214 Polyma (Netzsch, Germany) at a heating rate of 10 K/min and a sample weight of 5±0.1 mg. The temperature scale was calibrated against an indium standard (T_m = 156.6°C, ΔH = 28.44 J/g).

Electron Microscopy The average diameter of fibers was determined by electron microscopy using a Hitachi TM-1000 scanning electron microscope (Japan). A fiber sample was applied to a receiving cylinder with an electrically conductive scotch tape directly during fiber formation, and micrographs were taken at different degrees of magnification and solution flow rates. The fiber width was measured using the Photoshop.CS3.Extended program. At least 300 fibers were to be measured for constructing the size distribution and at least 100 fibers for determining the average diameter.

UV Irradiation To study the UV aging, fiber samples with a mass of ~50 mg were selected and placed in a chamber with a DRT-375 high-pressure mercury lamp, where they were kept to a state of brittle fracture. At certain time intervals (20–30 min), the samples were taken out and weighed on a Vibra AF-R220CE balance with an accuracy of 0.0001 g, and the mass loss was determined.

Hydrolytic Degradation The hydrolytic destruction of fibrous materials was studied in a phosphate buffer, which simulates the effect of a physiological medium. To measure the hydrolytic degradation of PHB fibrous materials, they were incubated in a 0.025 M phosphate buffer solution (pH 7.4) at 70°C for 21 days. The nonwoven fibrous materials were then taken out of the phosphate buffer at regular intervals, ashed with distilled water, placed in an incubator at 70°C for 3 h, and then weighed on a scale (the measurement error was 0.1 mg). In the experiments with the phosphate buffer, the buffer was replaced every 3 days.

Degradation in soil The microbial activity of soil was monitored with cotton along the extension of the experiment. The soil was maintained at approximately pH 7 and a relative humidity of 0.87 g water/g wet soil. Three specimens of each sample were extracted after certain times of burial in soil, cleaned and kept in a desiccator during 4 days in order to ensure water desorption before being analyzed.

Results and discussion

According to table 1, the crystal phase of PHB in the samples melts in a very narrow temperature range: 176–177°C. This means that the modification of PHB crystallites in the initial powder and fibers does not change. The initial temperature of decomposition (T_{init}) and the temperature of the maximum of the decomposition peak (T_{max}) in the fibers are much lower than in original PHB (286 °C); the difference is 11–14°C at the initial stage of thermal decomposition and 13– 18°C when the maximum is reached on the thermogram of destruction.

Table 1. Thermophysical parameters of PHB samples.

PHB content, wt %	T_m, °C	χ_c, %	T_{init}, °C	T_{max}, °C
5	176.5	70	244	273
7	177	73	239	270
9	176	72	236	268

Let us first consider the effect of UV radiation. Figure 1 shows the kinetic dependences of mass loss by the PHB nonwoven fibrous materials on the time of exposure to UV radiation. According to figure1, the intensity of the photooxidative destruction of the fibrous materials increases when the PHB concentration in the spinning solutions decreases. This occurs because fibers with a larger diameter are obtained at increased polymer concentrations in the spinning solution. As shown in [10], as the concentration of PHB increases during fiber formation, it undergoes more complete crystallization, and the amorphous region of the polymer phase becomes more compact, which, in turn, leads to increased fiber resistance to UV radiation.

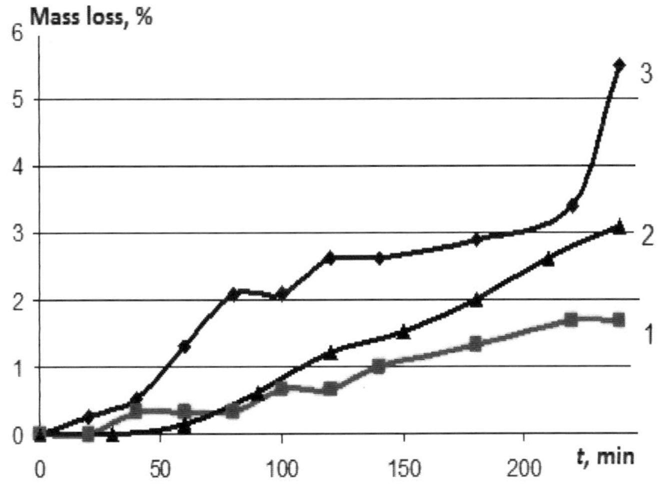

Figure 1. Dependences of mass loss of the nonwoven fibrous samples obtained from the (*1*) 9, (*2*) 7, and (*3*) 5 wt % PHB spinning solutions on the UV irradiation time.

Thus, the fiber formation from more concentrated PHB solutions increases their resistance to photooxidative destruction. According to figure 1, in the fibers obtained from the 7 and 9% PHB solutions, even an induction period appears, which increases more than twofold when passing from 7 to 9% solution.

Similar results were obtained in our study of the hydrolytic effect on the nonwoven fibrous materials of the phosphate buffer solution as a model of the physiological environment of a living organism. Figure 2 presents the microphotographs of fibrous materials after exposure for 14 and 21 days. According to figure 2, after 14 days of buffer action, the materials retained their original morphology regardless of the polymer concentration in the spinning solution; after 21 days, the samples were actively hydrolyzed, and the fibers became brittle and broke into fragments. The degree of degradation of the material obtained from the 5% PHB solution was higher; as the sample had thinner fibers, less time was needed for interaction with the buffer and degradation.

IOP Conf. Series: Materials Science and Engineering **921** (2020) 012026 doi:10.1088/1757-899X/921/1/012026

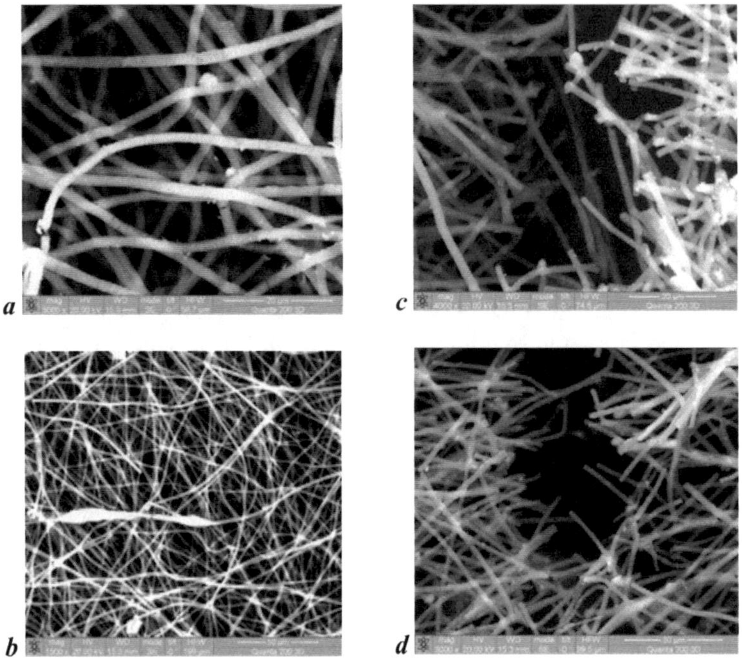

Figure 2. Micrographs of the samples fibrous material from the 5% solution (*a*), fibrous material from the 7% solution (*b*) and of the same samples after hydrolysis in a phosphate buffer solution: after 14 days of storage (*c*) and after 21 days of storage (*d*).

In the second part of the work we present the results of studies of nonwoven fibers PLA and PLA-PHB with 10 wt % of PHB content. Fibrous material can be used as a covering, for example, for germination of breeding seeds. During the application of agrofiber inevitably gets into the soil, so the study of biodegradation in the soil of these materials is necessary. Biodegradation of polymers usually takes place in two main steps: primary degradation, in which fragmentation of the polymer chain occurs due to hydrolysis or another oxidative reaction, and ultimate biodegradation, in which the microorganisms assimilate the low M_w chains formed. Both the PLA fibrous material and the PLA-PHB sample should be well degraded in the soil due to the combined action of moisture (hydrolysis) and microorganisms. Active biodestructors are fungi, which quickly affect polymeric materials based on natural polyesters [11]. The most aggressive micromycetes in relation to PLA and PHB are *Aspergillus Niger*, *Penicillium Chrysogenum* and *Trichoderma Viride*.

Figure 3. Photographs of the PLA-PHB sample after 60 days degradation in soil.

In our experiment, after 60 days of soil exposure, the samples of nonwoven fibers PLA and PLA-PHB lost about 30 wt.% and collapsed into fragments (figure 3). In figure 3 one can see the small fragments, pores and dirt, which is not removed and is the result of the action of microorganisms.

Thermophysical characteristics of the initial samples of non-woven material and after using them as agromaterials were determined by the method of DSC. The data are presented in the table 2.

Table 2. Thermophysical characteristics of the initial samples of non-woven fibers before and after the experiment.

Sample	T_m, °C		χ_c, %	
	initial	after degradation	initial	after degradation
PLA	164	152	40	25
PLA-PHB	163/171	150/–	39/56	23/–

It is noticeable that during the experiment the samples of nonwoven fibers undergo changes in the structure. The melting temperature of both samples is reduced by 11-12 °C, there is a significant decrease in the degree of crystallinity. Moreover, for the PLA-PHB sample, the change in the PLA χ_c, value is 16%, and the melting peak of PHB is generally absent, which indicates the fracture of the crystal structure and the processes of degradation.

Conclusion

Biopolymer nanofibers of PHB, PLA and PLA-PHB were produced by electrospinning considering different solution types. Based on the experimental data with PHB fibers, we can conclude that the polymer concentration in the spinning solution is one of the main factors responsible for the formation of the molecular and supramolecular structure of nonwoven fibrous materials by electrospinning.

Fibers with a content of 5 wt % PHB in solution are subjected to hydrolytic destruction in the buffer and UV radiation more quickly. The study of nonwoven agrofibers (PLA and PLA-PHB) showed that biodegradation in the soil is quite intense, which is confirmed by the DSC data.

It follows from the foregoing that fibers with varied structural organization and resistance to aggressive environmental factors can be obtained depending on the requirements to the material.

References

[1] Podzorova M V, Popov A A and Tertyshnaya Yu V 2018 *IOP Conf. Ser. Mater. Sci. Eng.* **347**, 12015
[2] Arrieta M P, López J, López D, Kenny J M and Peponi L 2016 *Ind. Crops. Prod.* **93**, 290-301.
[3] Tertyshnaya Yu V, Khvatov A V and Lobanov A V 2017 *Rus. J. Phys. Chem B.* **11**, 828–832.
[4] Carrasco F, Pagès P, Gámez-Pérez J, Santana O O and Maspoch M L 2010 *Polym. Degrad. Stab.* **95**, 116-125
[5] Vink E T H, Rábago K R, Glassner D A and Gruber P R 2003 *Polym. Degrad. Stab.* **80**, 403–419
[6] Ortiz R, Moreno – Flores S, Quintana I, Vivanco MdM, Sarasua J R and Toca – Herrera J L 2014 *Mater. Science and Engin.: C.* **37**, 241–250.
[7] Henton D, Gruber P, Lunt J and Randall J 2005 *Natural fibres, biopolymer, and biocomposites.* **16,** 527–78
[8] Tertyshnaya Yu V and Podzorova M V 2018 *Rus. J. Appl. Chem.* **91**, 417–423.
[9] Zhang L, Luo J, Menkhaus T J, Varadaraju H, Sun Y and Fong H 2011 *J. Membr. Sci.* **369**, 499–505
[10] Olkhov A A, Tertyshnaya Yu V, Chizhov A S, Karpova S G and Iordanskii A L 2018 *Rus. J. Phys. Chem. B.* **12**, 293–299
[11] Tertyshnaya Y V, Khvatov A A, Popov A A and Likhachev A N 2011 *International Polym. Sci. Tech.* **38**, 21-23

IOP Conf. Series: Materials Science and Engineering **921** (2020) 012027 doi:10.1088/1757-899X/921/1/012027

Chemical stability of modern jet fuels

L A Vlasenkova, K V Shatalov, N M Likhterova

FAE "The 25th State Research Institute of Chemmotology of Ministry of defense of Russian Federation", Moscow, 10 Molodogvardeyskaya street, Russia

vlasenkova.l@yandex.ru

Abstract. The article shows the need to study the chemical stability of jet engine fuels containing components of deep oil refining and offers a method for studying the chemical stability of jet engine fuels. Based on this method, research was conducted on jet engine fuels of various production technologies and the influence of the hydrocarbon composition on the chemical stability of a mixture of individual hydrocarbons was studied. The inhibitory effect of aromatic hydrocarbons in a mixture with different content of naphthenic hydrocarbons has been studied. Using artificial aging of jet engine fuels, a low level of chemical stability of jet engine fuels based on hydrocracking of vacuum oil strips is shown.

1. Introduction

The current task of modern refining is to improve the quality of petroleum products by developing destructive processes for refining operations heavy oils and oil residues [1]. One such process is the hydrocracking of vacuum oil, which has recently been actively used by manufacturers to produce jet fuels.

Among the technologies for obtaining jet fuels (crude distillation and hydrotreated kerosene fractions), hydrocracking is the only process during which there is a radical transformation of the group hydrocarbon composition. In addition to the reactions of removal of heteroatomic compounds and hydration of olefin hydrocarbons (characteristic reactions for the hydrotreatment technology) hydrocracking process involves hydration of aromatic hydrocarbons, cracking high-molecular and naphten compounds, dealkylation of cyclical structures and isomerization of paraffin hydrocarbons. In this way, isoparaphin-naphthene base with a low content of aromatic hydrocarbons is formed (5%) [2].

It is known that aromatic hydrocarbons inhibit the oxidation of naphthenes [3,4]. Accordingly, such a change in hydrocarbon composition would lead to increase the rate of fuel oxidation, reduce its chemical stability and storage life.

At the moment, the assessment of chemical stability in the scope of regulatory documentation requirements for the product is limited only to indirect physical and chemical indicators - the evaporation gum test, iodine test, acidity - which are point characteristics that do not allow us to assess the dynamics of changes in the chemical stability of jet fuels.

2. Chemical stability of jet fuels obtained by various production technologies

In FAE "The 25th State Research Institute of Chemmotology of Ministry of defense of Russian Federation" a method for assessing chemical stability was developed. The essence of method is to heating (100°C) fuel (75 cm^3) in a closed space at elevated oxygen pressure (400 kPa) in the presence of 0.125 g initiator of oxidation (dicumyl peroxide) with continuous recording of temperatures, pressures

IOP Conf. Series: Materials Science and Engineering **921** (2020) 012027 doi:10.1088/1757-899X/921/1/012027

and estimates (mass of oxygen absorbed Δm_{o_2}, maximum oxygen absorption rate v_{max}) [5]. The organization standard was developed for the method and its metrological certification was performed in accordance with GOST 8.563-2009 and GOST R ISO 5725-2002 [6].

The study of commercial fuels for jet engines of various production technologies using the developed method (table 1) made it possible to establish that fuels for jet engines based on kerosene fractions of hydrocracking have estimates of chemical stability (Δm_{o_2} =612.5…618.4 μg, v_{max}= 0.599…0.613 μg/min) is 2-3 times worse than jet fuels obtained by hydrotreating kerosene fractions (Δm_{o_2} =173.8…386.5 μg, v_{max}=0.186..0.354 μg/min), and 6 times worse than fuels based on kerosene fractions of direct distillation oil (Δm_{o_2}= 35.0…52.1 μg, v_{max}=0.058…0.066 μg/min).

Table 1. Chemical stability of jet fuels from various refineries

№	Sample, fuel processing technologies	Chemical stability	
		Δm_{o_2} (μg)	ϑ_{max} (μg/min)
1	TS-1 jet fuel, straight-run KF (kerosene fraction)	35.0	0.066
2	TS-1 jet fuel, straight-run KF	39.6	0.058
3	TS-1 jet fuel, straight-run KF	52.1	0.059
4	TS-1 jet fuel, demercaptanizated KF, antioxidant additive, anti-wear additive	24.0	0.058
5	TS-1 jet fuel, demercaptanizated KF, straight-run KF, antioxidant additive, anti-wear additive	42.2	0.058
6	TS-1 jet fuel, straight-run KF, hydrocracked KF, antioxidant additive, anti-wear additive	30.1	0.058
7	TS-1 jet fuel, straight-run KF, hydrotreated KF, antioxidant additive, anti-wear additive	121.5	0.109
8	RT jet fuel, hydrotreated KF, antioxidant additive, anti-wear additive	173.8	0.161
9	RT jet fuel, hydrotreated KF, antioxidant additive, anti-wear additive	361.5	0.203
10	RT jet fuel, hydrotreated KF, antioxidant additive, anti-wear additive	386.5	0.354
11	RT jet fuel, hydrotreated KF, antioxidant additive, anti-wear additive J	357.8	0.186
12	RT jet fuel, hydrotreated KF, antioxidant additive, anti-wear additive	287,1	0.301
13	TS-1 jet fuel, KF hydrocracked, antioxidant additive, anti-wear additive	612.5	0,613
14	RT jet fuel, KF hydrocracked, antioxidant additive, anti-wear additive	695.6	0,713
15	RT jet fuel, KF hydrocracked, KF hydrotreated, antioxidant additive, anti-wear additive	621.7	0.587
16	TS-1 jet fuel, KF hydrocracked, antioxidant additive, anti-wear additive	618.4	0.599

The high chemical stability of jet engine fuels based on straight-run kerosene fraction is explained by the increased content of heteroatomic compounds in their composition, which are natural inhibitors of oxidation [7]. Hydrotreated jet engine fuels, which have content of natural oxidation inhibitors, are stabilized using an antioxidant additive [8].

The low level of chemical stability of fuels for jet engines obtained by hydrocracking of vacuum oil strips is explained by a change in the hydrocarbon medium of fuels of a similar composition compared to "traditional" fuels of crude distillation of oil and hydrotreating of kerosene fractions.

Chromato-mass spectrometric analysis of fuel samples from various production technologies (table 2) showed that straight-run and hydrotreated fuels contain ≈30% of normal alkanes, ≈20% of isoalkanes and ≈17% of naphthenes protected from oxidation by a high content of aromatic hydrocarbons (≈20%). Hydrocracked fuels are characterized by a low content of normal alkanes (≈15%), a high content of isoalkanes that are more prone to oxidation (≈33%), and a high content of naphthenes (≈46%) with a low content of monoaromatic hydrocarbons (≈4%) and polycyclic aromatic hydrocarbons.

Table 2. Hydrocarbon composition of jet engine fuels

Groups of hydrocarbons (% wt.)	Fuel processing technologies		
	Straight-run jet fuel	Hydrotreatment technology	Hydrocracking of vacuum oil strips
Alkanes			
- *linear alkane*	33.47	29.88	15.04
- *nonlinear*	22.38	20.51	33.11
Naphthenes			
- *monocyclic*	17.00	15.65	35.92
- *polycyclic*	2.70	4.42	10.13
Aromatic hydrocarbons			
- *monocyclic*	16.19	21.87	3.99
- *polycyclic*	0.70	1.51	0
Unsaturated hydrocarbons	0.69	0.01	0.95

3. Chemical stability of mixtures of individual hydrocarbons

Using the developed method, mathematical models (1) and (2) were constructed, reflecting the relationship between chemical stability and individual hydrocarbon composition:

$$\Delta m_{o_2} = 459.6 + 176.4x_1 - 56.9x_2 - 22.0x_3 - 87.7x_1x_2 + 19.7x_1x_3 - 35.2x_2x_3 - 44.0x_1x_2x_3 \quad (1)$$

$$v_{max} = 0.291 + 0.116x_1 - 0.028x_2 - 0.003x_3 + 0.028x_1x_2 + 0.003x_1x_3 - 0.015x_2x_3 + 0.016x_1x_2x_3 \quad (2)$$

where x_1 is the content of naphthenes (from 30 to 70 % wt.), x_2 is the content of bicyclic hydrocarbons (from 1 to 4% wt.), x_3 is the content of monoaromatic hydrocarbons (from 5 to 15% wt.).

These mathematical models allowed us to study the inhibitory effect of aromatic hydrocarbons (figure 1, 2) and establish that, the chemical stability estimates are at a low level, when the content of naphthenic hydrocarbons in the mixture is more than 50%, regardless of the content of aromatic hydrocarbons, ($\Delta m_{o_2} > 400$ µg, $v_{max} > 500$ µg/min).

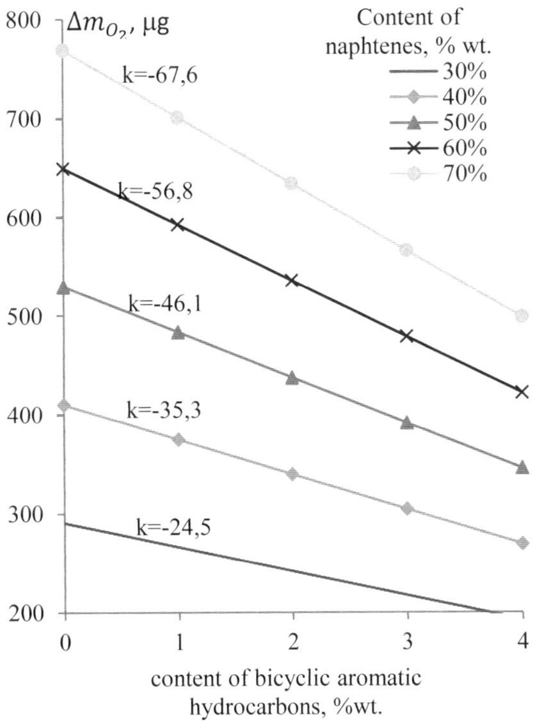

Figure 1. Influence of bicyclic aromatic hydrocarbons on the chemical stability of a mixture with 15% monoaromatic hydrocarbons

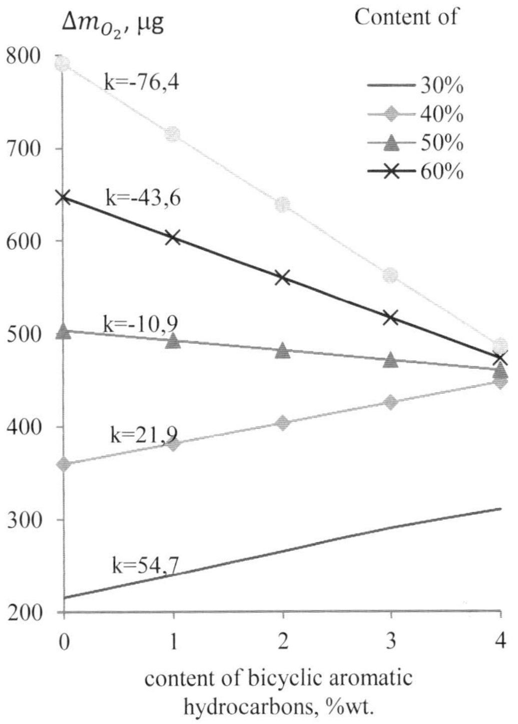

Figure 2. Influence of bicyclic aromatic hydrocarbons on the chemical stability of a mixture with 5% monoaromatic hydrocarbons

The inhibitory effect of bicyclic aromatic hydrocarbons is clearly manifested when the mixture contains 15% monoaromatic hydrocarbons. When the mixture contains 5% monoaromatic hydrocarbons, the inhibitory effect of polyaromatics appears only when the content of naphthenes is more than 50%.

4. Accelerated laboratory aging of jet fuels

The chemical stability of jet fuels has also been investigated under accelerated laboratory aging. Artificial storage in the laboratory was carried out for 30 days, keeping samples with fuel in a thermostat at a temperature of 100 °C (figure 3) and controlling the change in indicators "acidity" and "evaporation gum test" (table 3).

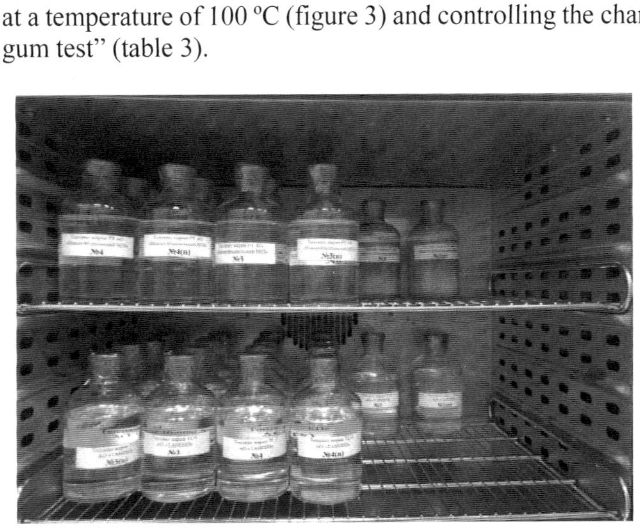

Figure 3. Accelerated laboratory aging of jet fuels

The rate of oxidation products accumulation for fuels based on kerosene fractions obtained by hydrocracking is 45 times higher in terms of "acidity" and 17 times higher in terms of "evaporation gum test" than the rate of oxidation products accumulation in fuels obtained by crude distillation and hydrotreating of kerosene fractions (table 3).

Table 3. Results of accelerated laboratory aging of jet fuels

Sample	Chemical stability		Acidity (GOST 5985)		Evaporation gum test (GOST 8485)	
	Δm_{O_2}, μg	ϑ_{max}, μg/min	Rate of oxidation products accumulation, μgKOH/ 100 sm^3	The estimate output beyond the limits set by GOST 10227, number of days	Rate of oxidation products accumulation, μg/100 см^3	The estimate exceeds the limits set by GOST 10227, number of days
- 100 % straight-run jet fuel	36.6	0.058	0.05	30	0.17	30
- 100 % hydrotreated kerosene fraction, antioxidant additive, anti-wear additive	193.9	0.179	0.03	21	0.12	>30
- 100 % hydrotreated kerosene fraction, antioxidant additive, anti-wear additive	675.1	0.637	1.37	6	2.05	3

The output of estimated fuel indicators based on kerosene hydrocracking fractions beyond the limits specified in GOST 10227 is 5 times faster in relation to crude distillation fuel and 3.5 times faster in relation to hydrotreated fuel.

5. Efficiency of antioxidant additive Agidol-1 introduction

The estimation of efficiency of antioxidant additive Agidol-1 introduction (table 4) in kerosene fractions of hydrogenation processes by developed method showed that a demercaptanizated kerosene fraction has high level of chemical stability without added antioxidant, at the level of straight-run fuel.

Table 4. Efficiency of antioxidant additive Agidol-1 introduction

Sample	ϑ_{max} (μg/min)	Additive efficiency coefficient
Kerosene fraction of hydrodemercaptanization	0,066	1,1
Kerosene fraction of hydrodemercaptanization + 0,0034 % Agidol-1	0,058	
Kerosene fraction of hydrotreatment	0,512	2,5
Kerosene fraction of hydrotreatment + 0,0040 % Agidol-1	0,203	
Kerosene fraction of hydrocracking	0,670	1,2
Kerosene fraction of hydrocracking + 0,0040 % Agidol-1	0,559	

The most effective is the introduction of Agidol-1 in the hydrotreated kerosene fraction. This allows to increase the chemical stability of the hydrotreated fraction by 2.5 times.

The introduction of Agidol-1 into the kerosene fractions of hydrocracking makes it possible to increase their chemical stability by only 1.2 times. Moreover, if you look at the curve of the dependence of the maximum oxygen absorption rate on the time of the test (figure 4), you can see that regardless of the introduction of an antioxidant, the sample is completely oxidized.

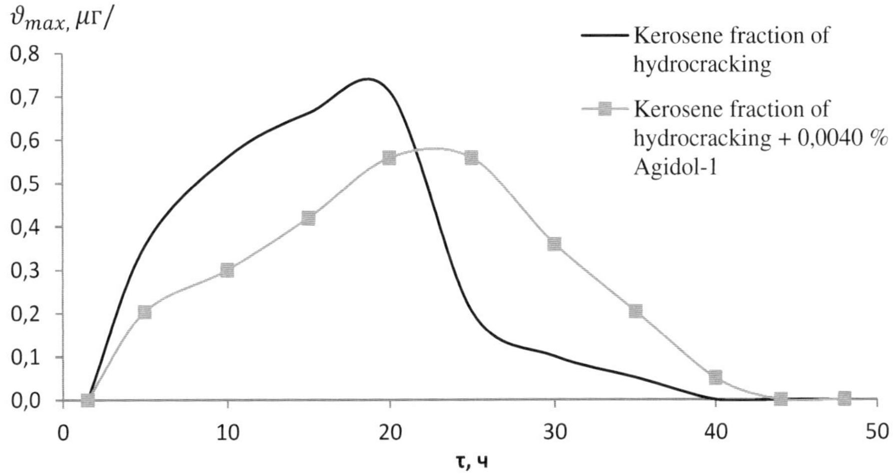

Figire 4. The effect of antioxidant additives on the chemical stability hydrocracked kerosene fraction

It is known that the maximum storage period for jet fuels based on the kerosene fraction of crude distillation is 10 years, and fuels based on hydrotreated kerosene fractions is 7 years [9]. The guarantee period for storage of jet fuel is 5 years. Given that the chemical stability of jet fuel based on kerosene fractions of hydrocracking is two times lower than the chemical stability of hydrotreated jet fuel, it can be assumed that the storage life of hydrocracking fuels will be two times lower than that of hydrotreated fuels, i.e. 3,5-4 years.

References

[1] Khavkin V, Gulyaeva L, Vinogradova N, Shmel'kova O 2009 *Mir nefteproduktov.* **3** 15-21
[2] Khavkin V, Belousov A 2014 Jet fuel production technologies *Neftepererabotka i neftekhimiya* **11** 5-7
[3] Veselyanskaya V, Radchenko E, Englin B 1983 *Chem Technol Fuels Oils* **12** 27-28
[4] Chernozhukov N I and Crane S E 1973 *Oxidability of mineral oils* (Moscow: Gostoptekhizdat) 370
[5] Vlasenkova L, Shatalov K, Likhterova N 2019 *Neftepererabotka i neftekhimiya* **4** 3-8
[6] Vlasenkova L, Shatalov K 2019 *Measurement Techniques* **62 (1)** 82-90
[7] Kovalev G, Gogitidze L, Kuranova V, Denisov E 1981 *Neftekhimiya* **21 (5)** 718-725
[8] Denisov E T and Kovalev G I 1983 *Oxidation and stabilization of jet fuels* (Moscow: Himiya) p 272
[9] Seregin E P 2018 *Development of chemotology* (Moscow: Izdatel'stvo pervyj tom) p 880

IOP Conf. Series: Materials Science and Engineering **921** (2020) 012028 doi:10.1088/1757-899X/921/1/012028

Effect of fertilization practice on biological properties of crude oil-polluted soil

X Wei[1], E A Bocharnikova[1,2]*, V V Matichenkov[2], D V Demin[2]

[1]Hunan University of Finance and Economics, Changsha 410205, China
[2]Institute Basic Biological Problems RAS, 142290, Pushchino, Russia

E-mail: nimedd@yandex.ru

Abstract. Hydrocarbon soil pollution is widespread problem in many regions of the world, where oil is extracted and transported. Pollution with crude oil has numerous adverse impacts on the soil-microorganism-plant system. Although adding fertilizers has been shown to be essential to enhance remediation rate, excessive nutrients may exert toxicity for microorganisms and plants. Field test was conducted in the area previously subjected to several accidental oil spills in the central European Russia. The effects of fresh pig manure and/or conventional NPK fertilizers on plant growth, hydrocarbon content and activities of catalase, dehydrogenase, urease, sucrose, and phosphatase were evaluated depending on an application rate. Fertilizer application resulted in restoration of plant cover and reduction of hydrocarbons by 13 to 46%. Oil pollution reduced activities of catalase, dehydrogenase, sucrase and phosphatase and increased urease activity. Manure application was more effective in regard to plant growth, hydrocarbon removal and enzymatic activity in comparison with mineral fertilizers. The maximum positive effect on contaminated soil-microorganism-plant system was obtained under combined application of manure and NPK.

1.Introduction

The annual world oil extraction is about 15.8 billion liters per day and keeps rising. Increasing oil production and construction of new oil pipelines and refineries lead to increased environmental pollution [1]. Summarizing land being negatively affected by oil and gas extraction and transportation was increased from 0.2 million ha in 2000 to about 3 million ha in 2012 [2]. Soil degradation is one of the most deleterious consequences of leakage of pipelines and storage tanks and accidental spills [3]. As reported by numerous studies, the presence of hydrocarbons disturbs physical, chemical and biological soil properties, including structure, water-holding capacity, redox conditions, nutrient availability, thus leading to deterioration of soil fertility and health [4; 5; 6]. In this regard, the elaboration of environmentally friendly and economically feasible approaches aimed at intensifying hydrocarbon degradation and restoration of soil fertility is an acute issue.

Because of varying physical and chemical properties of multiple components that compose oil, both the migration and transformation modes in soil and response to the remediation approaches differ depending on composition. Numerous clean-up technologies have been suggested such as bioremediation, bioaugmentation, biostimulation, ventilation, phytoremediation, phytodegradation, chemical oxidation and others [7; 8]. To successfully develop advanced technologies or enhance the efficiency of conventional strategies for soil remediation, comprehensive understanding of the physical, chemical and biological soil processes under oil contamination is critical.

Microbial-based remediation is considered to be nature-like, relatively cost-effective, and widely used approach. Different microorganisms possess the ability to utilize petroleum hydrocarbons as a source of carbon and energy. Among them, bacteria play a leading role. Numerous bacterial genera indigenous to soil have been reported to decompose crude oil and oil products, for example, *Pseudomonas*, *Bacillus*, *Phodococcus*, and *Micrococcus* [9]. The efficiency of biodegradation is affected by multiple factors including water regime, temperature, oxygen availability, and many others [10]. Insufficient nutrient supply limits the natural processes of oil degradation [11]. General strategy to enhance the biodegradation is aimed at maintaining optimal conditions for microbial growth.

Traditional mineral and organic fertilizers are widely used to mitigate a negative influence of crude oil on soil and stimulate microbial activity. Mineral fertilizers promote soil fertility and plant growth as well as provide balanced nutrient supply required for hydrocarbon (HC)-degrading microbes [12; 13]. Organic fertilizers improve soil properties, activate the microbial growth, and serve as a source of additional microorganisms and nutrients [14]. Some studies evidence that the combined application of inorganic and organic soil amendments is more efficient than their single application [15]. Excessive amounts of nutrients applied can be deleterious for soil health and microbial activity [10]. The type and application rate of soil amendments depend highly on soil properties.

Any biological process in soil involves the enzyme participation. Soil enzymes play a key role in converting and decomposition of organic pollutants. The main objective of the current investigation was to determine an influence of mineral fertilizers and manure on enzymatic activity and HC removal in crude oil-contaminated soil.

2. Material and Methods

Field test (Gray Forest Soil, Bashkiria, Russia) was conducted in an area previously subjected to several accidental crude oil spills. The upper layer of unpolluted soil had the following characteristic: 2.26±0.10 C%, pH=7.1±0.1, clay content 35.6±0.5%, CEC = 15.8±05 cmol(+) kg^{-1}; total phosphorus 1.24±0.15 g kg^{-1} P; total potassium 28.9±0.3 g kg^{-1} K; total nitrogen 4.5±0.3 g kg^{-1} N.

The oil had a high content of sulfur (2.5±0.2%), the content of paraffin 8.0 ±0.2%; naphthenes 6.4±0.1%; aromatic hydrocarbons 15.1±0.4%; density 0.938±0.015. Twenty four plots of 10 m^2 each were separated and the following treatments were conducted: (1) control without pollution; (2) pollution; (3) pollution + $N_{60}P_{90}K_{90}$; (4) pollution + $N_{120}P_{180}K_{180}$; (5) pollution + manure, 70 t ha^{-1}; (6) pollution + manure, 140 t ha^{-1}; (7) pollution + $N_{60}P_{90}K_{90}$ + manure, 70 t ha^{-1}, and (8) pollution + $N_{120}P_{180}K_{180}$ + manure, 140 t ha^{-1}. Fresh pig manure with a moisture of 50% was used. Mineral fertilizers were added as ammonium nitrate (Fosagro, Russia), potassium chloride (Uralkali, Russia) and superphosphate (Biomaster, Russia). After application of fertilizers and manure, barley (*H. vulgare* L., cv. *Suzdaletz*) was planted with a density of 1000 seeds per m^2. After the growing season, a percent of survived plants was evaluated.

Soil samples were collected from a 0-20 cm depth in a year. The HC content was analyzed by hexane (C_6H_{14}) and ethanol-benzole (1:1) extraction methods using ST 255 Soxtec system (FOSS). Ten (10) g of soil was placed into Soxtec flask and 100 mL of solvent was added. Samples were extracted at +70°C for 4 hours [16]. Total HC content in an extract was estimated by a gravimetric method [17].

The activities of catalase, dehydrogenase, urease, sucrose, and phosphatase were determined by the methods described by Achuba and Peretiemo-Clarke [19], Tabatabai [20], Guo [21], Ge et al. [22], and Khomutova et al. [23], accordingly. Absorbance of solutions was measured colorimetrically by a spectrofotometer Shimadzu UV–VIS 160A (Kyoto, Japan). Each treatment and each analysis were conducted in 4 replications. All data obtained was subjected to a statistical analysis based on comparative methods using Duncan's multiple range tests for mean separation at the 5% level of significance [24].

3.Results and Discussion

Uncontaminated soil was characterized by high content of organic matter and high level of fertility. However, the contamination with oil resulted in death of all barley plants in control soil (Figure 1). The application of mineral fertilizers and/or manure provided a significant increase in the plant survival. Manure had a greater effect as compared to mineral fertilizers.

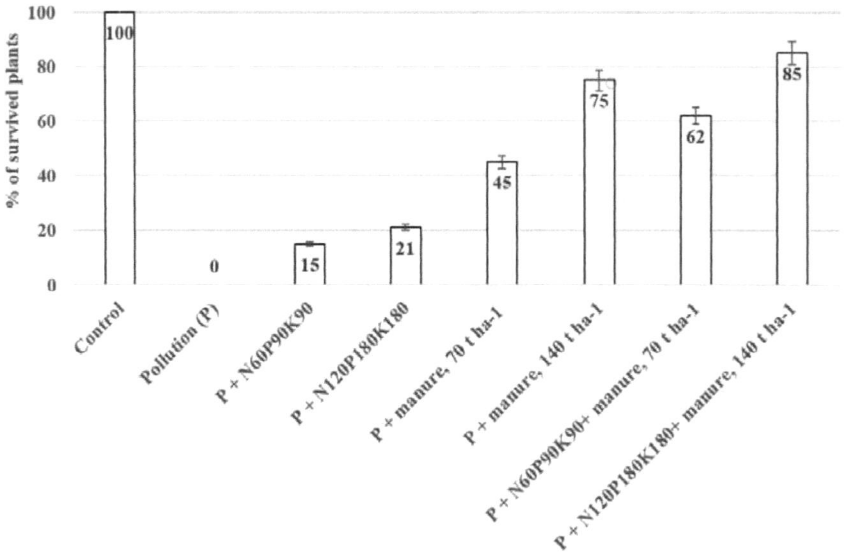

Figure 1. Amount of survived plants, %.

The HC content is presented in Table 1. The application of mineral fertilizer or manure considerably decreased the HC content in comparison with untreated soil (by 35.4-46.3%; 20.0-24.6%; and 25.3-46.2% with mineral fertilizers, manure, and combined fertilization, accordingly). Organic C in the polluted soil was slightly reduced under the application of mineral fertilizers, from 7.9 to 7.2 and 7.5% of C, accordingly at a low and high rate of fertilization. The addition of manure significantly increased the soil C up to 12.5%, however, a higher increase was observed at a lesser rate of manure. Manure+mineral fertilizers or manure at 140 t ha^{-1} provided a smaller raise in the total C, thus evidencing more intensive mineralization of organic compounds due to high microbial activity.

Table 1. Petroleum hydrocarbons and total C in soil.

	C_6H_{14}-extractable	C_2H_5OH+ C_6H_6 -extractable	Total C, %
	---------------- %	-----------------	
Control	0.08	0.14	2.26
Pollution (P)	5.20	6.29	7.94

P + $N_{60}P_{90}K_{90}$	5.04	5.18	7.20
P + $N_{120}P_{180}K_{180}$	3.36	3.38	7.50
P + manure, 70 t ha^{-1}	4.50	4.41	12.51
P + manure, 140 t ha^{-1}	3.92	5.03	11.01
P + $N_{60}P_{90}K_{90}$+ manure, 70 t ha^{-1}	3.66	5.08	11.53
P + $N_{120}P_{180}K_{180}$+ manure, 140 t ha^{-1}	2.80	4.70	11.85
LSD_{05}	0.10	0.12	0.10

Soil enzymatic activity is one of the most fast and sensitive features reflecting the microbial activity and the HC degradation potential [9; 11]. Different enzymes are related to different types of microbial activity. In our experiment, the oil pollution sharply reduced the catalase, sucrase, urease and phosphatase activities, while dehydrogenase did not change (Table 2). The dehydrogenase activity is considered as a measure of microbial oxidative processes in oil contaminated soil [25]. High level of dehydrogenase evidences the intensive oil degradation. Application of nutrients or organic matter significantly increased dehydrogenase activity, with a maximum effect by manure (an increase by 33-76% as compared with contaminated control).

Catalase is known to promote the decomposition of hydrogen peroxide, thereby preventing poisoning living organisms [26]. Reduced catalase activity is related to decreased soil respiration. In our experiment, the catalase activity reduced 2.3 times in the oil-polluted soil. The manure application benefitted the catalase activity better than mineral fertilizers, probably due to better air penetration to soil.

The sucrase activity reflects the intensity of the sucrose hydrolysis. A sharp reduction in this enzyme observed in the contaminated soil (by 41.8%) may indicate reduced content of sucrose and other sugar-based organic molecules to be released to soil with root exudates. Mineral fertilizers only slightly improved this parameter. The application of manure or manure with NPK enhanced the sucrase activity, but less than that of catalase, dehydrogenase and phosphatase. Perhaps, the restoration of the sucrase activity requires more time.

Urease is the enzyme responsible for the hydrolysis of urea to CO_2 and NH_3. Most authors report a decrease in soil urease activity under petroleum pollution [21]. In our study, the urease activity sharply increased in polluted soils. Similar significant increase in urease activity has been reported in soil contaminated with crude oil leaked from pipeline [27]. Such contradictory data may be caused by differences in oil composition, contamination level, exposure time and others. Mineral fertilizers had insignificant effect on the urease activity, while manure or manure with NPK considerably increased the urease activity by up to 46% as compared with contaminated control and 4.6 times compared with uncontaminated soil.

The activity of phosphatase is inversely dependent on the availability of mobile phosphorus in soil [28]. The phosphatase activity was reduced 2.4-fold under oil contamination. The application of NPK or/and manure further reduced the phosphatase activity, thereby indicating less P deficiency.

Table 2. Enzymatic activity of crude oil-polluted soil as influenced by organic and mineral fertilizers.

	Catalase, O_2 mL	Dehydrogenase, mg of formasana	Sucrase, glucose mg	Urease, NH_3 mg	Phosphatase, phenolphthalein mg
	g^{-1} of dry soil				
Control	5.41	3.52	20.06	0.35	7.56
Pollution (P)	2.30	3.60	11.98	1.10	3.10
P + $N_{60}P_{90}K_{90}$	2.92	3.74	12.87	0.88	2.51

P + $N_{120}P_{180}K_{180}$	2.92	4.19	12.31	1.00	2.10
P + manure, 70 t ha^{-1}	3.56	4.81	13.13	1.12	2.47
P + manure, 140 t ha^{-1}	4.11	6.36	13.40	1.29	2.03
P + $N_{60}P_{90}K_{90}$+ manure, 70 t ha^{-1}	3.05	4.71	13.91	1.29	2.37
P + $N_{120}P_{180}K_{180}$+ manure, 140 t ha^{-1}	4.61	5.87	15.14	1.61	2.63
LSD_{05}	0.05	0.04	0.15	0.02	0.05

Our data evidences that the contamination with crude oil altered the soil microbial activity. A main negative impact was observed on catalase and sucrase, while urease increased and dehydrogenase changed insignificantly. Reduced phosphatase activity can reflect reduced P deficiency as a result of contamination and fertilization.

Crude-oil pollution adversely affects the soil-microorganism-plant system and alters the microbial activity. Analysis of soil ferment activity demonstrates a reduction in catalase, dehydrogenase, sucrase and phosphatase, whereas urease activity increased. The data obtained evidences that organic fertilizer had more pronounced effect than only mineral fertilizers. A maximum positive effect was observed on contaminated soil-microorganisms-plant system under co-application of manure and NPK fertilizers.

4.Acknowledgements

This work was financially supported by the Ministry of Science and Higher Education of Russian Federation, theme AAAA-A17-117030110137-5 and AAAA-A17-117030110139-9 and Key Research and Development Program of Hunan Province, China (2019WK2031), Hunan Provincial Base for Scientific and Technological Innovation Cooperation (2018WK4013) and China Central Guide the Development of Local Science and Technology Special Funds (2017XF5042).

5.Reference

[1] Vedomosti.ru https://www.vedomosti.ru/business/articles/2018/09/13/780795-neftivpervie-100
[2] Allred B W, Smith W K, Twidwell D, Haggerty J H, Running S W, Naugle D E and Fuhlendorf S D 2015 *Science* **348(6233)** 401
[3] Villacís J, Casanoves F, Hang S, Keesstra S and Armas C 2016 *Sci Total Environ* **566** 761
[4] Essien O E and John I A 2010 *J. Appl. Sci. Environ Manage* **14(4)** 147
[5] Hewelke E, Szatyłowicz J, Hewelke P, Gnatowski T and Aghalarov R 2018 *Water Air Soil Pollut,* **229** 51
[6] Uzoije A P and Agunwamba J C 2011 *Environ Sci Technol* **4** 313-323.
[7] Lim M W, Von Lau E and Poh P E 2016 *Mar Pollut Bull* **109(1)** 14
[8] Stupin D J 2009 *Soil pollution and new technologies for its restoration* (St. Peterburg Lan)
[9] Das N and Chandran P C 2011 *Biotechnol Res Int* **2011** 13
[10] Ukpaka C P 2017 *Curr Sci* **3(2)** 67
[11] Shukry W M, Al-Hawas G H S, Al-Moaikal R M S and El-Bendary MA 2013 *Br J. Environ Clim Change* **3(1)** 103
[12] Agarry S E, Owabor C N and Yusuf R O 2010 *Bioremediat J* **14(4)** 189
[13] Ijah U J J, Safiyan U H and Abioye O P 2008 *Sci World J* **3(2)** 63
[14] Agarry S E 2018 *J Appl Sci Environ Manage* **22(4)** 587
[15] Nwaichi E O, Frac M, Nwoha P A and Eragbor P 2015 *Int J Phytorem* **17(12)** 1253
[16] Bocharnikova EA 1990. *The effect of oil pollution on the properties of Gray-Brown Soil of Apsheron and Gray Forest soils of Bashkiria*: diss. …cand. biol. sci.: Moscow State University

[17] Okoh A, Ajisebutu S, Babalola G and Trejo-Hernandez M 2001 *Inter Microb* **4(2)** 83
[18] Orlov D S and Grishina L A 1981 *Practical Manual on Humus Chemistry* (Moscow State University, Moscow)
[19] Achuba F I and Peretiemo-Clarke B O 2008 *Int Agrophys* **22(1)** 1
[20] Tabatabai M A 1982 In: *Methods of Soil Analysis. Part 2. Chemical and Microbiolgical Properties* (ASA and SSSA, Madison, WI)
[21] Guo H et al. 2012 *Chemosphere* **87(11)** 1273
[22] Ge G, Li Z, Fan F, Chu G, Hou Z and Liang Y 2010 *Plant Soil* **326(1-2)** 31
[23] Khomutova T E, Demkina T S, Kashirskaya N N and Demkin V A 2012 *Eurasian Soil Sci* **45(4)** 423
[24] Duncan D B 1957 *Biometrics* **13(2)** 164
[25] Namkoong W, Hwang E Y, Park J S and Choi J Y 2002 *Environ Pollut* **119(1)** 23
[26] Wang J, Zhang Z, Su Y, He W, He F and Song H 2008 *Petrol Sci* **5(2)** 167
[27] Trasar-Cepeda c, Leiros M C, Seoane S, Gil-Sotres F 2000 *Soil Biol Biochem* **32** 1867
[28] Alrumman S A, Standing D B and Paton G I 2015 *J King Saud Univ Sci* **27(1)** 31

IOP Conf. Series: Materials Science and Engineering **921** (2020) 012029 doi:10.1088/1757-899X/921/1/012029

Purification of hydrocarbon-contaminated soils by silicon-rich substances

X Wei[1], V V Matichenkov[1,2], E A Bocharnikova[2], S M Sevostianov[2]

[1]Hunan University of Finance and Economics, Changsha 410205, China
[2]Institute Basic Biological Problems RAS, 142290, Pushchino, Russia

E-mail: Sevost2000@rambler.ru

Abstract. Hydrocarbon (HC) contamination of soil occurs during oil extraction, transportation and treatment. To enhance the efficacy of HC-polluted soil restoration is an important challenge. In greenhouse and field experiments, two silicon (Si)-rich substances were tested for reduction of HC content in diesel + used motor oil contaminated soil. Over 3 months the HC content decreased from 3 to 2.8; 1.1 and 0.8% in control, calcium silicate slag- and amorphous Si dioxide-treated soils, respectively. Silicon-rich substances remarkably increased the catalase and dehydrogenase activities in the contaminated soil. Several mechanisms underlying the Si-mediated purification of HC-contaminated soil are suggested: a) changing soil physical properties; b) capsulation of HC molecules by polysilicic acid; c) adsorption of monosilicic acid by HCs leading to enhanced hydrophilicity; d) direct effect on soil microbial activity. Many types of Si-rich materials are cheap and available in large volumes (industrial by-products) that potentially could be used for the purification of HC-polluted environment after ecological expertise of the absence of inorganic pollutants.

1.Introduction

Soil contamination with crude oil and oil products is widespread problem because of rapidly rising production, refining, and transportation [1]. Increasing use of petroleum products globally leads to an immense increase in environmental contamination by diesel, spent engine oil or jet fuel [2]. Soil contamination with hydrocarbon (HC)-based products has a negative impact on public health, being a source of acute and chronic diseases, such as: poisoning, cardiovascular diseases, cancer, and others. [3].

Selection of the strategy applicable for soil restoration primary depends on a level of pollution. If the content of HCs is higher than 3%, oil and oil products are recommended to remove mechanically from soil [4]. If soil contains 0.3-3% of HCs, it is classified as contaminated and special techniques are required to reduce their content. Various remediation strategies have been elaborated, for instance, application of soil amendments such as peat, fertilizers, manure, microbial preparations, specific chemicals (enzymes), special plant cultivation and others [5; 6; 7].

Microorganisms possess the ability to transform or mineralize naturally occurring organic molecules, and this can be utilized in the *in situ* remediation of oil-contaminated areas [2; 8]. Numerous remediation-related studies are aimed at the selection of microorganisms able to degrade petroleum HCs [7; 9]. Because of inhibitory effect of HC on microbial community, the efficiency of bioremediation processes are often low [4; 8; 10].

Over the last decades, beneficial impacts of Si-rich compounds on various soil properties and fertility level have been well documented [11; 12]. Silicon fertilizers and soil amendments provide increased adsorption and water holding capacity; improved plant availability of P, K, Fe, Zn and other nutrients; decreased toxicity of organic and inorganic pollutants; strengthened texture stability and others [13; 14]. It should be noted that some types of industrial wastes can be classified as Si fertilizers [11]. Due to numerous Si-mediated favourable effects on soil, plant and biota, Si-rich substances could be a promising agent for involvement in the HC-polluted soil restoration. There is data that application of Si-rich substances increases the soil microbial activity [13].

A lot of studies are available regarding soil contamination by diesel or oil-products, while their combined presence is rarely studied [7; 9]. The aim of the current study was to determine the effect of Si-rich substances on HCs content and enzymatic activity in diesel+spent motor oil contaminated soil.

2.Materials and Methods

Two Si-rich substances (calcium silicate slag and amorphous silicon dioxide) were tested in greenhouse and field experiments. Calcium silicate slag was obtained from the Kosogorsky metallurgical enterprise (Tula region, Russia). This slag was mostly comprised of $CaSiO_3$. Heavy metals or other inorganic pollutants (As, Cd, Hg, Pb et al.) were not detected. Amorphous silicon dioxide (SiO_2) (Aerosil A-300) (Ltd Silica, Russia) is chemically pure substance contained 99.8% of SiO_2. The content of water-soluble Si was 4.6 ± 0.5 and 48.5± 1.5 mg kg^{-1} of Si, accordingly in $CaSiO_3$ and SiO_2.

Greenhouse experiment was conducted on Grey Forest Soil in the Institute Basic Biological Problems Russian Academy of Sciences. The field test was conducted on Soddy Podzolic Soil in the north of Moscow region. Soil chemical properties are presented in Table 1. The scheme of both experiments was the following:

1) Control;
2) HC pollution;
3) HC pollution + $CaSiO_3$ 5 t ha^{-1};
4) HC pollution + $CaSiO_3$ 10 t ha^{-1};
5) HC pollution + SiO_2 5 t ha^{-1};
6) HC pollution + SiO_2 10 t ha^{-1}.

Table 1. Selected properties of experimental soils.

	pH	Total C, g kg^{-1}	CEC cmol(+) kg^{-1}	Total nitrogen, g kg^{-1}	Total phosphorus, g kg^{-1}	Total potassium, g kg^{-1}
Grey Forest Soil	7.2	24.7	12.8	3.08	0.91	22.3
Soddy Podzolic Soil	6.1	9.8	8.5	2.51	0.68	15.4

Both experiments were performed at the same level of pollution. In greenhouse pot test, 30 mL of diesel and spent motor oil (1:1) mixture was added to 1 L of soil and agitated. In field test, 3 L of diesel and spent motor oil (1:1) per m^2 was added and mixed in a 10-cm layer of soil. After this, the Si substances were applied and soil was mixed again. Then barley (*Hordeum vulgare* L. cv Elf) was planted in depth of 2-3 cm. In greenhouse test, 10 seeds were planted in each pot. In field test, 1000 seeds per 1 m^2 were planted. Before seeding all seeds were soaked in distilled water for 24 hours. The air temperature in the chamber was kept at 20 ± 2^{0}C during the day time and 12 ± 2^{0}C during the night. The light period was 12 h; the light intensity was 130 μmol photons m^{-2} s^{-1}. The relative air humidity was 45 ± 5% during the day and 70 ± 5% during the night. The field test was conducted from June to July, the

IOP Conf. Series: Materials Science and Engineering **921** (2020) 012029 doi:10.1088/1757-899X/921/1/012029

daytime temperature was about 20-25°C and night temperature was 15-20°C. The natural precipitation was 200-300 mm.

In greenhouse test, soil samples were collected weekly over a 4-week period; in field test - every 2 weeks during 8 weeks from 0-5 and 5-10 cm depth. The HC content was analyzed by hexane (C_6H_{14}) extraction method using ST 255 Soxtec system (FOSS). Ten (10) g of soil was placed into Soxtec flask and 100 mL of hexane was added. The extraction temperature was +70°C and time – 24 hours [15].

The biomass of barley was measured 4 weeks after seeding in pot test and in 8 weeks in field test. Considering that microorganisms play a key role in the HC degradation, the soil microbial activity was evaluated by testing the activity of catalase and dehydrogenase.

The catalase and dehydrogenase activities were determined using the methods described by Achuba and Peretiemo-Clarke [16] and Tabatai [17]. Absorbance of solutions was read spectrophotometrically using Shimadzu UV–VIS 160A (Kyoto, Japan). Each treatment and each analysis were conducted in 4 replications. All data obtained was subjected to a statistical analysis based on comparative methods using Duncan's multiple range tests for mean separation at the 5% level of significance [18].

3. Results and Discussion

The dynamics of HCs in greenhouse experiment is presented in Table 2. In the Si-untreated soil, the HC content insignificantly decreased from 3.1 to 2.8% for 4 weeks. Application of Si-rich substances substantially reduced the soil HCs. The maximum effect was observed under SiO_2 application at 10 t ha^{-1} (from 3.1% to 0.4%). The minimum reduction was obtained at lower application rate of $CaSiO_3$ (from 3.1% to 1.8%).

Table 2. Effect of Si-rich substances on HCs in greenhouse experiment.

	Hydrocarbons, %			
Treatment	1week	2 weeks	3 weeks	4 weeks
Control	0.2	0.2	0.2	0.1
Hydrocarbon pollution (HP)	3.1	3.0	2.7	2.8
HP+ $CaSiO_3$, 5 t ha^{-1}	3.0	2.9	2.5	1.8
HP+ $CaSiO_3$, 10 t ha^{-1}	3.0	2.7	2.4	1.5
HP+SiO_2, 5 t ha^{-1}	2.8	2.5	1.3	0.8
HP+SiO_2, 10 t ha^{-1}	2.7	2.3	1.0	0.4
LSD_{05}	0.2	0.2	0.2	0.2

The used oil products had a toxic effect on barley plants in both greenhouse and field tests. In the contaminated control, all plants have died. The treatment with $CaSiO_3$ or SiO_2 promoted a reduction in the toxic effect of contaminants on tested plants. Application of the Si-rich substances resulted in increased biomass from 0.5 to 1.8 g of dry weight $plant^{-1}$ in greenhouse test and from 1.7 to 2.8 g of dry weight $plant^{-1}$ in field test. Silicon dioxide applied at 10 t ha^{-1} demonstrated a higher efficacy, whereas $CaSiO_3$ at 5 t ha^{-1} was lesser efficient.

The dynamics in the HC soil content is presented in Table 3. Application of the Si-rich substances significantly reduced the content of oil-based products in 0-5 and 5-10 cm soil layers. The maximum effect was obtained at a higher rate of SiO_2 (decreased HCs from 2.5% in control to 0.3%). Additionally, the Si-rich substances reduced the leaching of used motor oil and diesel.

Table 3. Effect of Si-rich substances on HCs in field experiment.

Treatment	Depth, cm	Weeks 2	4	6	8
Control	0-5	0.2	0.2	0.2	0.2
	5-10	0.1	0.1	0.1	0.1
Hydrocarbon pollution	0-5	2.9	2.9	2.6	2.5
(HP)	5-10	0.9	1.1	1.2	1.2
HP+ $CaSiO_3$, 5 t ha^{-1}	0-5	2.8	2.7	2.2	1.8
	5-10	0.9	1.0	0.8	0.6
HP+ $CaSiO_3$, 10 t ha^{-1}	0-5	2.8	2.5	2.0	1.4
	5-10	0.7	0.6	0.6	0.4
HP+ SiO_2, 5 t ha^{-1}	0-5	2.5	2.0	1.8	1.0
	5-10	1.0	0.8	1.0	0.7
HP+ SiO_2, 10 t ha^{-1}	0-5	1.6	1.3	1.1	0.3
	5-10	0.6	0.4	0.3	0.2
LSD_{05}		0.4	0.3	0.4	0.3

Dehydrogenase and catalase together are considered as a sensitive indicator of viable microorganisms and soil quality. Dehydrogenase enzymes take part in oxidation of organic compounds and correlate with HC degradation [5]. Catalase is an antioxidant enzyme providing decomposition of hydrogen peroxide to water and oxygen [16]. The HC pollution led to substantial reductions in the activities of catalase and dehydrogenase (Table 4). Silicon substances, especially SiO_2, promoted a significant increase in the activities of both enzymes. Increased soil dehydrogenase activity evidences enhanced HCs oxidation. The effect of SiO_2 was much higher than that of $CaSiO_3$, perhaps due to a higher solubility on Si.

Table 4. Effect of Si substances on catalase and dehydrogenase activities in HC-polluted soil.

	Catalase, O_2 mL	Dehydrogenase, mg of formasana	Sampling depth, cm	Catalase, O_2 mL	Dehydrogenase, mg of formasana
	-------------	------------- g^{-1} of dry soil	-------------	-------------	-------------
	Greenhouse test			Field test	
Control	6.52	4.87	0-5	5.83	4.11
			5-10	4.22	2.21
Hydrocarbon pollution (HP)	2.30	3.60	0-5	1.54	1.12
			5-10	1.11	0.34
HP+ $CaSiO_3$, 5 t ha^{-1}	2.95	3.94	0-5	2.14	1.98
			5-10	2.43	0.95
HP+ $CaSiO_3$, 10 t ha^{-1}	3.93	4.12	0-5	5.12	2.21
			5-10	3.84	1.34
HP+ SiO_2, 5 t ha^{-1}	5.16	4.93	0-5	5.34	4.02
			5-10	4.11	2.13
HP+ SiO_2, 10 t ha^{-1}	5.14	5.65	0-5	6.34	4.23
			5-10	5.44	2.36
LSD_{05}	0.05	0.04		0.06	0.05

The obtained data has shown that Si-rich substances had a positive influence on the HC-contaminated soil, accelerated HC degradation, and promoted microbial activity. Among two tested Si substances, SiO_2, as having a higher solubility of Si, was more efficient. It is possible to suppose that soluble forms of Si (mono- and polysilicic acids) play an important function in the HC-polluted soil remediation.

Several mechanisms responsible for Si-mediated purification of the HC-contaminated soil can be supposed: a) changing physical soil properties; b) capsulation of HC molecules by polysilicic acid; c) adsorption of monosilicic acid by HCs leading to enhanced hydrophilicity; d) direct effect on soil microbial activity. Some types of Si-rich materials are cheap and available in large volume as being industrial by-products. Therefore, followed ecological expertise on the content of inorganic pollutants, such materials could be involved in soil restoration strategies.

4.Acknowledgements
This work was financially supported by the Ministry of Science and Higher Education of Russian Federation, theme AAAA-A17-117030110137-5 and AAAA-A17-117030110139-9; the Key Research and Development Program of Hunan Province, China (2019WK2031) and Science and Technology Program of Changsha City (kq1706153).

5.Reference

[1] Bekuzarova S A, Weisfeld L I and Alexandrov E N 2017 In: *Heavy Metals and Other Pollutants in the Environment* (Apple Academic Press) p. 129
[2] Nwaogu L A, Onyeze G O and Nwabueze R N 2008 *Afr J Biotechnol* 7**(12)** 1939
[3] Ramirez M I, Arevalo A P, Sotomayor S and Bailon-Moscoso N 2017 *Environ Pollut* **231** 415-425
[4] Stupin D J 2009 *Soil pollution and new technologies for its restoration* (St. Peterburg Lan)
[5] Dindar E, Şağban F O T and Başkaya H S 2015 *Int Biodeter Biodegr* **105** 268
[6] Fingas M 2016 *Oil spill science and technology* (Gulf professional publishing)
[7] Oyewole O A, Zobeashia S L T, Oladoja O E, Musa I O and Terhemba I T 2020 *J Biosci* ***28*** 33
[8] Namkoong W, Hwang E Y, Park J S and Choi J Y 2002 *Envir Pollut* **119(1)** 23
[9] Sprocati A R et al. 20121. *Process Biochem* **47(11)** 1649
[10] Lapinskienė A, Martinkus P and Rėbždaitė V 2006 *Environ Pollut* **142(3)** 432
[11] Liang Y, Nikolic M, Bélanger R, Gong H and Song A 2015 *Silicon in agriculture: From theory to practice* Springer
[12] Matichenkov V V and Bocharnikova E A 2001 *Studies in Plant Science* **8** 209
[13] Kulikova A C 2012 *Silicon and silicon-rich rocks in system of fertilization for agricultural plants* (Ulyanovsk State Agricultural Academy press)
[14] Ma J F and Takahashi E 2002 *Soil, fertilizer, and plant silicon research in Japan* Elsevier
[15] Bocharnikova E A 1990. *The effect of oil pollution on the properties of Gray-Brown Soil of Apsheron and Gray Forest soils of Bashkiria*: diss. ...cand. biol. sci.: Moscow State University
[16] Achuba F I and Peretiemo-Clarke B O 2008 *Int Agrophys* **22(1)** 1
[17] Tabatabai M A 1982 In: *Methods of Soil Analysis. Part 2. Chemical and Microbiolgical Properties* (ASA and SSSA, Madison, WI)
[18] Duncan D B 1957 *Biometrics* **13(2)** 164

AUTHOR INDEX